Achieving sustainable production of eggs

Volume 2: Animal welfare and sustainability

It is widely recognised that agriculture is a significant contributor to global warming and climate change. Agriculture needs to reduce its environmental impact and adapt to current climate change whilst still feeding a growing population, i.e. become more 'climate-smart'. Burleigh Dodds Science Publishing is playing its part in achieving this by bringing together key research on making the production of the world's most important crops and livestock products more sustainable.

Based on extensive research, our publications specifically target the challenge of climate-smart agriculture. In this way we are using 'smart publishing' to help achieve climate-smart agriculture.

Burleigh Dodds Science Publishing is an independent and innovative publisher delivering high quality customer-focused agricultural science content in both print and online formats for the academic and research communities. Our aim is to build a foundation of knowledge on which researchers can build to meet the challenge of climate-smart agriculture.

For more information about Burleigh Dodds Science Publishing simply call us on +44 (0) 1223 839365, email info@bdspublishing.com or alternatively please visit our website at www.bdspublishing.com.

Related titles:
Achieving sustainable production of eggs Volume 1: Safety and quality
Print (ISBN 978-1-78676-076-0); Online (ISBN 978-1-78676-078-4, 978-1-78676-079-1)

Achieving sustainable production of poultry meat Volume 1: Safety, quality and sustainability
Print (ISBN 978-1-78676-064-7); Online (ISBN 978-1-78676-066-1, 978-1-78676-067-8)

Achieving sustainable production of poultry meat Volume 2: Breeding and nutrition
Print (ISBN 978-1-78676-068-5); Online (ISBN 978-1-78676-070-8, 978-1-78676-071-5)

Achieving sustainable production of poultry meat Volume 3: Health and welfare
Print (ISBN 978-1-78676-072-2); Online (ISBN 978-1-78676-074-6, 978-1-78676-075-3)

Chapters are available individually from our online bookshop: https://shop.bdspublishing.com

climate•SMART•publishing
IN AGRICULTURAL SCIENCE

BURLEIGH DODDS SERIES IN AGRICULTURAL SCIENCE

NUMBER 17

Achieving sustainable production of eggs

Volume 2: Animal welfare and sustainability

Edited by Professor Julie Roberts,
University of New England, Australia

burleigh dodds
SCIENCE PUBLISHING

Published by Burleigh Dodds Science Publishing Limited
82 High Street, Sawston, Cambridge CB22 3HJ, UK
www.bdspublishing.com

Burleigh Dodds Science Publishing, 1518 Walnut Street, Suite 900, Philadelphia, PA 19102-3406, USA

First published 2017 by Burleigh Dodds Science Publishing Limited

Library of Congress Control Number: 2016961789

British Library Cataloguing in Publication Data
A catalogue record for this book is available from the British Library

ISBN 978-1-78676-080-7 (print)
ISBN 978-1-78676-082-1 (online)
ISBN 978-1-78676-083-8 (online)
ISSN 2059-6936 (print)
ISSN 2059-6944 (online)

Typeset by Deanta Global Publishing Services, Chennai, India
Printed by Lightning Source

Contents

Series list

Title | Series Number

Achieving sustainable cultivation of maize - Vol 1 001
From improved varieties to local applications
Edited by: Dr Dave Watson, CGIAR Maize Research Program Manager, CIMMYT, Mexico

Achieving sustainable cultivation of maize - Vol 2 002
Cultivation techniques, pest and disease control
Edited by: Dr Dave Watson, CGIAR Maize Research Program Manager, CIMMYT, Mexico

Achieving sustainable cultivation of rice - Vol 1 003
Breeding for higher yield and quality
Edited by: Prof. Takuji Sasaki, Tokyo University of Agriculture, Japan

Achieving sustainable cultivation of rice - Vol 2 004
Cultivation, pest and disease management
Edited by: Prof. Takuji Sasaki, Tokyo University of Agriculture, Japan

Achieving sustainable cultivation of wheat - Vol 1 005
Breeding, quality traits, pests and diseases
Edited by: Prof. Peter Langridge, The University of Adelaide, Australia

Achieving sustainable cultivation of wheat - Vol 2 006
Cultivation techniques
Edited by: Prof. Peter Langridge, The University of Adelaide, Australia

Achieving sustainable cultivation of tomatoes 007
Edited by: Dr. Autar Mattoo, USDA-ARS, USA & Prof. Avtar Handa, Purdue University, USA

Achieving sustainable production of milk - Vol 1 008
Milk composition, genetics and breeding
Edited by: Dr Nico van Belzen, International Dairy Federation (IDF), Belgium

Achieving sustainable production of milk - Vol 2 009
Safety, quality and sustainability
Edited by: Dr Nico van Belzen, International Dairy Federation (IDF), Belgium

Achieving sustainable production of milk - Vol 3 010
Dairy herd management and welfare
Edited by: Prof. John Webster, University of Bristol, UK

Ensuring safety and quality in the production of beef - Vol 1 011
Safety
Edited by: Prof. Gary Acuff, Texas A&M University, USA & Prof.James Dickson, Iowa State University, USA

Ensuring safety and quality in the production of beef - Vol 2 012
Quality
Edited by: Prof. Michael Dikeman, Kansas State University, USA

Achieving sustainable production of poultry meat - Vol 1 013
Safety, quality and sustainability
Edited by: Prof. Steven C. Ricke, University of Arkansas, USA

Achieving sustainable production of poultry meat - Vol 2 014
Breeding and nutrition
Edited by: Prof. Todd Applegate, University of Georgia, USA

Achieving sustainable production of poultry meat - Vol 3 015
Health and welfare
Edited by: Prof. Todd Applegate, University of Georgia, USA

Introduction

The welfare of laying hens is dependent on the provision of appropriate housing, excellent management, high-quality feed, and prevention and appropriate treatment of diseases. This volume, 'Achieving sustainable production of eggs Volume 2: Animal welfare and sustainability', provides a detailed account of laying hens' nutritional requirements and a practical guide to maintaining their health. The welfare of hens is addressed by examining their welfare standards, identifying and discussing welfare issues affecting free-range laying hens, reviewing the welfare costs and benefits of beak trimming, and investigating the management of laying hen flocks with intact beaks. The sustainability of the egg industry is addressed from two perspectives: waste management in egg production and an assessment of the sustainability of organic egg production.

Part 1 Animal health and welfare

The major economic cost in commercial egg production is the provision of an appropriate diet for hens both during the rearing and during the laying stages. The scientifically formulated diets of hens should meet all their nutritional requirements, at the same time making use of cost-effective feed ingredients. Chapters 1 and 2 provide a detailed account of the nutritional requirements of laying hens, with an emphasis on how to meet the industry goal of keeping flocks in production for a longer period – up to 100 weeks of age. **Chapter 1** reviews ways of optimising energy intake to achieve high production and the desired egg size and weight. The importance of nutrition during the rearing period is stressed as this has a lasting impact on subsequent hen performance. Egg production commences when the pullet is still undergoing its final stages of growth, including the development of the gastrointestinal tract. This competition for energy expenditure between growth and egg production requires careful balancing of the hens' diet. Breeder company's recommendations should be followed for the different phases during rearing, pre-lay and production, in order to optimise egg production, persistency and longevity. Attention must be paid to body weight and its uniformity at all hen ages. Modern hens tend to eat too little at the onset of lay but can be trained to eat more by including 5–6% crude fibre in the developer diet and by varying particle size. During the laying period, and in particular in extended production cycles, the challenge is to supply feed adjusted to energy and protein requirements to optimise egg production, and in particular egg weight, for commercial profitability, without increasing body weight and fatness. During the laying period, feed intake can be manipulated by adjusting energy concentration, particle size and diet texture. Egg weight is influenced by energy concentration, as well as dietary lipids, proteins and amino acids. Hens' diet can be based on a range of cereals and other protein sources including corn, soya bean, wheat, barley, sorghum, dried distiller's grains and solubles, peas, faba beans, white lupin, sunflower meal, cottonseed meal, rapeseed and even insects. Alternative, non-cage production systems, including organic production, pose particular requirements with respect to diet.

Chapter 2 concentrates on optimising hen's performance, bone quality and eggshell quality, with particular attention being paid to calcium, phosphorus, vitamins and minerals, fatty acids and yolk pigments, as well as dietary means of reducing the incidence of metabolic diseases (hepatic steatosis, osteoporosis, keel bone disorders, feather picking).

Calcium and phosphorus are critical requirements in the diets of laying hens, and the ratio of these two minerals is also of importance. Eggshells contain a high percentage of calcium, and bone contains both calcium and phosphorus. Providing sufficient phosphorus in diets devoid of meat/bone or fishmeal can be enhanced by the use of phytase enzymes, which make available more of the phosphorus contained in plant ingredients. The size of calcium particles is a critical factor during egg production, with at least two-thirds of large particles (1–2.5 mm for low solubility particles, 2–4 mm for highly soluble particles) being recommended in the diet. Vitamin D is important in calcium metabolism, and the use of these metabolites may enhance shell quality. Vitamins K, E, C and A also play vital roles. Magnesium and trace minerals such as copper, zinc, manganese and boron are required for optimal performance. Proprietary additives may, under some circumstances, provide beneficial effects. As birds cannot synthesise carotenoids, yolk colouration depends directly on the level of carotenoids consumed by the hen, its intestinal transfer efficacy, and the chemical composition of the carotenoid source, which determines its colouring capacity. In commercial production, carotenoids, natural or synthetic, are commonly added to the feed. The role of nutrition in the occurrence and prevention of metabolic diseases such as hepatic steatosis, osteoporosis, keel bone disorders and feather picking is discussed.

Chapter 3 provides an overview of the welfare of laying hens. Societal concerns about hen welfare have resulted in a trend away from conventional cage systems towards enriched cages and non-cage production systems in many countries. This has been driven mainly by concerns about confinement of hens and about the ability of hens to carry out natural behaviours such as nesting and dust bathing. However, all types of production systems represent a trade-off between welfare costs and benefits. Evaluation of the welfare of laying hens requires definitions and methods of measuring welfare. The authors point out that the scientific assessment of animal welfare involves multiple measures that capture different viewpoints on what constitutes a good quality of life for animals. These include measures of health and biological fitness (e.g. physical condition, mortality, production and indicators of stress), the emotional or subjective experiences of animals (e.g. conditions leading to pain, fear, discomfort, reward and pleasure) and the ability to perform natural or species-typical behaviour patterns. Conceptual frameworks for animal welfare are continuously evolving. The welfare trade-offs inherent in the different types of production systems are discussed. Conventional cages result in high levels of health and hygiene and low levels of mortality, but prevent birds from moving around freely and engaging in some natural behaviours. Non-cage systems permit birds to move around more freely, but increase the risks of injury and poorer health. Behavioural requirements of laying hens are assessed in a range of ways including preferences, strength of motivation and the extent to which a hen exhibits signs of negative mental state if she cannot access a resource or perform a behaviour. Four commonly agreed-upon behavioural needs of laying hens are nesting, foraging, dust bathing and perching. Injurious pecking can be influenced by environmental factors, nutrition, genetic predisposition, physiological states and rearing environments. Once it occurs, it can be controlled by two main means – beak trimming and dim lighting. However, beak trimming has its own welfare concerns, as discussed in Chapter 6. Poor skeletal health is a serious welfare issue for laying hens. The high rate of egg production means that laying hens have one of the highest rates of calcium turnover of any animal. Optimising calcium availability and metabolism and facilitating load-bearing exercise, as well as genetic selection, are key to prevention of osteoporosis in hens.

Chapter 4 focuses on welfare standards for laying hens. Laying hen production systems are categorised differently in different countries. In the European Union (EU), Council Directive 1999/74/EC categorises the common laying hen systems into three groups: alternative, un-enriched cage and enriched cage systems. This directive required the adoption of enriched or colony cages ('colony' cages in the United Kingdom house between 40 and 80 birds) in place of conventional cages. The European Food Safety Authority Scientific Opinion Statement reports on animal-based measures to assess the welfare of animals. In the United Kingdom, welfare standards include the Farm Animal Welfare Council, the British Egg Industry Council 'Lion Code', the Royal Society for the Prevention of Cruelty to Animals Freedom Foods Laying Hen Standard and AssureWel and Welfare Quality®. Global standards include the World Organisation for Animal Health (OIE) (Office International des Epizooties) Terrestrial Animal Health Code. Different mechanisms are in place to enforce implementation of the welfare standards. Consumers and supermarket chains have considerable influence in determining welfare standards. However, consumers are not always consistent in making purchases in line with the welfare standards they say that they support. AssureWel has identified the following as important indicators of hen welfare: feather loss, bird dirtiness, beak trimming, antagonistic behaviours, flightiness, birds needing further care and mortality. Welfare Quality® collects welfare data under the headings: good feeding (absence of prolonged hunger and absence of prolonged thirst); good housing (comfort around resting, thermal comfort and ease of movement); good health (absence of injuries, absence of disease and absence of pain induced by management procedures); appropriate behaviour (expression of social behaviours, expression of other behaviours, good human–animal relationship and positive emotional state); and classifies these categories as excellent, enhanced, acceptable and not classified. Welfare Quality® specifies what indicators need to be used and the order in which they are assessed on-farm.

Chapter 5 reviews welfare issues affecting free-range laying hens. Definitions of 'free range' vary among and within countries. They may involve static sheds with an outdoor area or mobile sheds that are rotated around pasture. Regulations or guidelines vary between countries in relation to indoor/outdoor stocking density, flock size, age of first access, daily hours of available access and so on, and in some cases, independent auditing bodies will have further stipulations (e.g. shelter on range) for certification labels. However, the international consensus is that free-range hens need daily access to outdoor open-air runs. The areas of major welfare concern for free-range laying hens reviewed in this chapter include individual differences in range use, behavioural expression on the range, correlations between ranging and health variables, parasite loads, adequate nutrition, grass impaction, feather pecking and cannibalism, and predation. Each of these issues were addressed in detail and, where possible, preventative measures suggested.

Chapter 6 reviews the costs and benefits of the commercial practice of beak trimming, which is a procedure routinely applied in the egg industry to prevent feather pecking and cannibalism. The traditional method of trimming by application of a hot blade (HB) has been superseded by infrared (IR) beak trimming. Both methods are associated with acute pain, but there is evidence that IR trimming has welfare advantages over HB trimming. In the past, HB beak trimming in early life was sometimes followed by a second trim in adulthood, but IR trimming is only applied to chicks. Bird-to-bird pecking has been categorised into five types: aggressive pecking, gentle feather pecking, severe feather pecking, tissue pecking and vent pecking. Although repeated aggressive pecks can cause feather and tissue damage, various studies have demonstrated that feather pecking and aggressive pecking have different underlying mechanisms; however, there is evidence of a genetic correlation between the two. There is general agreement that damaging feather pecking is related to foraging. Injurious pecking in laying hens remains a complex and intractable problem, and unpredictable pecking outbreaks are more likely and more severe in intact beak flocks. Beak trimming is still the most effective preventative strategy available, while it remains the case that injurious pecking cannot be reliably controlled under commercial conditions, and the author concludes that the welfare costs imposed by IR beak trimming are probably justified. Considerable research has been conducted into the welfare impacts of beak trimming. HB trimming causes initial acute pain followed by a period of more chronic pain. It also results in the formation of scar tissue and neuromas, which may result in chronic pain. IR trimming is conducted by application of a localised, non-contact IR heat source, which results in necrosis of the affected area causing loss of the beak tip within 2–3 weeks following treatment. The author summarises the advantages of IR over HB as: the beak tip gradually erodes, giving the bird time to adjust to the new beak shape, which may improve pecking accuracy and feeding efficiency; the process eliminates the presence of open wounds following trimming, preventing bleeding and reducing the risk of inflammation and infection; the automation and standardisation of the process reduces variability and the risk of human error or inappropriate handling. The effects of HB or IR on production parameters are not always consistent. Beak trimming usually results in reductions in damaging pecking and improved plumage but may also increase feed wastage and reduce the effectiveness of preening in removing ectoparasites. Research is continuing into alternative strategies to reduce the need for beak trimming.

Chapter 7 provides an excellent practical guide to maintaining the health of laying hens. Biosecurity is a critical factor in maintaining healthy flocks and the author categorises factors into separation, people, traffic control, equipment and vehicles, air, feed and water. Where possible, hens of different ages should be housed separately from one another and this can be facilitated by having breaks or buffer zones between housing facilities. Staff should enter the 'dirty' side of the farm or building, shower and then enter the 'clean' side before changing into coveralls, boots and hair coverings that are specific to that location. Birds should be reared separately from pets and pests that may be capable of transmitting diseases. Keeping the areas around poultry houses free from overgrown vegetation and use of rodent bait stations are standard precautions. All movements of people and vehicles should be strictly controlled and disinfection procedures followed. Detection of disease within a flock can be challenging, so staff need to be familiar with the symptoms of possible diseases in the flock. Common poultry diseases are respiratory diseases (e.g. infectious bronchitis, Newcastle disease, infectious laryngotracheitis, *Mycoplasma gallisepticum*, collibacillosis and infectious coryza), diseases of the nervous system (e.g. Marek's disease and Newcastle disease) and gastrointestinal diseases (e.g. coccidiosis, focal duodenal necrosis and intestinal parasites such as tapeworms and nematode worms). Other diseases of concern are the skeletal system disorder osteomalacia, urogenital system diseases such as gout and urolithiasis, fatty liver syndrome, vent prolapse and cannibalism, mites and egg production drops caused by avian encephalomyelitis. Vaccination is a critical tool in preventing disease in layer flocks and can be applied by a variety of methods: spray, in drinking water or by subcutaneous injection.

Chapter 8 reviews ways of managing flocks with intact beaks and complements Chapter 6. Chapter 8 further discusses the nature and causes of feather pecking as a good understanding of the underlying causes that are critical to identification of ways to prevent this potentially damaging activity. The author points out that beak trimming will not necessarily prevent pecking but rather reduces the amount of damage that pecking can inflict on other birds. Some clues can be obtained from observations of chickens in semi-wild situations. A genetic association appears to exist between fearfulness and feather pecking and there is evidence of an association for stress and feather damage between parents and offspring. Good management focuses on the prevention of the onset of feather pecking as, once it starts, it is very difficult to control. Rearing conditions are of vital importance with factors such as presence or absence of litter, stocking density, the complexity of the environment, climatic conditions, light intensity and feed (e.g. nutritional completeness and particle size). The transition period from the rearing facility to the layer house is also critical. Factors affecting feather pecking in the laying period are similar to those for rearing but the laying period is of a much longer duration. Type of housing, environmental enrichment, access to a free range, feed characteristics, management (particularly changes in management), light intensity, health status and ventilation all influence the incidence of negative pecking behaviours and need to be carefully controlled and monitored.

Part 2 Sustainability

Chapter 9 reviews waste management associated with commercial egg production. Wastes that need to be managed include manure, bird mortalities, egg washing and processing wastewater. Most of the nitrogen in poultry manure is contained in urea, uric acid and undigested protein; when these compounds are broken down, they release ammonia. Ammonia production from manure is influenced by strain, production system, ventilation, feed consumption, body weight, duration of egg production and egg size. Frequency of removal of manure from the housing system also plays a role. Manure may be stored for a period prior to distribution or further processing. Mortalities can occur as a result of diseases, accidents or natural disasters, and they need to be treated carefully in order to avoid spread of disease. Many countries practice egg washing, and different practices are reviewed and characteristics of waste wash water summarised in this chapter. Waste and wastewater can be treated by anaerobic digestion, or by aerobic processes such as composting and aerobic stabilisation of wastewater prior to use for irrigation. Thermochemical processes such as incineration, pyrolysis and gasification may also be used for treatment of waste. Dead hens may be burnt, composted or rendered. Land application of laying hen manure can be a controversial issue and it is important to consider the nutrient status of the land, possible run-off into water sources and the presence of contaminants such as antibiotics, pesticides, larvicides, endocrine-disrupting compounds and coccidiostats. An interesting case study is provided for anaerobic digestion of laying hen manure and mortalities. There is considerable scope for improved technologies for processing and utilisation of waste arising from egg production.

The last chapter, **Chapter 10**, assesses the sustainability of organic egg production. Organic food production has increased in popularity in the last few decades, although it still remains as a small portion of global food production. Because of concerns about animal welfare, there has been a movement away from the caged-rearing systems used in conventional egg production. Some would argue that organic production systems increase the welfare of the birds and, at the same time, reduce environmental impact. In the EU, organic farming regulations are set out under Council Resolution 2029/91 and its amendments and organic poultry production is dictated by EC 834/2007. For organic egg production in the EUS, poultry must be housed in free-range systems with stipulated furnishing and stocking density for both the poultry house and range, and a maximum flock size of 3000 hens. The US regulations do not restrict the size of the flock. In the EU, synthetic amino acids cannot be used, and both the EU and the United States ban the use of synthetic medications such as antibiotics, dewormers, insecticides and coccidiostats. Some areas allow vaccinations if they were not produced biotechnologically. In all countries, organic regulations require the feeding of diets made with organic feedstuffs. The authors discuss the sustainability of organic production using a life cycle assessment approach. The authors point out that 75% of conventional crop producers in the United Kingdom believed that organic farming methods are better for the environment but only 13% thought that such techniques could produce sufficient food and fibre for society. Conversely, only 73% of organic crop producers in the United Kingdom interviewed believed that organic production systems could produce the necessary food and fibre for the growing world population. Organic egg production will most likely remain as a niche market.

Summary

The chapters contained in Part 1 of Volume 2 review the relevant aspects of animal health and welfare. Nutrition is of crucial importance to hen welfare and the profitability and sustainability of the commercial egg industry. Concerns about hen welfare centre on the type of housing system, the costs and benefits of beak trimming and the maintenance of good health in the flocks. As outlined in Part 2, the sustainability of the industry relies not only on the provision of housing and suitable feed ingredients but also on the management of waste arising from the industry. The final chapter reviews the sustainability of organic egg production and concludes that such production is likely to remain at a relatively small scale.

Part 1

Animal health and welfare

Chapter 1

Laying hen nutrition: optimizing energy intake, egg size and weight

Y. Nys, Institut National de la Recherche Agronomique (INRA), France

1 Introduction

The priority for producers, breeding and feed companies is to increase egg production. Continuous improvement in egg number has been initially obtained by selecting hens for early sexual maturity and improved peak of production. However, nowadays, the priority of breeding companies is to increase persistency in lay and maintain egg quality in older hens (Bain et al., 2016). Selection of egg persistency, combined with improvement in eggshell and internal egg quality in hens at older ages, have enabled egg production to be extended to more than 80 weeks of age (Dunn, 2013; Pottguetter, 2015). Currently, the aim is to extend the laying cycle of commercial flocks to 100 weeks to get more than 500 eggs in a single cycle without moult. This corresponds to exportation of more than 26 kg of eggs (about 14-fold the hen's body weight) and more than 1 kg of calcium (3 kg of eggshell).

This goal challenges the metabolism and homeostasis of the hen and some organs, in particular the liver where yolk protein and lipid synthesis takes place and the oviduct where the eggwhite, eggshell membranes and eggshell are made. Optimal eggshell quality is the main reason to terminate a layer flock, rather than the persistency of egg

http://dx.doi.org/10.19103/AS.2016.0012.20

production. One strategy to extend the laying period for commercial egg production might be to sell eggs from older hens as egg product. In such a situation, internal quality (Haugh unit and strength of the vitelline membrane) will be the limiting factors. Nowadays, producers should also take hen welfare and health more into consideration. Hen nutrition is crucial, not only in optimizing the excellent genetic potential of modern hens in terms of productivity and high egg quality but also for avoiding bone demineralization or fatty liver syndrome in hens producing high numbers of eggs for more than 80 weeks.

Hen nutrition influences egg composition, especially that of minor components. The possibility of modifying global egg composition (protein, energy) is rather limited because of the stability in egg composition for major components and the limited possibility of modifying the proportion of egg white and yolk. In contrast, addition of certain feed constituents directly influences their levels in the egg but will not be discussed in this chapter. Egg composition can be modified by hen nutrition as shown for fatty acids (Sirri and Meluzzi, 2011; chapter 16) and for nutrients that are present in small quantities such as vitamins and trace elements (Schiavone and Barroeta, 2011; chapter 16).

The early period before sexual maturity influences subsequent hen performance and egg quality. The early onset of sexual maturity in hens and extension of the production cycle requires management of the growth curve of the pullets, their feed intake capacity and their weight at the onset of lay. The objective is to avoid any competition between pullet growth and requirements for egg production and to optimize initial egg weight, then to control egg weight changes, shell quality and bone mineralization. Energy and protein supply influence hen performance and egg quality.

During the laying period, the most important nutritional tool supporting calcium metabolism and eggshell quality is high-quality limestone with fine and coarse structure. Extension of the laying period also reinforces the need to preserve by nutrition liver health and intestinal tract function by adjusting the feed composition or presentation for optimizing egg mass and avoiding metabolic or physiological dysfunction such as hepatic steatosis, osteoporosis, keel bone and feather picking. This latter risk might be reinforced in the near future by the expected ban on beak trimming.

The aim of this chapter is to update knowledge on hen nutrition taking into account the extension of the laying period and the need to prevent any unexpected metabolic pathologies of organs highly utilized by the process of egg formation. This chapter focuses on the role of nutrition in optimizing energy intake, egg size and weight. An accompanying

Table 1 Estimation of egg mass, yolk, shell and calcium exported by layers through their commercial laying life expressed in kg or multiples of bodyweight (xBW), based on ISA Brown Standard

		Egg mass		Yolk		Shell		Calcium	
		kg	xBW	kg	xBW	kg	xBW	kg	%BW
Weeks of age	60	15.7	8	5.2	2.8	1.89	1	0.7	37
	70	19.3	10	6.4	3.4	2.35	1.2	0.9	46
	80	22.6	12	7.5	4	2.82	1.5	1	55
	90	25.7	14	8.6	4.5	3.29	1.7	1.2	65

Chapter 2, Vol. 2 discusses the role of nutrition in optimizing hen performance, bone strength and eggshell quality.

2 Feeding the pullet and the pre-laying phase

The rearing of the pullet, including its feeding, will influence the growth curve, live weight and body composition at sexual maturity, parameters which will impact egg production and quality. It is well established that the body weight of hens at the onset of lay is highly correlated with initial egg weight and therefore influences the egg mass ($r^2 = 0.85$, $P < .01$; 0.7 g for each 100 g gain in live weight of the pullet calculated by meta-analysis; Bouvarel et al., 2011; Bouvarel and Nys, 2013). A similar correlation is also observed but to a lesser extent when considering the egg weight in mature hens (Bouvarel et al., 2011), demonstrating the long-term effect of pullet feeding on egg mass. This is partly explained by the competition in energy requirement for the late growth of the pullet and the initiation of egg production, combined with the limited capacity for feed intake in young modern hens. Reducing the need for pullet growth at the onset of lay will save dietary protein and energy for the process of egg formation.

An additional strategy will be to reinforce, during the intermediary period (9–16 weeks), the development of the gastrointestinal tract and the hen's capacity to increase feed intake. A higher feed intake is needed during the transition phase (16–20 weeks) at sexual maturity when the protein, energy and calcium requirements are greatly increased for production of eggs. The pullet diets should therefore be adjusted throughout the rearing phase, initially to ensure the development of muscle and bone until about 8 weeks, then at sexual maturity for the development of the reproductive organs and bone reserve (>15–16 weeks). During the intermediary phase (9–16 weeks), the nutritional requirements for growth are lower, and this period can be used for developing the gastrointestinal tract and the capacity of birds to eat more feed for optimizing the transition phase to egg laying.

The initial growth of internal organs, muscles and the active development of the skeleton until about 8 weeks are heavily dependent on protein and amino acid intake as well as calcium and phosphorus supply. Studies exploring sequence of dietary level at different ages have been reported for proteins. Some authors demonstrated an increase of 0.32 g of body weight/supplemental protein point when supplying 11 to 20% dietary proteins throughout the pullet phase (Keshavarz and Jackson, 1992; Summers and Leeson, 1994). During the first period (< 8 weeks, starter and grower diet; starter being slightly more dense in energy and protein than grower diet) an optimum level of energy (2950–2800 kcal of AMEn/kg corresponding to 12.33–11.7 kJ), protein (20–18%), digestible amino acids (0.48–0.4 methionine; 1.2–1.0 lysine) and mineral (1% calcium) should be provided to ensure growth of the different tissues. Levels depend on breed and should be adapted following breeder recommendations to reach optimum pullet weight. The supply of phosphorus should be high enough (>0.4 NPP) during the starting and growing period to ensure bone growth. A low phosphorus level during the first 10 weeks of age increases mortality of mature hens and will increase incidence of osteoporosis in adult birds (Nys, 2014).

The switch to developer diet should be carried out when body weight target is reached for a genetic line, rather than at a fixed age of the pullets as recommended by breeding companies. During the intermediary phase (9–16 weeks), the protein and mineral requirement are lower and the ratio of protein to energy should be optimized to avoid fattening of the birds as increased energy content of the diet enhances the fattening scores. However, pullets fed on variable dietary energy levels (2850 to 3000 kcal/kg) will decrease feed intake when fed higher energy level as a crumble diet, but this ability will be limited when fed on a mash diet (Guzman et al., 2015a). Guzman et al. (2015a) also demonstrated an improvement in feed conversion ratio when the average feed intake was reduced (crumble diet). There is evidence that the relationship of body development at 12 to 13 weeks of age with egg weight will be stronger than that at 18 weeks because nearly 95% of the body frame is developed at this pullet age (Thiele, 2012). Lower body weight at this stage of development is likely to be still present at 18 weeks and if so, will have a negative effect on egg weight.

It will be more difficult to compensate for a lower body size after 12 weeks by increasing energy feed density because overconsumption of feed will favour the development of fat and such change in body composition has been associated with a lower egg weight from 20 to 64 weeks of age (Cheng et al., 1991). In addition, pullets grown on high-energy developer diets might exhibit poor feed consumption at the onset of lay (Elliot, 2002), possibly because such diets poor in fibre will not facilitate digestive organ development.

The developer diet should contain 5 to 6% crude fibre to stimulate the development of the intestinal tract, and this period will be used for training birds to eat more as modern hens show difficulties in consuming enough feed at the onset of lay (Bouvarel and Nys 2013; Pottgueter, 2015). Detailed studies explored the optimum period, level and type of fibre supply between 0 and 17 weeks of age. Guzman et al. (2015b) showed that a continuous supply of fibre (2 or 4%, cereal straw or sugar beet pulp (SBP)) neither affected pullet performance (except for a negative effect of 4% SBP) nor did it affect the flock uniformity for body weight, and confirmed a slight increase in gizzard weight with fibre inclusion. Guzman et al. (2015a) showed that the inclusion of fibre, particularly at 2%, slightly improved pullet performance.

The stimulation of the gastrointestinal weight depends on the duration and period of fibre supply. A positive effect is observed when pullets are fed for at least ten weeks (0–10 weeks of age) with insoluble fibre (1%, Yokhana et al., 2016). Yokhana et al. (2016) also clearly demonstrated the stimulation of gastrointestinal proteases (pepsin, trypsin, chymotrypsin) by dietary fibre allocated for 10 weeks and at a lower magnitude when only fed for the first 5 weeks. When the fibre was introduced only at 17 weeks of age, a positive effect on liver weight was observed three weeks later but only after 12 weeks at the gizzard level.

The particle size is also a convenient way to stimulate feed intake and digestive capacity. During the initial period of pullet rearing (<45 days of age), small particle size obtained by grinding of main cereals (screen size <8 mm) is more suitable to beak size and favours feed intake to optimize weight gain (Frikha et al., 2009, 2011). In contrast, during the intermediary period before sexual maturity, use of whole cereals promotes feed intake and develops digestive tract capacity. This can also be reinforced by feeding management, limiting the amount of feed distribution to force the pullet to empty the feed trough and to stimulate their appetite. A crumble diet distributed for the first 5 weeks of age improves feed conversion ratio (5 to 8%) relative to a mash diet (Guzman et al., 2015a). This was confirmed when diets were allocated for 17 weeks by the same research group (Saldana et al., 2015) who observed an higher average body weight and lower FCR (3–4%) with crumble compared to a mash diet, but a slightly lower gastrointestinal tract development. Bone growth is limited during the intermediary period, and there is no evidence that increasing dietary phosphorus or calcium in the developer diet improves future hen performance or bone mineralization; therefore, dietary phosphorus can be lowered to 0.25% NPP and calcium to 0.8%. The stimulation of sexual maturity through the lighting programme is important to optimize pullet weight when the first eggs are laid. There is a strong relationship in individual hens between their early body weight and their initial egg weight, as shown when sorting hens on body weight (Galea, 2014). This correlation was observed throughout the laying period, underlining the importance of the combination of lighting programme and nutritional factors for reaching the correct body weight at sexual maturity.

After 16 weeks, the energy–protein ratio must be adjusted to supply sufficient feed and cope with growth of reproductive organs and the onset of egg production. Protein levels should be increased to 17–18%, keeping similar energy content to that in the developer period to stimulate feed consumption (Thiele, 2012). Sufficient early nutrient intake is essential for a successful start and to reach high peak rates of production. In addition, management of calcium supply between the pullet and laying stages is important for the development of medullary bone and for supplying calcium needed for eggshell formation. It also affects egg weight through its effects on feed ingestion during this period.

It is recommended to use a pre-laying diet containing 2.5% Ca for a maximum of 10 days (1 kg/pullet) at the very beginning of egg production (15–16 weeks of age) to ensure a transition between the developer and layer diets. A smooth transition to the high calcium amounts of a layer diet by using a pre-layer diet with only 2–2.5% calcium helps to maintain feed intake and egg weight, will improve the eggshell strength and is a tool to increase the uniformity of a flock. The pre-layer diet should be introduced at 15–16 weeks when the pullet is developing the medullary bone under the control of sex steroids at sexual maturity. The duration and amount of pre-layer diet should be adapted to the onset of egg production. If there is any delay in pullet transfer or earlier onset of lay, it is recommended to shorten the period of distribution of the pre-layer diet and introduce the layer diet with the risk, however, of slightly decreased feed consumption. Use of large particle size of calcium distributed on top of the formulated diet can limit this risk especially when the initiation of egg production is heterogeneous in a flock. Evidence for a positive effect of increasing dietary phosphorus is scarce in the early literature (Sauveur, 1987), even if the medullary bone formation induces a higher requirement for phosphorus. However, Rao et al. (1995) clearly demonstrated increased bone breaking strength and density when total phosphorus was increased from 0.4 to 0.55% TP (0.18 to 0.32% NPP). In addition, there is an increased retention of Ca and phosphorus by the pullets during the 2–3 week period before onset of lay as the result of renal physiological stimulation of the active metabolite of vitamin D3. It is, therefore, recommended to slightly increase the dietary phosphorus (0.3% NPP) when considering the physiological demand for medullary bone formation, to limit the resorption of cortical bone (Nys, 2014).

3 Feeding the laying hen

During the egg laying period, in particular when considering extended production cycles, the challenge will be to supply feed adjusted to energy and protein requirements to optimize egg output. It is crucial, to control egg weight by nutrition for commercial profitability but without increasing body weight and fatness or favouring hepatic steatosis and bone demineralization. In addition, nutrition system should maintain good feathering by limiting feather picking. Feed formulation needs to take into account the expected egg production and feed intake to ensure the daily requirement for protein, digestible amino acid (lysine, methionine, methionine plus cysteine, arginine, isoleucine, tryptophan, threonine, valine), calcium, non-phytic phosphorus, sodium, linoleic acid, unsaturated fatty acids and other additives. Feed composition will be adjusted continuously to daily feed intake and for controlling egg weight or quality. Another difficulty in old flocks results from the increased heterogeneity in individual egg production with age. Nutritionists should adjust the feed composition to keep in lay the most highly productive hens.

The energy requirement of laying hens is mainly due to maintenance and egg production. It is relatively stable throughout the laying period because of the increased maintenance and lowered egg production requirements after 35 weeks of age when hens have completed their body growth. The requirement tends to decrease slightly as egg production progresses. It is influenced by body weight, egg production and to a lesser extent by environmental temperature, activity and hen feathering. It has been estimated by Joly (2005) in brown egg layers:

$$\text{Daily energy requirement (kcal)} = 130 \text{ body weight (kg)}^{0.75} + 1.8 \text{ egg mass (g)} + 6 \text{ daily weight gain (g/day)} \text{ (temperature at 22°C)}$$

For white leghorns, the coefficient relative to body weight is 137. The energy cost per gram of egg is therefore estimated to be 1.8 kcal. Environmental temperature also affects the energy requirement. At temperatures lower than 22°C, it increases by 2 kcal per degree and kg body weight, decreases by about 1 kcal between 24 and 28°C but increases at higher temperatures depending on thermoregulation (Joly, 2005). In alternative systems, the requirement for maintenance is higher by about 1.2 kcal/d due to locomotor activity (free-range compared to conventional system). Feed consumption can be increased by 10–20% in free-range systems compared to indoor systems (MacLeod and Bentley, 2012). The energy requirement in an organic system, relative to a conventional one, increases by 15% (Leinonen et al., 2012). Feather loss is frequently observed in flocks of old hens; this increases the maintenance energy requirement and decreases the feed efficiency of layers at normal temperature (Joly, 2005). A lower feather covering affects the insulation of hens, favouring heat loss and increasing energy requirement. A lower feather coverage induces an increase in feed consumption, the magnitude of which depends on environmental temperature: +20 g at 12.8°C if the feather coverage corresponded to 50% normal or +7 g at 23.9°C (Peguri and Coon, 1991, 1993).

4 Control of feed consumption in laying hens

4.1 Energy concentration

Feed intake is crucial in determining the daily energy consumption that influences egg mass and hen fattening. Hens have the ability to adjust their feed consumption based on dietary energy concentration (Joly, 2005; Thiele, 2012; Bouvarel et al., 2011). Energy concentration can therefore be modified within a relatively large range, and the choice of a dietary energy level will be largely influenced by economic considerations. The meta-analysis carried out by Bouvarel et al. (2011), on more than 15 papers collected between 1991 and 2009, quantified the linear relationship between the changes in feed consumption induced by modification of dietary metabolizable energy (± 30%):

$$\text{Difference in consumption (\%)} = 1.452–0.685 \times \text{difference ME (\%)};\ R2 = 0.875$$

A large change in dietary ME was experimentally induced by introducing sand or adding fat. It is noteworthy that this adjustment is linear but is only partial as hens cannot totally compensate a high dilution of the diet or, conversely, decrease their feed intake of high-density diet to reach the same energy consumption. In brown egg-laying hens, the decrease in energy intake was estimated to be on average 1.2% when dietary energy concentration was decreased from 3000 to 2200 kcal/kg (Joly, 2005). It is also possible to increase the daily energy intake by enriching the diet with fat (1.3%/100 kcal; Joly and Bougon, 1997; Grobas et al., 1999a,b; Leeson et al., 2001; Valkomen et al., 2008; Van Krimpen et al., 2009) and this option is used for supplying enough energy at onset of egg production or for increasing the egg weight. The adjustment of feed intake to lower energy level (between 2950 and 2650 kcal/kg; other nutrients being similar per unit of energy) was confirmed by Perez-Bonilla et al. (2012) in brown egg-laying hens whatever their initial body weight (1.73 or 1.3 kg): the feed intake increased when energy levels were below 2850 kcal. However, the compensatory increase in feed intake did not alleviate a reduction in egg production when energy level was 2650 kcal/kg. In addition, the FCR was optimized when the hens were fed with at least 2850 kcal. In leghorn hens, this capacity of adjusting feed intake to energy dilution (until 10% less AME; other nutrient density, protein, amino acids, Ca and P being also similar by energy unit) was not observed (de Persio et al., 2015). de Persio et al. (2015) observed a linear reduction in egg production and weight when diets were diluted by 5 or 10%.

4.2 Particle size and diet texture

The feed consumption of laying hens not only depends on energy requirement but is also influenced by their preferences and ability to recognize particles in the diet. Poultry feed particles must be large enough to be recognized and picked up (Rogers, 1995). Laying hens clearly select their food intake according to the relative size of the particles in relation to the beak. In addition, distribution of large particles allowed hens to express specific appetite as demonstrated for calcium during the period preceding eggshell formation (Mongin et Sauveur, 1979; Wilkinson et al., 2011). Calcium is preferentially eaten during the late afternoon either as particles when available or whole diet intake is stimulated when calcium cannot be discriminated from other feed components. By using feeds presented as crumbs (72% of particles >1.18 mm), Portella et al. (1988) demonstrated an immediate preference of laying hens for the largest particles.

On the other hand, hens do not demonstrate a preference for the largest particles when the feed is presented as a fairly homogeneous meal in which 'mean-sized' particles are mainly between 0.60 and 2.36 mm. Small particles (< 0.60 mm) disappear slowly from the trough whatever the feed presentation (Portella et al., 1988). These flour particles contain the micronutrients such as vitamins, minerals and amino acids, and percolate to the bottom of the feeding trough (Tang et al., 2006). This problem is alleviated by addition of water or fat to adhere the small particles (Tang et al., 2006). Distribution of a fine rather than a more coarse meal (31% vs 9% particles < 0.5 mm) reduced feed consumption, laying rate and egg weight (−0.9 g) (Joly, 2008). Safaa et al. (2009) also showed that feed intake increased when coarser particle size was slightly enlarged (13–16% vs 21–22% particles <0.63 mm), without any effect on hen productivity. The grinding method of the cereal of the diet might impact feed consumption depending on the cereal type. The consumption in hens fed wheat or maize was slightly higher than in hens fed barley, especially when the cereals were ground by a roller mill compared to a hammer mill (Perez-Bonilla et al., 2014).

This ability of hens to eat more food when it is presented as larger particle can be used to stimulate feed consumption when diets are low in energy and enriched in fibre. Hens will eat more of feeds diluted with fibre (45% wheat bran) when it is presented as pellet compared to meal (Vilariño et al., 1996). Moistened feed is rarely used in laying hens, as it might be a risk for egg hygiene. It has been shown to stimulate intake of dry matter (Tadtiyanant et al., 1991; El Kaseh and Forbes, 1995) but, if fermented and consequently showing an acidic pH, reduces ingestion of dry matter in hens 18–37 weeks of age (110 vs 125 g/day) (Engberg et al., 2009).

It is noteworthy that birds need to be trained to a new type of diet. They are able to learn to recognize their feed and their ability to adapt can be developed to better control their feeding behaviour. A sudden change in feed, for example, from flour meal to large particles (>2.35 mm), can result in an immediate reduction in consumption (−28%) which, in turn, will affect egg weight (Portella et al., 1988). Some authors advised exposing pullets during rearing to all the grains that they may be offered later in life (Forbe et Covasa, 1995). It has also been demonstrated that the characteristics of the previous diet affect feed intake and performance of choice-fed birds.

Crumbled or pelleted diets will stimulate feed consumption in hens, avoid consumption of fine particles but will reduce the duration of time taken for feed consumption, which can favour feather picking. Ileal digestibility of starch was also slightly improved by feeding hens coarse compared to fine feed (Ruhnke et al., 2015). Therefore, meals with larger particles of cereals are more frequently used than crumbles. The physical form of the feed therefore interacts with the ability of the hen to match its energy intake to the energy concentration of the feed.

5 Dietary effects on egg weight

5.1 Effect of energy concentration and dietary lipids

Dietary energy concentration affects egg mass mainly through its effect on egg weight, as its effect on egg production is rather limited (about 1%, Joly, 2005). It mainly affects feed consumption, which affects egg mass, through egg weight and feed conversion ratio (+0.8% per 100 kcal/kg when using diluted diet). Dietary energy concentration can be modified throughout the laying period by supplying dietary lipid or fibre to adjust the egg weight to commercial demand or to improve welfare parameters in hens. The impact on egg weight of decreasing energy concentration was estimated by Joly (2005) to be 0.4% per 100 kcal/kg diet. When considering the daily energy intake, a mean variation of 0.96 g in egg weight was calculated when the daily energy intake was increased by 10 kcal, based on a meta-analysis carried out on 15 references (Bouvarel et al., 2011):

$$\text{Egg weight (g)} = 32.58 + 0.096 * \text{energy consumption (kcal/day; R2} = 0.328)$$

This was observed on hens with different genetic background and level of production as data were collected between 1991 and 2008. In recent studies, an effect of energy concentration was observed mainly when energy concentration was reduced to 2650 kcal/kg (Perez- Bonilla et al., 2012) or lower than 2800 kcal/kg (de Persio et al., 2015). A high-energy diet during the first part of the laying period is therefore used to stimulate initial egg weight. This might be partly due to the enrichment of lipids that are used for increasing energy concentration; however, Perez-Bonilla et al. (2012) did not observe change in egg weight when supplying 3.6% fat compared to 1.8% in an isoenergetic diet. Conversely, a decrease in lipid and energy level is used to reduce egg weight in older hens, which might partly explain the tendency of hens to progressively increase their feed intake with increasing age.

Dietary fatty acid profile influences egg weight. It was demonstrated that linoleic acid has a positive effect on egg weight when added at 1% in the diet (Whitehead, 1981; Grobas et al., 1999a,b). The response is amplified when both linoleic and oleic acids are combined in the diet (Whitehead, 1981; Dänicke et al., 2000). These fatty acids are suspected to increase the retention of other fatty acids and are particularly efficient at the onset of egg production for stimulating egg weight (Whitehead et al., 1991). However, Joly (2005) attributed this positive effect to lipids in general rather than specifically to linoleic acid, in agreement with Grobas et al. (1999b), who demonstrated no change in egg weight when supplying similar levels of lipid with a variable amount of linoleic acid. The incorporation of lipids stimulates energy consumption by improving texture and appetence of diets in particular by aggregation of fine particles. In older hens (94–106 weeks), there is, however, evidence that linoleic acid did not affect egg weight (Yousefi et al., 2006). Buitendach et al. (2015) also observed no change in egg weight at the end of lay (58–74 weeks of age) in hens fed 3% lipids throughout the whole laying period (from 20 to 74 weeks), in isoenergetic diets enriched either with polysaturated fatty acids (linseed and fish oil), oleic acid (sunflower oil) or saturated fatty acid (tallow).

5.2 Effect of dietary proteins and amino acids

Egg weight and egg mass are also influenced by protein and key amino acid supply to the hens. A daily additional gram of protein increases average egg weight by 1.4 g when compiling numerous references (Bouvarel et al., 2011):

$$\text{Egg weight (g/day)} = 35.55 + 1.347 * \text{crude protein consumption (g/day)}, R2 = 0.58$$

In brown egg-laying hens, the increased protein level stimulated the egg weight between 16.5 and 18.5% without affecting egg production and the effect was larger in heavier hens (Pérez-Bonilla et al., 2012).

Amino acid dietary level affected the number and size of eggs. Any deficiency in amino acids will, at first, negatively impact egg production and therefore cannot be used for decreasing egg weight. However, egg weight responds to the concentration of first limiting amino acids in the diets, the response following an asymptotic exponential curve; therefore, egg size can be controlled by precise supply of amino acids (methionine, cysteine, lysine, isoleucine, tryptophan, valine and threonine). Several studies have evaluated the effects of methionine content on egg weight at peak production by adding DL-methionine to diets with sufficient protein content (Schutte et al., 1994; Bertram et al., 1995a,b; Dänner and Bessei, 2002; Narvaez-Solarte et al., 2005). Egg weight follows a curvilinear relationship

Table 2 Laying rate at 80 weeks of age of 1366 pure line layers hatched in 2010 – ISA-Hendrix internal data

Quintile	Laying rate (%)	Daily egg mass (g/d)
1st	97.17	60.22
2nd	91.19	55.26
3rd	86.30	51.90
4th	79.03	47.77
5th	55.64	34.04
Average	81.87	49.84
% hen above the average	65.80	62.08

with methionine content, with a plateau at 0.36–0.38% of methionine in the feed (Bouvarel et al., 2011). Previously, Joly (2002, 2008) estimated the methionine requirement to be 7.5 mg/g egg when compiling 15 papers published between 1990 and 2005 and underlined that methionine requirement was preponderant relative to that of cysteine. Using a similar approach, Joly (2002, 2008) estimated the requirement in lysine at 15 mg of total lysine/g egg, corresponding to 13.3 mg digestible lysine.

An estimation of the daily requirement, relative to the exported egg weight, facilitates the adjustment of protein and amino acid supply to the change in feed intake and hen requirement at a given age. In hens, it is important to adjust diet composition to the expected feed consumption and to record the daily feed intake and control its level by management of feed distribution, particle size or sequential feeding. Moreover, feed formulation is calculated for a flock, not on individual hens. There is an increased heterogeneity in the flock with increasing age (Table 2) (a mean egg production of 82% can correspond to only 60% of hens laying between 86 and 97% and 40% laying between 40 and 80%). That raises questions about the nutritional strategy. The best strategy might be to feed high producing hens to reach their potential, assuming that low egg-producing hens will reduce their feed intake sufficiently to limit the risk of additional cost induced by consumption of higher density diets.

The proportion of albumen and yolk in the egg depends mainly on hen's age. The yolk can increase from 23 to 28% at about 30 weeks after initiation of egg production (Zita et al., 2009). However, it is also affected by the dietary protein content but to only a small extent. A reduction to 13–14% dietary protein slightly reduced the albumen proportion ($< 1\%$) and did not change yolk percentage (Novak et al., 2006). When the egg weight was increased by methionine or lysine, it affected both egg components similarly (Shafer et al., 1996).

6 Effect of cereals on egg quality

6.1 Cereals

Corn and soya bean are most frequently used in laying hen diets but numerous cereals or protein sources can be substituted in order to utilize local sources. Total substitution of corn by wheat or barley in a laying hen diet (Safaa et al., 2009; Perez-Bonilla et al., 2011) did not influence egg weight, proportion of compartments within the egg, eggshell quality or internal quality (Haugh unit) as long as protein, energy, methionine, lysine and linoleic acid are adjusted and when the diets containing barley and wheat are supplemented with enzymes (Lázaro et al., 2003). Frikha et al. (2009, 2011) also demonstrated that wheat can be substituted in pullet diets with, however, a slight decrease in body weight.

6.2 Dried distiller's grains with solubles

The increased use of corn grain for ethanol production has impacted the corn price and has greatly increased the availability of its co-product, the dried distiller's grains with solubles (DDGS). DDGS contains, on average, 27% CP, 10% oil and 8% starch (Parsons et al., 2006; Salim et al., 2010). Corn DDGS can be fed to laying hens up to 15–20% as a substitute for corn and soyabean with no adverse effect on egg production (Salim et al., 2010; Masa'deh et al., 2011). Substitution of 20% DDGS for corn and soyabean meal has been shown to reduce daily NH_3 emissions by 24% and H_2S by 58% (Wu-Haan et al., 2010) and also improved N retention (Masa'deh et al., 2011). DDGS, however, showed a high variability in composition and nutritional value for metabolizable energy (2500 to 2900 kcal) due to its lipid (7–12%), fibre (5–11%) and starch composition (4 to 12%). High variability was also observed for CP (23 to 32%) and for the level and digestibility of the limiting amino acids for poultry, in particular methionine and lysine (Parsons et al., 2006; Salim et al., 2010). These variations are due to differences in proportion of grain versus solubles, owing to the production process and to residual starch content depending on fermentation efficiency.

High dietary levels of DDGS, up to 69% (experimental diets being isocaloric and adjusted for digestible methionine and lysine), stimulate feed and protein consumption and egg weight, but decrease egg production, without any effect on egg mass (Pineda et al., 2008). However, the incorporation of levels higher than 20%, in isocaloric and isonitrogenous (including lysine and methionine) diets, can drastically reduce egg production and egg weight (Sun et al., 2012) but the negative effect was alleviated when diets were adjusted with additional methionine. Masa'deh et al. (2011) also underlined the interest in adjusting methionine levels in DDGS diets to avoid lower egg weight.

In Europe, wheat distiller's grains with solubles (wheat DDGCs) are also available and can be used in poultry even if they are more suitable in ruminant feeds than pigs or poultry that have particular needs for amino acids. Wheat DDGS contained 36% CP (32 to 39%), 4.4% lipids (3 to 5%), 7.6% crude fibre (6–9%) and 5% starch (2.5 to 10%) but their composition was also quite variable (Noblet et al., 2012). Most amino acids of wheat DDGS are considered to be slightly less digestible than those in corn by about 5–10% when measured in caecectomized cockerels or forced fed roosters (Cozannet et al., 2010, 2011). These authors observed similar egg mass (140 hens at 25 weeks of age, 11 diets, 12 birds per replicate) and AMEn value of 9.6 Mj/kg for layers (7.8 to 10.5) when comparing 10 European wheat DDGS and proposed a prediction equation ($AMEn = 3.32 + 0.12\ L^*$, L being luminance of the sample or $AMEn = 12.52 - 0.24ADF$ (%), ADF being the acid detergent fibre). Globally, wheat DDGS nutritional values are expected to be quite similar to corn DDGS because of their numerous similarities in composition even if its energy value is lower, but there is little experimental work exploring the use of wheat DDGS in poultry (Noblet et al., 2012; Oryschak et al., 2010).

In broilers, wheat DDGS are equivalent to corn DDGS for energy and nutrient apparent ileal digestibility up to 15% of dietary supply and can be incorporated into diets to at least until 10% without any detrimental effect on broiler performance (Oryschak et al., 2010). However, in hens, the literature is scarce.

7 Alternative protein sources and egg quality

Soyabean is the most commonly used protein source for poultry but its price instability and its fluctuant availability has stimulated interest in alternative protein sources, especially in Europe which imports more than 95% of soyabean meal for poultry production. The main characteristics and nutritional value for monogastrics of legume seeds has been summarized by numerous authors (Lacassagne, 1988a; Gatel, 1994; Perez-Maldonado et al., 1999, Peyraud et al., 2015, Heuzé et al., 2016).

7.1 Peas

Peas have a high starch content (48–54% DM) and are relatively low in fibre (NDF 10–18% DM), have low levels of anti-tryptic activity (2–6 TIU/mg) and very low levels of tannins (Feedipedia, 2016). Peas can be an alternative to soyabean meal. Their amino acid profile is well balanced in lysine but deficient in tryptophan and methionine. Incorporation of peas at levels of 10, 15, 20 and even 30% does not affect the total number of eggs but can decrease egg weight up to 3.1% (Lacassagne, 1988a). This is in agreement with Koivunen et al. (2015) who observed similar egg quality and hen performance throughout the laying cycle (21–70 weeks), a slight increase in FCR for one period (41–57 weeks) and in final body weight (+3%) for hens fed peas. In contrast, Fru-Nji et al. (2007) observed a decreased body weight gain in proportion to pea inclusion (15%) but no negative effect on egg quality and production until 50% of dietary pea inclusion.

7.2 Faba beans

Faba bean seed composition is quite similar to peas, rich in protein (25–33%) and starch (40–48%) with moderate content of fibre (7–11%) and high lysine content (5.4–6.8%), but low level in sulphur-containing amino acids (0.6–1.0% methionine). Faba beans contain not only a high proportion of linoleic and linolenic acids but also condensed tannins (5 to 10 g/kg), vicine and convicine (6 to 14 g/Kg but more than 10 fold less in low V+C cultivars; Duc et al., 1999). Dehulled faba beans contain more protein and starch and much less fibre (about 50% less). Faba beans decreased egg weight in proportion to the level of inclusion when supplied at levels higher than 7%, because of the presence of anti-nutritional factors such as vicine–convicine (Lacassagne, 1988a; Lessire et al., 2016), and greatly reduced egg mass and production when supplied from 24 to 40% dietary level (Fru-Nji et al., 2007). Egg weight was greatly reduced when 20% of faba bean containing high levels of V+C was incorporated but was similar to the control diet when varieties with low V and C were used (Lessire et al., 2016). Koivunen et al. (2014) also observed a decreased egg weight when providing 5 or 10% faba beans, a similar egg mass but feed conversion tended to be higher at 10% inclusion. Dietary levels up to 30% did not, however, affect egg weight when the faba bean variety contained low vicine and convicine concentrations (Dänner, 2003) or when dehulled-micronized faba beans (24%) were used (Laudadio and Tufarelli, 2010).

7.3 White lupin

White lupin is suitable for poultry nutrition because of its high protein level, low alkaloid level and high level of unsaturated fatty acids when its amino acid imbalance is corrected by an adequate supplementation of limiting amino acids (lysine and sulphur amino acids). However, it contains high non-starch polysaccharide concentrations (Lacassagne, 2008a; Laudadio and Tufarelli, 2011; Nalle et al., 2012; Krawczyk et al., 2015). Blue lupin meal (Hammershøj and Steenfeldt 2015) supplied up to 15% had no negative effect on hen performance or egg quality but depressed these parameters when included at a higher level because of amino acid imbalance. Perez-Maldonado et al. (1999) also demonstrated that inclusion of sweet lupin seeds (25%) in laying hen diets did not depress egg production and quality. That was confirmed by Krawczyk et al. (2015), who supplied 10, 20 or 30% of lupin seed meal as a substitute for soyabean meal in isocaloric and isonitrogenous diets. These authors demonstrated the absence of adverse effects on performance and egg quality and, in particular, on taste and texture of the eggs. Dehulling of lupins increased its nutritional value by increasing its protein and amino acid content and by decreasing its fibre level (Laudadio and Tufarelli, 2011). Dehulled-micronized lupin seeds contain no alkaloids, contain concentrated protein and, at 18% dietary level, can totally substitute for soyabean. This inclusion slightly reduced feed intake but had no effect on feed conversion, egg mass or egg weight, and egg quality parameters, while increasing slightly yolk colour and the amount of albumen (Laudadio and Tufarelli, 2011).

7.4 Sunflower meal

Sunflower meal showed a high variability in level of protein depending on the processing method, low lysine level but with high availability of amino acids and without any anti-nutritional factors (Senkoylu and Dale, 1999). Numerous studies demonstrated that it can be used to substitute 50 to 100% soyabean protein corresponding to a dietary level of 30%, without adverse effect on laying hen performance. However, sunflower meal contained

Table 3 Composition of protein sources and DDGS from Feedipedia (Heuzé et al., 2016). Composition sorted by variety of a large range of feedstuffs and nutritional values for animal production is regularly updated on the Feedipedia site

Protein sources	Soyabean 48	Peas	Faba beans	White lupin	Corn DDGS	Wheat DDGS
AME MJ/kg DM	10.9	13.4	12.2	10.4	12.6	9.9
CP	51.6	25.2	31.1	29	29.5	37.5
Lysine %CP	6.1	6.5	6.2	4.7	3.0	2.3
Methionine %CP	1.4	1.4	0.8	0.7	2.0	1.5
Threonine %CP	3.9	3.8	3.5	3.4	3.7	3.0
Tryptophan %CP	1.3	1.1	0.8	0.8	0.8	1.0

relatively high fibre, diluting energy concentration and contributing to wet manure and dirty eggs (Brenes et al., 2008).

7.5 Cottonseed meal

Cottonseed meal (CSM), resulting from oil extraction of cottonseeds is highly available in China, Africa and other tropical countries. CSM is a rich but high variable source of protein (30 to 50%) with low concentration of lysine, high level of fibre (12 to 28%) and contains anti-nutritional factors such as gossypol (Swiatkiewicz et al., 2016). Gossypol is an antagonist of pepsin and reduces protein and amino acid digestibility of CSMs. Amino acid digestibility was evaluated at 74 to 91% in CSM containing 1.5% gossypol but was higher when the level of gossypol was only 0.02% in glandless CSM (Salas et al., 2013). CSM also contained phytin, which reduced iron availability by iron binding and might negatively affect haematological parameters in birds, although Adeyemo (2008) observed no effect in hens fed a diet containing up to 13% CSM and 0.57g gossypol/kg. The results of the majority of recent studies with laying hens indicated that CSM can be safely used in the nutrition of laying hens up to 10% dietary level, as reviewed by Swiatriewicz et al. (2016) or even 15% when using CSM with low gossypol levels (He et al., 2015). A negative effect on egg weight was observed only at the higher level of inclusion (18.9%) but other egg quality parameters (eggshell, yolk colour) as well as liver and kidney tissue were not affected. However, CSM with a high level of gossypol supplied at 6% negatively affected egg yolk colour and egg weight, in contrast to expanded CSM with low level of gossypol (−68%) which can be supplied up to 10% without any negative effect on hen performance and egg quality (Yuan et al., 2014).

7.6 Rapeseed

Rapeseed (canola for modern varieties) is one of the major oilseeds produced throughout the world. Its composition has been altered by selection, particularly in Canada and Europe, to reduce its anti-nutritional factors: the erucic acid, an unpalatable toxic fatty acid and glucosinolates, goitrogenic compounds (Newkirk, 2009). It was previously demonstrated that the inclusion of rapeseed in feed reduces egg weight with a decrease proportional to the dietary level of rapeseed meal up to 8–9%, with more marked effects when the rapeseed contains high levels of glucosinolates (Lacassagne, 1988b). However, breeding for rapeseed with low anti-nutritional compounds and the application of some feed processing technology to rapeseed have allowed its use in poultry at higher levels of up to 20% (Newkirk, 2009; Khajali and Slominski, 2012; Koslowski and Jeroch, 2014). Modern rapeseeds or canola have very low content of glucosinolates, less than 10 μmol/g, compared to old varieties (120 μmol/g) (Khajali and Slominski, 2012; Koslowski and Jeroch, 2014). Canola seed meals have an high protein content (34–39% DM) and a good amino acid profile with high level of methionine compared to plant feedstuffs but contain less lysine than soyabean (Newkirk, 2009; Feedipedia, 2016). The mean digestibility of amino acids in caecectomized hens, evaluated in 9 rapeseed meals, varied between 0.78 and 0.84, being higher for methionine (0.87) but lower for lysine (0.74) (Rezvaniy et al., 2012). Its level of crude fibre is, however, higher than other oilseeds. Fibre, but also tannins, limits incorporation of rapeseed in poultry diets but selection of triple zero varieties or hulling lowered fibre and tannin content (Koslowski and Jeroch, 2014) and improved AMEn and amino acid digestibility. Eggs from brown egg-laying strains were reported to develop a fishy taste if rapeseed levels are higher than 5%, due to the presence of sinapine, the precursor of trimethylamine in the yolk (Lacassagne et al., 1988b). Some hens are genetically unable to convert trimethylamine into odourless components. This unusual feature has been related to a polymorphism of the gene coding for the enzyme-degrading trimethylamine present in some brown egg hens (Honkatukia et al., 2005). Its incidence has now been reduced through selection for this criterion, associating this polymorphic gene and undesirable egg taste (Newkirk, 2009). In conclusion, rapeseed meal can be partly substituted for soyabean and incorporated up to 20% in diets without any negative effect on hen performance but, in sensitive hens, its incorporation might need to be more limited. In practical conditions, the substitution of soyabean is often partial and is conducted using diverse alternative sources of protein to avoid any risk on egg performance and quality, for favouring amino acid balance and for economical optimization.

7.7 Insects

Free-range birds are highly attracted by insects which constitute an interesting source of protein when birds have access to outdoor areas. Insects are a rich source of protein, essential amino acids and fat but legislation, high cost of production and limited availability has until now restricted their use in animal production including poultry. Limited knowledge of their nutritional value has further limited their use. However, many efforts are currently being carried out to analyse the potential of insect supply in poultry feed and to improve the production of insects as a source of animal protein. Experimental work has focused on the mealworm (larvae of the beetle *Tenebrio molitor*), the maggot and pupae of the housefly (*Musca domestica*), the black soldier fly (*Hermetia illucens*), and insect families belonging to the order Orthoptera, including locusts, grasshoppers, crickets and katylids, and has been reviewed by Makkar et al. (2014) and Jozefiak et al. (2016). Insects are a rich source of protein (38 to 64%), essential amino acids (5–8 g/kg lysine but lower content of methionine, 1–2 g/kg) and fat (9 to 40%) with 19–28 Mj/kg gross energy. The nutritional value is, therefore, quite variable between species, physiological stage of development (adult or larvae) but can also vary within species (Jozefiak et al., 2016). Protein and fat concentration changes also with the rearing conditions and the composition of the growth media used for insect production (Makkar et al., 2014). Live insects have a high water content (average of 30% dry matter). The components of dried insect meal are crude protein and fat. Chitin (a polysaccharide component) can also be present in insect at level of 3 to 16% in larvae of aquatic insects (Cauchie, 2002) or 18 to 60% in arthropods in general (Klasing, 1998). The relative proportion of fat and chitin largely explained the large range of ME values found in maggot meal (Bovera et al., 2015). The nutritional value of insects has been evaluated in broilers. Housefly larvae may be added up to 25% DM, without any negative effects on weight gain (BWG), feed intake (FI) and feed efficiency (Pretorius, 2011; De Marco et al., 2015). The AMEn of the two insect larval meals (*Tenebrio molitor* and *Hermetia illucens*) were, respectively, 16.02 and 16.60 MJ/kg DM for broiler chickens and the average amino acid apparent ileal digestibility coefficients of 17 analysed amino acids were, respectively, 0.86 and 0.68 (De Marco et al., 2015). Maggot meal included at a level of 16% did not alter feed intake and growth of young pullets up to 56 days of age (Enberg et al., cited in Jozefiak et al., 2016) but no information seems to be available currently for laying hens. A high level of chitin might limit insect meal use because of its negative effect on feed consumption (Longvah et al., 2011). Its low digestibility in chickens (Hossain and Blair, 2007) contributed also to lowering the ME energy in poultry. Insect are, however, a promising source of animal protein and optimization of some insect production and expected experimental work in this area should contribute to a more general use in poultry in the near future.

8 Feed management of cereal grains and protein sources

8.1 Mode of distribution

Several studies have sought to evaluate the use of whole cereal grains in feeds directly on the farm to reduce feed costs. Hens have the ability to select diet components when particles such as grains are present. Whole grains of wheat or barley are of a size suitable to be eaten by poultry and can, therefore, be used as low protein feeds which can be fed in free choice with high-protein concentrate pellets (Forbes and Covasa, 1995). Choice feeding permit individual birds in a flock to select from various feed ingredients offered and enables them to compose their own diet according to their actual needs. However, hens exhibit marked preferences for types of cereals which, in turn, can affect their ability to compose a suitable diet. In general, hens offered free choices of different cereals tended to eat larger amounts of grain as the energy content of the grain decreased and, in the early literature, it was considered that production performance of birds was not affected by the type or form of the grain used in free-choice feeding, even if it might influence the protein intake (Forbes and Covasa, 1995). Keshavarz (1998a,b) evaluated sequential supply of dietary proteins throughout the day (positive control 16% CP, negative control 13% CP, two sequential diets combining 13% and 16% CP fractions), and concluded that optimum performance may be expected as long as the daily protein intake is adequate and satisfies the daily needs of laying hens. That is in agreement with the observation that hens are not able to fit their feed intake based on CP level in the diet (Summers and Leeson, 1985) but regulate their feed intake mainly on ME level of the diet (Plavnik et al., 1997; Nahashon et al., 2005). Whole grains supplying mainly energy can also be provided along with a complete feed, concentrating the supply of proteins, minerals and vitamins as a mixture, or sequentially during the day. Blair et al. (1973) reported that the sequential supply of cereals (wheat, barley and corn) in the morning and a balanced diet (diet enriched in protein, minerals and vitamins to provide the daily requirement of hens) in the afternoon, did not result in a deterioration in egg weight. Umar-Faruk et al. (2010) obtained similar performance when sequentially distributing whole wheat in the morning and a balanced diet fed in a limited quantity in the afternoon (adjusted at 60g/hen to provide the daily hen protein and mineral requirement), or when mixing the two fractions. In this sequential feeding with whole wheat, intake was reduced due to lower wheat intake, while a similar level of egg weight and production was observed. This consequently improved the feed conversion ratio compared to complete diet and loose mix (respectively, 5 and 10%). First, it was probably due to the positive effects of whole grains by enlarging gizzard size and improving digestive capacity of hens, and then by a lower gain and fat deposition in body weight. Use of ground wheat in sequential feeding was less efficient than whole wheat (Umar Faruk et al., 2011), but it was possible to substitute, at least partially, whole wheat by ground wheat or ground corn, allowing more flexibility and economic optimization (Traineau et al., 2013).

8.2 Alternative systems

Organic schemes are the most demanding nutritionally. There is a range of additional production system, including certification schemes or label production, which are also associated with nutritional constraints due to the use of particular feedstuffs or to the ban of some nutrients such as synthetic products or GMO. Alternative systems apply existing knowledge used for conventional systems, rather than settling novel nutritional principles (Hammershoj, 2011; MacLeod and Bentley, 2012). Organic food is governed by certifications run by some large retailers or by national or European directives (EU, 2007, 2012, 2014). Diets should be formulated without animal proteins, genetically modified organisms or synthetic amino acids. Therefore, organic and other alternative diets can be more difficult to optimize and more costly. It is possible to formulate diets without animal protein, genetically modified organisms and synthetic amino acids, but it is difficult to attain nutritional optima with a risk of lower performance in case of nutritional imbalance. Organic diets may not supply sufficient levels of certain essential amino acids, and a substandard level of methionine which cannot be compensated by the synthetic form may influence egg production and stress laying hens, initiating unwanted behaviour such as feather pecking (Van Krimpen et al., 2005; Steenfeld and Hammershoj, 2015). It is possible to use alternative sources of protein to soyabean as outlined in part 7, but lupins, peas and fava beans supply lower levels of sulphur amino acids. Consequently, sufficient levels of amino acid are provided by feeding diets with higher protein content. This strategy increases nitrogen output in excreta and the risk of wet litter and higher ammonia concentration in the house (Dekker et al., 2012, 2013); however, the negative environmental impact can be partly alleviated by the lower density of organic flocks. Reducing protein level of the diet in organic production elicited lower performance and egg quality (Steenfeldt and Hammershøj, 2015; Hammershøj and Steenfeldt, 2015). Access to the outside (free-range hen husbandry with pasture) allows consumption of pasture with herbs and insects, but their contribution as essential amino acid sources is not well quantified, even if insects are nutritionally better defined (Jozefiak et al., 2016). On the other hand, the free-range system requirement for maintenance is higher due to locomotor activity (by about 5 kJ/d in free-range system compared to conventional system) and to the thermal environment (+1kj per degree of temperature, 1–1.5% of feed consumption in well-feathered hens, double if low feathering hens). Therefore, feed consumption can be increased by 10–20% in free-range systems compared with indoor systems (MacLeod and Bentley, 2012). This will increase the intake of other components of the diet unless their concentrations are adjusted to the feed intake. A pasture or the availability of forage plants can be a source of nutrient, protein and energy, but also of minor nutrients such as carotenoids which will reinforce the yolk colouration (Hammershøj and Steenfeldt, 2012; Hammershøj and Steenfeldt, 2015). Hens expressed more appetite for maize silage and carrots than for alfalfa and their consumption can be large at 104 g/kg/d and 57 g/hen/d, respectively (Steenfeldt and Hammershøj, 2015). Providing additional foraging material can, however, contribute to lower consumption of the hen's main diet. Egg weight decreased with lowering of protein level and in hens fed maize – carrot compared with alfalfa forage. Feeding the alfalfa silage increased egg yolk colour and its carotenoid content (Hammershøj and Steenfeldt, 2015).

It is noteworthy that outdoor access, as well as poultry rearing in family backyards may cause abnormal contamination of eggs. Laying hens are likely to be exposed to a wide range of persistent organic pollutants (POPs), including polychlorinated biphenyls (PCBs). PCBs can be released from soil particles in the digestive tract of laying hens (Fournier et al., 2012) and soil ingestion is believed to be the main source of PCB exposure in domestic poultry reared outdoors. Hens, in a free-range system, can consume 10 g of dry soil, 7 g of dry vegetation and 20 g of insects and earthworms a day (De Vries et al., 2006), but in cases of an imbalanced diet, this can increase to 30 g of soil daily (Waegeneers et al., 2009; Jondreville et al., 2011). Soil contamination can issue from traffic at close proximity to the free range or from fire even when carried out years ago. The POPs are preferentially stored in fat and subsequently will be transferred into the egg yolk because of their lipophilic properties (Fournier et al., 2015).

9 Conclusions

Hen nutrition must address a number of different goals, including maximizing egg production, maintaining egg quality, extending the laying cycle as well as ensuring hen health and welfare. This chapter has reviewed many variables involved in hen nutrition, including differing dietary requirements in the pre-laying and laying phases. It has also reviewed ways feed intake can be controlled to manage energy consumption affecting parameters such as egg weight, including particle size in the diet and the use of dietary lipids and proteins. The final part of the chapter has discussed the use of cereals and protein sources in hen nutrition, feed management and particularities of alternative systems.

Hen nutrition is a complex process as numerous nutrients can affect hen performance and egg quality. There are considerable pieces of evidence indicating that the rearing period is important for the future production of hens. Any deviation from optima might have long-term effects. Much precise knowledge is, however, available on hen nutrition but the difficulty might be due to interactions between factors that are less easily controlled. Future development in modelling might help in integrating such complexity. The extension of the production period relies on liveability and reduced mortality of the birds and maintenance of an acceptable level of egg quality to be commercially viable. The exportation of large amounts of energy, protein and calcium are required from the liver for yolk formation and the oviduct for eggwhite and shell synthesis. There is a need for more robust hens adaptable to variable conditions of production throughout different seasons especially in alternative systems. It is obvious that genetic selection is an efficient means for developing further egg production, hen performance and egg quality. Breeding companies select hens which fit for more diverse environments throughout the world, rather than developing lines specific to one climatic condition.

However, hen nutrition remains a crucial parameter for sustainability of the egg production system to ensure high performance and profitability for producers. The limit of extension of the laying period is still determined by controlling, through nutrition, the quality of the final product and avoiding any deterioration in hen heath. In terms of sustainability of the production system in different parts of the world, nutrition remains the crucial factor because of the similarities of genetic background used throughout the world and because of the large economical contribution of nutrition to egg production. The differential cost of production between different areas in the world mainly results from the nutritional cost and the availability and cost of feedstuffs which remain the limiting factors for developing egg production.

10 Where to look for further information

A second chapter of this book described additional aspects of hen nutrition. Numerous reviews have been mentioned throughout the text and described more exhaustively the information we did summarize. The site feedipedia is of interest as it describes in detail feedstuff composition, animal requirements and is continuously updated. WPSA proceedings and the WPSJ are useful sources of information in particular the reviews published in the proceeding of the European Symposium on Poultry Nutrition which takes place every two years.

11 References

Abdelqader, A., Al-Fataftah, A. R. and Das, G. (2013), Effects of dietary Bacillus subtilis and inulin supplementation on performance, eggshell quality, intestinal morphology and microflora composition of laying hens in the late phase of production. *Anim. Feed Sci. Rech.* 179, 103–11.

Adeyemo, G. O. (2008), Effects of cottonseed cake-based diets on hematology and serum 277 biochemistry of egg-type chickens. *Int. J. Poult. Sci.* 7, 23–7.

Atteh, J. O., and Leeson, S. (1985), Response of laying hens to dietary saturated and unsaturated fatty acids in the presence of varying dietary calcium levels. *Poult. Sci.* 64, 520–8.

Bain, M., Nys, Y. and Dunn, I. C. (2016), Increasing persistency in lay and stabilising egg quality in longer laying cycles. What are the challenges? *Br. Poult. Sci.* 57, 330–8.

Bertram, H. L., Dänner, E., Jeroch, J. and Jeroch, G. (1995a), Effect of DL-Methionine in a cereal-pea diet on the performance of brown laying hens. *Arch. Geflügelk* 59, 103–7.

Bertram, H. L., Schutte, J. B. and De Jong, J. (1995b), Influence of DL-Methionine supplements on the performance of laying hens under heat-stress. *Arch. Geflügelk* 59, 314–18.

Blair, R. W., Dewar, W. A. and Downie, J. N. (1973), Egg production responses of hens given a complete mash or unground grain together with concentrate pellets'. *Br. Poult Sci.* 14, 373–7.

Bouvarel, I., Nys, Y. and Lescoat, P. (2011), Hen nutrition for sustained egg quality. In *Improving the Safety and Quality of Eggs and Egg Products*, Vol. 1, Cambridge: Woodhead Publishing, pp. 261–99.

Bouvarel, I., and Nys, Y. (2013), Optimizing egg mass and quality traits in modern laying hens through nutrition. In Proceedings 19th European Symposium on Poultry Nutrititon. Presented at 19th European Symposium on Poultry Nutrition (ESPN), Potsdam, DEU. (2013-08-26 - 2013-08-29).

Bovera, F., Loponte, R., Maronos, S., Piccolo, G., Parisi, G., Laconisi, V., Gaco, L. and Nizza A. (2015), Use of Tenebrio molitor larvae meal as protein source in broiler diet: Effect on growth performance, nutrient digestibility, and carcass and meat traits. *J. Anim. Sci.* 94: 639–47.

Brenes, A., Centeno, C., Viveros, A. and Arija I. (2008), Effect of enzyme addition on the nutritive value of high oleic acid sunflower seeds in chicken diets. *Poult. Sci.* 87, 2300–10.

Buitendach, G. C., de Witt, F. H., Hugo, A., Van Der Merwe, H. J. and Fair, M. D. (2015), Effect of dietary fatty acid saturation on egg production at end-of-lay. *South Afr. J. Anim. Sci.* 2013, 43 (Issue 5, Supplement 1), S126–31.

Butler, E. J. (1976), Fatty liver diseases in the domestic fowl-a review. *Avian Pathol.* 5, 1–14.

Cauchie, H. M. (2002), Chitin production by arthropods in the hydrosphere. *Hydrobiologia* 470, 63–96.

Cheng, T. K., Peguri, A., Hamre, M. L. and Coon, C. N. (1991), Effect of rearing regimens on pullet growth and subsequent laying performance. *Poult. Sci.* 70, 907–16.

De Vries, M., Kwakkel, R. P. and Kijlstra A. (2006), Dioxins in organic eggs: a review, *Wageningen J. Life Sci.* 54, 207–21.

Cozannet, P., Lessire, M., Gady, C., Métayer, J. P., Primot, Y., Skiba, F. and Noblet, J. (2010), Energy value of wheat dried distillers grains with solubles in roosters, broilers, layers and turkeys. *Poult. Sci.* 89(10), 2230–41.

Cozannet, P., Primot, Y., Gady, C., Métayer, J. P., Lessire, M., Skiba, F. and Noblet, J. (2011), Standardized amino acids digestibility of wheat distillers dried grains with solubles in force-fed roosters. *Br. Poult. Sci.* 52(1), 72–81.

Dänicke, S., Halle, I., Jeroch, H., Bottcher, W., Ahrens, P., Zachmann, R. and Gotze, S. (2000), Effect of soy oil supplementation and protein level in laying hen diets on precaecal nutrient digestibility, performance, reproductive performance, fatty acid composition of yolk fat, and on other egg quality parameters. *Eur. J. Lipid Sci. Tech.* 102, 218–32.

Dänner, E., and Bessei, W. (2002), Effectiveness of liquid DL-Methionine hydroxylanalogue-free acid (DL-MHA-FA) compared to DL-Methionine on performance of laying hens. *Archiv für Geflügelkunde* 66, 97–101.

Dänner, E. E. (2003), Use of low-vicin/convicin faba beans (Vicia faba) in laying hens. *Archiv für Geflügelkunde* 67, 249–52.

De Marco, M., Martínez, S., Hernande, F., Madrid, J., Gai F., Rotolod, L., Belfortid, M., Bergeroa, D., Katze, H., Dabboud, S., Kovitvadhid, A., Zoccaratod, I., Gascoc, L. and Schiavonea, A. (2015), Nutritional value of two insect larval meals (Tenebrio molitorand Hermetia illucens) for broiler chickens: apparent nutrient digestibility, apparent ileal amino acid digestibility and apparent metabolizable energy. *Anim. Feed Sci. Tech.* 209, 211–18.

Dekker, S. E. M., Aarnink, A. J. A., de Boer, I. J. M. and Groot Koerkamp, P. W. G. (2012), Total loss and distribution of nitrogen and phosphorus in the outdoor run of organic hens. *Br. Poult. Sci.* 53, 731–40.

Dekker, S. E. M., de Boer, I. J. M., Van Krimpen, M., Aarnink, A. J. A. and Groot Koerkamp, P. W. G. (2013), Effect of origin and composition of diet on ecological impact of the organic egg production chain. *Livest. Sci.* 151, 271–83.

De Persio, S., Utterback, P. L., Utterback, C. W., Rochell, S. J., O'Sullivan, N., Bregendahl, K., Arango, J., Parsons, C. M. and Koelkebeck, K. W. (2015), Effects of feeding diets varying in energy and nutrient density to Hy-Line W-36. laying hens on production performance and economics. *Poult. Sci.* 94, 195–206.

Duc, G., Marget, P., Esnault, R., Le Guen, J. and Bastianelli, D. (1999), Genetic variability for feeding value of faba bean seeds (Vicia faba L.). Comparative chemical composition of isogenics involving zero-tannin and zero-vicine genes. *J. Agric. Sci.* 133, 185–96.

Dunn, I. (2013), Long Life Layer; genetic and physiological limitations to extend the laying period. *Proceeding of the 19th Symposium on Poultry Nutrition.* WPSA. Postdam.

El Kaseh, R., and Forbes, J. M. (1995), Effect of wet food on the performance of laying hens, *Br. Poult. Sci.* 36, 839–40.

Elliot, M. A. (2002), Applied pullet and layer nutrition with an emphasis on problems commonly encountered in the field. *Cornell Poult. Pointers* 52(3), 16–32.

Engberg, R. M., Hammershoslashj, M., Johansen, N. F., Abousekken, M. S., Steenfeldt, S. and Jensen, B. B. (2009), Fermented feed for laying hens: effects on egg production, egg quality, plumage condition and composition and activity of the intestinal microflora *Br. Poult. Sci.* 50, 228–39.

E.U. (2007), Commission regulation N.834/2007 of 28 June 2007 on organic production and labeling of organic products and repealing Regulation (EEC) No. 2092/91. *Off. J. Eur. Communities* (L189), 1–23.

E.U. (2012), Commission implementing regulation (EU) No 505/2012 of 14 June 2012 amending and correcting Regulation (EC) No. 889/2008 laying down detailed rules for the implementation of Council regulation (EC) N0 834/2007 on organic production and labelling of organic products with regard to organic production, labelling and control. *Off. J. Eur. Union* (L154), 12–19.

E.U. (2014), Council Regulation No. 836/2014 amending Regulation No.889/2008 laying down detailed rules for the implementation of Council Regulation (EC) No. 834/2007 on organic production and labelling of organic products with regard to organic production, labelling and control. http://eur-lex.europa.eu/legal-content/EN/TXT/?uri=uriserv%3AOJ.L_.2014.230.01.0010.01.ENG.

Feedipedia – Animal Feed Resources Information System – INRA CIRAD AFZ and FAO. © 2012-2016. . http://www.feedipedia.org/

Forbes, J. M., and Covasa, M. (1995), Application of diet selection by poultry with particular reference to whole cereals. *World Poult. Sci. J.*, 51, 149.
Fournier, A., Feidt, C., Travel, A., Le Bizec, B., Venisseau, A., Marchand, P. and Jondreville, C. (2012), Relative bioavailability to laying hens of indicator polychlorobiphenyls present in soil. *Chemosphere* 88, 300–6.
Fournier, A., Martin, O., Travel, A., Puillet, L., Feidt, C. and Jondreville, C. (2015), Modeling PCB transfer into hen eggs: influence of physiological characteristics of the animal. *Environ. Toxicol. Chem.* 34, 173–83.
Frikha, M., Safaa, H. M., Serrano, M. P., Arbe, X. and Mateos, G. G. (2009), Influence of the main cereal and feed form of the diet on performance and digestive tract of brown-egg laying pullets. *Poult. Sci.* 88, 994–1002.
Frikha, M., Safaa, H. M., Serrano, M. P., Jiménez-Moreno, E. and Mateos, G. G. (2011) Influence of the main cereal of the diet and particle size of the cereal on productive performance and digestive traits of brown-egg laying pullets. *Anim. Feed Sci. Technol.* 164, 106–15.
Fru-Nji F., Niess, E. and Dfeffer, E. (2007), Effect of graded replacement of soyabean meal by faba beans (Vicia faba L.) or field peas (Pisum sativum L.) in rations for laying hens on egg production and quality. *J. Poult. Sci.* 44, 34–41.
Galea, F. (2014), Nutritional optimization of laying hens for improved health and egg quality. Nordic Symposium. On Bone Health. 23 May 2013 – Copenhagen. ISA-Hendrix genetic.
Grobas, S., Mendez, J., De Blas, C. and Mateos, G. G. (1999a), Influence of dietary energy, supplemental fat and linoleic acid concentration on performance of laying hens at two ages. *Br. Poult. Sci.* 40, 681–7.
Grobas, S., Mendez, J., De Blas, C. and Mateos, G. G. (1999b), Laying hen productivity as affected by energy, supplemental fat, and linoleic acid concentration of the diet. *Poult. Sci.* 78, 1542–51.
Guzmán, P., Saldaña, B., Mandalawi, H., Pérez-Bonilla, A., Lázaro, Q. and Mateos, G. G. (2015a), Productive performance of brown-egg laying pullets from hatching to 5 weeks of age as affected by fiber inclusion, feed form, and energy concentration of the diet. *Poult. Sci.* 94, 249–61.
Guzmán, P., Saldaña, B., Kimiaeitalab, M. V., García, J. and Mateos, G. G. (2015b), Inclusion of fiber in diets for brown-egg laying pullets: effects on growth performance and digestive tract traits from hatching to 17 weeks of age. *Poult. Sci.* 94, 2722–33.
Hammershøj, M. (2011), Organic and free-range egg production. In Y. Nys, M. Bain Van and F. Immerseel (eds), *Improving the Safety and Quality of Eggs andEgg Products: Egg Chemistry, Production and Consumption*, Vol. 1, 1st ed. Woodhead Publishing Limited, Cambridge, pp. 463–86.
Hammershøj, M., and Steenfeldt, S., (2012), The effects of kale (Brassica oleracea ssp acephala), basil (Ocimum basilicum) and thyme (Thymus vulgaris) as forage material in organic egg production on egg quality. *Br. Poult. Sci.* 53, 245–56.
Hammershøj, M., and Steenfeldt, S. (2015), Organic egg production. II: the quality of organic eggs is influenced by hen genotype, diet and forage material analyzed by physical parameters, functional properties and sensory evaluation. *Anim. Feed Sci. Technol.* 208, 182–97.
He, T., Zhang, H. J., Wang, J., Wu, S. G., Yue, H. Y. and Qi, G. G. (2015), Application of low-gossypol cottonseed meal in laying hens diet. *Poult. Sci.* 94, 2456–63.
Hetland, H., Choct, M. and Svihus, B. (2004), Role of insoluble non-starch polysaccharides in poultry nutrition. *World Poult. Sci. J.* 60, 415–22.
Heuzé, V., Tran, G., Delagarde, R., Lessire, M. and Lebas, F. (2016), Faba bean (Vicia faba). Feedipedia a programme by INRA, CIRAD., AFZ. and FAO. http://www.feedipedia.org/node/4926 Last updated on 20 May 2016.
Honkatukia, M., Reese, K., Preisinger, R., Tuiskula-Haavisto, M., Weigend, S., Roito, J., Maki-Tanila, A. and Vilkki, J. (2005), Fishy taint in chicken eggs is associated with a substitution within a conserved motif of the FMO3 gene. *Genomics*, 86, 225–32.
Hossain, S. M., and Blair, Q. (2007), Chitin utilisation by broilers and its effect on body composition and blood metabolites. *Br. Poult. Sci.* 48, 33–8.
Joly, P. (2002), Estimation of the amino-acid requirements of laing hens in relation to the genetic evolution. Bull Tech, ISA Joly P. (2005), Energy levels in feeds for laying hens: effects on

performance and intake. XIX. Congresso Latino Americano de Avicultura Panama 5–7 Octubre 2015.

Joly, P. (2008), Nutrition Management guide ISA., Institut de selection Avicole. Saint Brieuc France.

Joly, P., and Alleno, C. (2001), Heures d'oviposition avec ou sans éclairement nocturne et influence sur la qualité de l'œuf en France. Quatrième journées de la recherche avicole. Nantes.

Joly, P., and Bougon, M. (1997), Influence du niveau énergétique sur les performances de la pondeuse à œufs roux et évolution de l'ingéré en fonction de l'âge. 2èmes Journées de la Recherche Avicole. Tours.

Jondreville, C., Travel, A., Besnard, J. and Feidt, C. (2011), Ingestion de sol et de végétaux par des poules élevées en plein air nourries avec un aliment complet ou à base de blé et de coquilles marines. *Neuvièmes Journées de la Recherche Avicole*, 29–30 Mars, Tours, France, pp. 91–5.

Jozefiak, A., Kieronczyk, B., Rawski, M., Swiatkiewicz, S., Dlugosz, J. and Engberg, R. M. (2016), Insects-a natural nutrient source for poultry- a review. *Ann. Anim. Sci.* 16(2), 297–313.

Keshavarz, J. (1998a), Investigation on the possibility of reducing protein, phosphorus, and calcium requirements of laying hens by manipulation of time of access to these nutrients. *Poult. Sci.* 77, 1320–32.

Keshavarz, J. (1998b), Further investigations on the effect of dietary manipulation of protein, phosphorus, and calcium for reducing their daily requirement for laying hens. *Poult. Sci.* 77, 1333–46.

Keshavarz, J., and Jackson, M. (1992), Performance of growing pullets and laying hens fed low-protein, amino acid-supplemented diets. *Poult. Sci.* 71, 905–18.

Khajali, F., and Slominski, B. A. (2012), Factors that affect the nutritive value of canola meal for poultry. *Poult. Sci.* 91(10), 2564–75.

Klasing, K. C. (1998), *Comparative Avian Nutrition*. New York, CABH Publishing.

Koivunen, E., Tuunainen, P., Valkonen, E. and Valaja, J. (2015), Use of semi-leafless peas (Pisum sativum L) in laying hen diets. *Agric. Food Sci.* 24, 84–91.

Koivunen, E., Tuunainen, P., Valkonen, E., Rossow, L. and Valaja, J. (2014), Use of faba beans (Vicia faba L.) in diets of laying hens. *Agric. Food Sci.* 23, 65–172.

Kozłowski, J., and Jeroch, G. (2014), Enhancing the nutritional value of poultry feedstuffs using the example of rapeseed products – a review. *Ann. Anim. Sci.* 14, 245–56.

Krogdahl, A. (1985), Digestion and absorption of lipids in poultry. *J. Nutr.* 115, 675–85.

Krawczyk, M., Przywitowski, M. and Mikulski, D. (2015), Effect of yellow lupine (L. luteus) on the egg yolk fatty acid profile, the physicochemical and sensory properties of eggs, and laying hen performance. *Poult. Sci.* 94, 1360–7.

Lacassagne, L. (1988a), Alimentation des volailles: substituts au tourteau de soja 1- Les protéagineux, INRA. *Productions Anim.* 1, 47–57.

Lacassagne, L. (1988b), Alimentation des volailles: substituts au tourteau de soja 2-Le tourteau de colza INRA. *Productions Anim.* 1, 123–8.

Laudadio, V., and Tufarelli, V. (2010), Treated fava bean (Vicia faba var. minor) as substitute for soyabean meal in diet of early phase laying hens: Egg-laying performance and egg quality. *Poult. Sci.* 89, 2299–2303.

Laudadio, V., and Tufarelli, V. (2011), Influence of substituting dietary soyabean meal for dehulled micronized lupin (Lupinus albus cv. Multitalia) on early phase laying hens production and egg quality. *Livest. Sci.* 140, 184–8.

Lázaro, R., García, M., Araníbar, M. J and Mateos, G. G. (2003), Effect of enzyme addition to wheat, barley, and rye-based diets on nutrient digestibility and performance of laying hens. *Br. Poult. Sci.* 44, 256–65.

Leeson, S., Summers, J. D. and Caston, L. J. (2001), Response of layers to low nutrient density diets. *J. Appl. Poult. Res.* 10, 46–52.

Leinonen, I., Williams, A. G., Wiseman, J., Guy, J. and Kyriazakis, I. (2012), Predicting the environmental impacts of chicken systems in the United Kingdom through a life cycle assessment: broiler production systems. *Poult. Sci.* 91, 8–25.

Lessire, M., Gallo, V., Prato, M., Akide-Ndunge, O., Mandili, G., Marget, P., Arese, P. and Duc, G. (2016), Effects of faba beans (Vicia faba L.) with different concentrations of vicine and convicine on egg production, egg quality and red blood cells in laying hens. *Animals*, on line 29 December 2016.

Longvah, T., Mangthyak, J. and Ramulu, P. (2011), Nutrient composition and protein quality evaluation of eri silkworm (Samia ricinii) prepupae and pupae. *Food Chem.* 128, 400–3.

MacLeod, M. G., and Bentley, J. S. (2012), Nutritional challenges of alternative production systems. In V. Sandilands and P. M. Hocking (eds), *Alternative Systems for Poultry. Poultry Science Symposium 30*, CAB. International, Oxford, UK, pp. 297–315.

Makkar, P. P. S., Tran, G., Heuzé, V. and Ankers, P. (2014), State-of-the-art on use of insects as animal feed. *Anim. Feed Sci. Tech.* 197, 1–33.

Masa'deh, M. K., Purdum, S. E. and Hanford, K. J. (2011), Dried distillers grains with solubles in laying hen diets. *Poult. Sci.* 90, 1960–6.

Mateos, G. G., Jimenez-Moreno, E., Serrano, M. P. and Lazaro, R. P. (2012), Poultry response to high levels of dietary fiber sources varying in physical and chemical characteristics. *J. Appl. Poult. Res.* 21, 156–74.

Mongin, P., and Sauveur, B. (1979), The specific appetite of domestic fowl. In K. N. Boorman and B. M. Freeman (eds), *Food Intake Regulation in Poutry*. Edinburgh, British Poultry Science, pp. 171–89.

Nahashon, S. N., Adefope, N., Amenyenu, A. and Wright, D. (2005), Effect of dietary metabolisable energy and crude protein concentrations on growth performance and carcass characteristics of French guinea fowl broilers. *Poult. Sci.* 84, 337–44.

Nalle, C. L., Ravindran, V. and Ravindran, G. (2012), Nutritional value of white lupins (Lupinus albus) for broilers: apparent metabolisable energy, apparent ileal amino acid digestibility and production performance. *Animal* 6, 579–85.

Narvaez-Solarte, W., Rostagno, H. S., Soares, P. R., Silva, M. A. and Uribe Velasquez L. F. (2005), Nutritional requirements in Methionine + Cystine for white-egg laying hens during the first cycle of production. *Int. J. Poult. Sci.*, 4, 965–8.

Newkirk, Q. (2009), Canola meal: feed industry guide. In R. Newkirk (ed.), 4th Ed., Canadian International Grains Institute, Canola Council, Winnipeg, MB, Canada.

Noblet, J., Cozannet, P. and Skiba, F. (2012), Nutritional value and utilization of wheat dried distillers grain with soluble in pigs and poultry in FAO. Biofuel co-products as livestock feed- opportunities and challenges, edited by P. S. Harinder, Makka Rome, Chapt 9, pp. 163–74.

Novak, C., Yakout, G. M. and Scheideler, S. E. (2006), The effect of dietary protein level and total sulfur amino acid:lysine ratio on egg production parameters and egg yield in Hy-Line W-98 hens. *Poult. Sci.* 85, 2195–2206.

NRC. (1994) National Research council. Nutrient Requirements of Poultry: Ninth Revised Edition, 1994.

Nys, Y. (2014), Phosphorus utilization in laying hens. In Proceedings of Associazione Scientifica di Avicoltura – Sezione Italiana della World's Poultry Science Association 50° Convegno Nazionale 'Il fosforo nell'alimentazione del pollame: un caso di sostenibilità', Fiera di Forlì, 15 Aprile 2014.

Oryschak, M., Korver, D., Zuidhof, M., Meng, X. and Beltranena, E. (2010), Comparative feeding value of extruded and nonextruded wheat and corn distillers dried grains with solubles for broilers. *Poult. Sci.* 89, 2183–96.

Parsons, C. M., Martinez, C., Singh, V., Radhakrishman, S. and Noll, S. L. (2006), Nutritional Value of Conventional and Modified DDGS for Poultry. Presented at the Multi State Poultry Nutrition and Feeding Conference, Indianapolis, IN. [Online] URL: http://www.ddgs.umn.edu/articlespoultry/

Peguri, A., and Coon, C. (1991), Effect of temperature and dietary energy on layer performance. *Poult. Sci.* 70, 126–38.

Peguri, A., and Coon, C. (1993), Effect of feather coverage and temperature on layer performance. *Poult. Sci.* 72, 1318–29.

Penz, A .M. Jr. and Jensen, L. S. (1991), Influence of protein concentration, amino acid supplementation, and daily time of access to high- or low-protein diets on egg weight and components in laying hens. *Poult. Sci.*, 70, 2460–6.

Pérez-Bonilla, A., Frikha, M., Mirzaie, S., Garcia, J. and Mateos, G. G. (2011), Effects of the main cereal and type of fat of the diet on productive performance and egg quality of brown-egg laying hens from 22 to 54 weeks of age. *Poult. Sci.* 90, 2801–10.

Pérez-Bonilla, A., Jabbour, C., Frikha, M., Mirzaie, S., Garcia, J. and Mateos, G. G. (2012), Effect of crude protein and fat content of diet on productive performance and egg quality traits of brown egg-laying hens with different initial body weight. *Poult. Sci.* 91, 1400–5.

Pérez-Bonilla, A., Frikha, M., Lázaro Q. P. and Mateos, G. G. (2014), Type of grinding of the main cereal of the diet affects production of brown egg-laying hens. *Anim. Feed Sci. Tech.* 194, 121–30.

Perez-Maldonado, R. A., Mannion, P. F. and Farrell, D. J. (1999), Optimum inclusion of field peas, faba beans, chick peas and sweet lupins in poultry diets. I. Chemical composition and layer experiments. *Br. Poult. Sci.* 4, 667–73.

Peyraud, J. L., Dourmad, J. Y., Lessire, M., Medale, F. and Pryronnet, C. (2015), Conséquence zootechniques d e l'introduction des légumineuses françaises dans les systèmes de productions animales. chap 4, in Les légumineuses pour de systèmes agricoles et alimentaires durables; Schneider A., Huyghe ed Quae, Paris., 225–61.

Pineda, L., Roberts, S., Kerr, B., Kwakkel, R., Verstegen, M. and Bregendahl, J. (2008), Maximum dietary content of corn dried distiller's grains with solubles in diets for laying hens: effects on nitrogen balance, manure excretion, egg production, and egg quality. Iowa State University Animal Industry Report 2008. AS. leaflet R2334. Iowa State University, Ames. Renewable Fuels Association. 2011.

Plavnik, H., Wax, E., Sklan, D., Bartov, H. and Hurwitz, S. (1997). The response of broiler chickens and turkey poults to dietary energy supplied either by fat or carbohydrates. *Br. Poult. Sci.* 76, 1000–05.

Portella, F. J., Caston, L. J. and Leeson, S. (1988), Apparent feed particle size preference by laying hens. *Can. J. Anim. Sci.* 68, 915–22.

Pottguetter, R. (2015), Nutrition of hens in extended production cycles – as a practical approach. In Proceeding of XVH. European Symposium on the Quality of Eggs and Egg Products May, 10 to 13 2015. Nantes – France.

Pretorius, P. (2011), The evaluation of larvae of Musca domestica (common house fly) as protein source for broiler production, Doctoral dissertation, Stellenbosch University.

Qaisrani, S. N., Van Krimpen, M. M. and Kwakkel, R. P. (2013), Effects of dietary dilution source and dilution level on feather damage, performance, behavior, and litter condition in pullets. *Poult. Sci.* 92, 591–602.

Rao, K. S., Roland, D. A., Orban, J. H., Rabon, H. W. and Bryant, M. M. (1995), Age at sexual maturity influences the response of single white leghorn pullets to marginal and low levels of dietary phosphorus. *J. Nutr.* 125, 1342–50.

Rezvaniy, M., Kluth, H., Bulang, M. and Rodehutscord, M. (2012), Variation in amino acid digestibility of rapeseed meal studied in caecectomised laying hens and relationship with chemical constituents. *Br. Poult. Sci.* 53, 665–74.

Rogers, P. J. (1995), *The Development of the Brain and Behavior in the Chicken*, CAB. Int., Wallingford, UK, pp. 95–110.

Safaa, H. M., Jiménez-Moreno, E., Valencia, D. G., Frikha, M., Serrano, M. P. and Mateos, G. G. (2009), Effect of main cereal of the diet and particle size of the cerealon productive performance and egg quality of brown egg-laying hens in early phase of production. *Poult. Sci.* 88, 608–14.

Salas, C., Ekmay, R., England, J., Cerrate, S. and Coon, C. N. (2013), The TMEn, proximate analysis, amino acid content and amino acid digestibility of glandless and commercial cottonseed meal for broilers. *Int. J. Poult. Sci.* 12 (405), 212–16.

Saldaña, B., Guzmán, P., Cámara, L., García, J. and Mateos, G. G. (2015), Feed form and energy concentration of the diet affect growth performance and digestive tract traits of brown-egg laying pullets from hatching to 17 weeks of age. *Poult. Sci.* 94, 1879–93.

Salim, H. M., Kruk, Z. A. and Lee, B. D. (2010), Nutritive value of corn distillers dried grains with solubles as an ingredient of poultry diets: a review. *World's Poult. Sci. J.* 66, 411–31.

Sauveur, B. (1987), Effect of prelaying dietary calcium and phosphorus levels on growth and subsequent laying performance of pullets. Proceedings of the 6th European Symposium on Poultry Nutrition, Königslutter. WPSA(D). E18–E25.

Schutte, J. B., De Jong, J. and Bertram, H. L. (1994), Requirement of the laying hen for sulfur amino acids. *Poult. Sci.*, 73, 274–80.

Sengor, E., Yardimci, M., Cetingul, S., Bayram, H., Sahin, G. and Dogan, H. (2007), Effects of short chain fatty acid (SCFA) supplementation on performance and egg characteristics of old breeder hens. *S. Afr. J. Anim. Sci.* 37, 158–63.

Senkoylu, N., and Dale, N. (1999), Sunflower meal in poultry diets: a review'. *World's Poult. Sci. J.* 55, 153–74.

Shafer, D. J., Carey, J. B. and Prochaska, J. F. (1996), Effect of dietary methionine intake on egg component yield and composition. *Poult. Sci.* 75, 1080–5.

Steenfeldt, S., and Hammershøj, M. (2015), Organic egg production I: effects of different dietary protein contents and forage material on organic egg production, nitrogen and mineral retention and total tract digestibility of nutrients of two hen genotypes. *Anim. Feed Sci. Technol.* 209, 186–201.

Summers, J. D., and Lesson, S. (1985), *Commercial Poultry Nutrition*, 3rd edition, Nottingham University Press Editions, University of Guelph, Ontario, Canada.

Summers, J. D., and Leeson, S. (1994), Laying hen performance as influenced by protein intake to sixteen weeks of age and body weight at point of lay. *Poult. Sci.* 73, 495–501.

Sun, H., Lee, E. J., Samaraweera, H., Persia, M., Ragheb, G. and D. U. Ahn (2012), Effects of increasing concentrations of corn distillers dried grains with solubles on the egg production and internal quality of eggs. *Poult. Sci.* 91, 3236–46.

Swiatkiewicz, S., Arczewska-Wlosek, A. and Jozefiak, D. (2016), The use of cottonseed meal as a protein source for poultry: an updated review. *World's Poult. Sci. J. S.* 72, 473–84.

Tadtiyanant, C., Lyons, J. J. and Vandepopuliere, M. (1991), Influence of wet and dry feed on laying hens under heat stress. *Poult. Sci.* 70, 44–52.

Tang, P., Patterson, P. G. and Puri, V. M. (2006), Effect of feed segregation on the commercial hen and egg quality. *J. Appl. Poult. Res.* 15, 564–73.

Thiele, H. H. (2012), Management tools to influence egg weight in commercial layers. *Lohman Inf.* 47(1), April 2012, 21–31.

Traineau, M., Bouvarel, H., Mulsant, C., Roffidal, L., Launay, C. and Lescoat, P. (2013), Effects on performance of ground wheat with or without insoluble fiber or whole wheat in sequential feeding for laying hens. *Poult. Sci.* 92, 2475–86.

Umar-Faruk, M., Bouvarel, H., Même, N., Rideau, N., Roffidal, L., Tukur, G. M., Bastianelli, D., Nys, Y. and Lescoat, P. (2010), Sequential feeding using whole wheat and a separate protein-mineral concentrate improved feed efficiency in laying hens, *Poult. Sci.* 89, 785–96.

Umar Faruk, M., Bouvarel, H., Mallet, S., Ali, M. N., Tukur, H. M., Nys, Y. and Lescoat, P. (2011), Is sequential feeding of whole wheat more efficient than ground wheat in laying hens? *Animal* 5, 230–8.

Van Krimpen, M. M., Kwakkel, R. P., Reuvekamp, B. F. J. and Verstegen, M. W. A. (2005), Impact of feeding management on feather pecking in laying hens. *World Poult. Sci. J.* 48, 389–98.

Van Krimpen, M., M., Kwakkel, R. P., Van Der Peet-Schwering, C. M., Den Hartog, L. A. and Verstegen, M. W (2009), Effects of nutrient dilution and nonstarch polysaccharide concentration in rearing and laying diets on eating behavior and feather damage of rearing and laying hens. *Poult. Sci.* 88, 759–73.

Waegeneers, N., De Steur, H., De Temmerman, L., Van Steenwinkel, S., Gellynck, X. and Viaene, J. (2009), Transfer of soil contaminants to home-produced eggs and preventive measures to reduce contamination. *Sci. Tot. Environ.* 407, 4438–46.

Whitehead, C. C. (1981), The response of egg weight to the inclusion of different amounts of vegetable oil and linoleic acid in the diet of laying hens. *Br. Poult. Sci.* 22, 525–32.

Whitehead, C. C., Bowman, A. S. and Griffin, H. D. (1991), The effects of dietary fat and bird age on the weights of eggs and egg components in the laying hen. *Br. Poult. Sci.* 32, 565–74.

Wilkinson, S. J., Selle, P. H., Bedford, M. R. and Cowieson, A. J. (2011), Exploring calcium-specific appetite in poultry nutrition. *World Poult. Sci. J.* 67, 587–95.

Wu-Haan, W., Powers, W., Angel, R. and Applegate, T. J. (2010), The use of distillers dried grains plus solubles as a feed ingredient on air emissions and performance from laying hens. *Poult. Sci.* 89, 1355–9.

Yokhana, J. S., Parkinson, G. and Frankel, T. L. (2016), Effect of insoluble fiber supplementation applied at different ages on digestive organ weight and digestive enzymes of layer-strain poultry. *Poult. Sci.* 95, 550–9.

Yousefi, M., Shivazad, M. and Sohrabi-Haghdoost, I. (2006), Effects of reducing dietary methionine, linoleic acid, choline and increasing energy on performance and eggshell quality in aged laying hens. EPC 2006–12th European Poultry Conference, Verona Italy, 10–14 September, 2006, World's Poultry Science Association (WPSA).

Yuan, C., Song, H. H., Zhang, X. Y., Jiang, Y. J., Zhang, A. T., Azzam, M. M. and Zou, X.T. (2014), Effect of expanded cottonseed meal on laying performance, egg quality, 422 concentrations of free gossypol in tissue, serum and egg of laying hens. *J. Anim. Sci.* 423 85, 549–54.

Chapter 2

Laying hen nutrition: optimising hen performance and health, bone and eggshell quality

Y. Nys, Institut National de la Recherche Agronomique (INRA), France

1 Introduction

Hen nutrition must address a number of different goals including maximising egg production, maintaining egg quality, extending the laying cycle and ensuring hen health and welfare. Chapter 1 reviewed the differing dietary requirements in the pre-laying and laying phases. It also reviewed ways feed intake can be controlled to manage energy consumption affecting parameters such as egg weight. This chapter reviews dietary influences on eggshell quality and bone health in particular. It looks first at the role of phosphorus and calcium in the diet. It then goes on to review other nutritional factors such as fatty acids, vitamins, magnesium and trace elements. After a review of dietary influences on egg yolk colouration, the chapter concludes by discussing the ways nutrition can promote hen health in such areas as the prevention of hepatic steatosis, osteoporosis, keel bone disorders and feather picking.

2 Hen phosphorus nutrition

Phosphorus is an essential nutrient for numerous metabolic functions in laying hens, from the cell level to the whole organism. Phosphorus is a constituent of nucleotides (ADN,

http://dx.doi.org/10.19103/AS.2016.0012.33

ARN), ensures transfer of energy (ATP, ADP, GTP), is part of phospholipids which form the cell membranes and participates in numerous regulations of cell pathways through protein phosphorylation (Nys, 2014). At the level of the organism, phosphorus contributes to the acid–base and water balance and buffers cells and external fluids. It also contributes to regulation of feed intake in cases of low and marginal supply of phosphorus but its impact is low in laying hens compared with broilers. Indeed, P deficiency is known to induce many metabolic disorders resulting in deterioration of bone status, appetite and reproductive functions. The major part of body phosphorus is present in bone (around 85%) as part of the Ca hydroxyapatite, the mineral phase of bone that is deposited on the organic bone matrix. A particular form of bone, the medullary bone, develops in hens at sexual maturity. It will be used as a calcium reservoir throughout the laying period to compensate for any inadequate dietary Ca supply for shell mineralisation, which occurs mainly during the night (Whitehead, 2004; Nys and Guyot, 2011). It is noteworthy that the medullary bone is a priority relative to the cortical bone for the reproductive hen and that sex steroids can favour its formation even by using the cortical bone as a calcium source.

The daily contribution of bone Ca can reach 40% of the eggshell (about 1 g) and concomitantly will liberate 0.42 g of phosphorus, which will be mainly excreted in the urine. The egg formation exports in contrast a limited amount of phosphorus (about 0.12 g/egg). The yearly exportation of phosphorus in the egg during 60 weeks of egg laying corresponded to about 36 g. It reaches more than 120 g in the urine in hens fed with the recommended level of dietary ground calcium when feed management does not take into account dietary options favouring the synchronisation of Ca intake with the nocturnal need of Ca for the shell formation. The intense mobilisation of bone associated with egg production can be at the origin of osteoporosis and create a risk of bone breakage particularly in aged hens (Whitehead, 2004). Hens also require phosphorus for body maintenance and to ensure cell metabolism. A particular demand for phosphorus occurred in immature hens for the initial bone growth of the pullet then, at sexual maturity, for the medullary bone formation (Sauveur, 1987). The effects of dietary level of phosphorus within a frequently used range (0.35–0.75% total phosphorus), on egg quality (egg weight, eggshell quality), are rather limited. Egg production and hen mortality are more sensitive to low supply of phosphorus, and can be affected when dietary phosphorus reaches low level (less than 1 g added non-phytic phosphorus, NPP). These are, consequently, the main parameters taken into account to optimise dietary P supply to hens. Bone analysis has been largely used in hens for quantifying hen osteoporosis at the end of the laying period (Fleming et al., 1998).

Defining phosphorus requirement is, therefore, complex in hens because of the duration of the experiments needed to record discontinuous variables such as egg production or mortality recorded as a percentage on large groups of hens and because of the interaction of phosphorus with supply of dietary calcium (level and form). These interactions are not limited to the intestinal availability of phosphorus and calcium but occur also through the precise regulation of their metabolism involving PTH (parathyroid hormone) and the active metabolite of vitamin D (1,25 dihydroxycholecalciferol). These regulations are mainly driven by ionic calcium (sensors) regulating the production of the phosphocalcic hormones effective on target tissues, the intestine, kidney and bone for both Ca and P (Nys, 1993; Bar, 2008, 2009; Nys and Leroy, 2017).

The literature on phosphorus requirements in hens remains relatively limited, but has been stimulated over the past thirty years by the constraints of reducing the dietary supply of inorganic phosphorus to decrease its negative impact on the environment.

Excessive levels of phosphorus contribute to eutrophication of groundwater. The high cost of inorganic phosphorus sources and the development of dietary microbial phytases have also generated the majority of recent experimental data. It is also noteworthy that the geologic sources of phosphorus and sites of extraction are limited in the world and that the ban on alternative phosphorus sources of animal origin has increased the risk for the shortage of the phosphorus supply. The main question for poultry nutritionists is, therefore, to know how far it is possible to reduce the supplementation of inorganic phosphorus without disturbing hen performance and bone integrity.

The subgroup on mineral nutrition from the European WPSA (Vogt et al., 1984) initially developed a global factorial method to recommend a large reduction in dietary phosphorus (0.3% available phosphorus (aP) during the laying period). This recommendation has since been studied for mature hens using global performance of the flock (egg production and mortality) for a whole laying period in diets most frequently supplemented with microbial phytases. It is noteworthy that the majority of studies on P requirement and phytase were carried out using ground limestone as the Ca source and, therefore, corresponded to a daily loss of P in urine. Since 1980, the recommendation has been to reduce the phosphorus concentration of layer diets (0.3% aP which corresponded to NPP) and to progressively increase the dietary calcium content.

However, numerous experiments demonstrated that suboptimal concentrations of dietary phosphorus (<0.25% NPP) induce lower egg production and high mortality (Hartel, 1990; Sohail and Roland, 2002). A corn–soya bean diet without any supplementation of inorganic phosphorus, 0.32% total phosphorus (TP) or slightly supplemented with an inorganic P reaching 0.42% TP, elicited 40% (or even up to 90% in one study) or more than 15% hen mortality, respectively, when low (2%) or high dietary Ca (>3.5%) were provided. The negative effect of low phosphorus (≤0.2% NPP) is also observed at a lower magnitude in negative controls of some studies exploring phosphorus phase feeding or phytase supplementation in hens fed 3.5–4% Ca (Keshavarz, 2003b; Lei et al., 2011). Feed intake, feed conversion index and egg weight are negatively affected at such low values, but these parameters are clearly less sensitive to dietary P than egg production or mortality. Excess of dietary phosphorus (>0.4% NPP) also negatively affected egg weight and shell quality, but can stimulate hen body weight (Sohail and Roland, 2002).

The majority of authors have established the P requirement for hens at a level of 0.25–0.35% of NPP. The NPP is calculated as the sum of NPP of the diet (TP–PP) and inorganic P, and corresponded to a level slightly higher than the available phosphorus used in the studies before 1990. Numerous studies, and the current recommendations of breeding companies, propose the progressive decrease of P level with age of hens, considering that the risk of osteoporosis due to the shortage of dietary P is decreased. The daily rate of shell formation and concomitant bone mobilisation remain high in aged hens, especially when the current tendency is to extend the laying period. Some experimental evidence in aged hens (65–90 wks) shows that dietary phosphorus should, rather, be increased to reduce mortality when dietary Ca is increased (4–5%) for improving shell quality (Bar et al., 2002).

In practical layer diets, phytase is nowadays used routinely and numerous studies have demonstrated the beneficial effect of microbial phytase on P retention, allowing a decrease in inorganic phosphorus supplementation and a lower phosphorus excretion. The majority of plant phosphorus is present as phytic phosphorus and is not hydrolysed in the bird's intestine. Supplementation of microbial phytase in hen diets efficiently hydrolyses phytic phosphorus as demonstrated by numerous publications in this area

(Keshavarz, 2003b; Nahm, 2007; Hughes et al., 2008; Lei et al., 2011) and might also improve utilisation of other nutrients in the diet (Cowieson et al., 2009). A meta-analysis (Ahmadi and Rodehuscord, 2012) has been used to explore the dose response of phytase in diets containing 0.1–0.4% NPP and shows that egg production and feed conversion ratio are optimised at 0.18% and 0.15% NPP when hens are supplemented with 150 or 300 FTU, respectively. The majority of poultry diets contain 2.2–2.5g/kg phytic phosphorus and it is considered that more than 70% of PP might be hydrolysed by recent batches of commercially available phytase. Plant phytases were initially explored and hydrolysed phytic phosphorus but their lower efficiency 40% less efficient compared to microbial phytases and high variability in activity due to plant genetic origin or dietary technological treatments have limited their use (Nys et al., 1996). It might, however, be of interest to take into account their contribution in alternative feeding systems using whole cereals with high phytase activity (wheat, triticale or rye).

It is noteworthy that calcium level and form influence bone mobilisation during the laying cycle and that all nutritional management favouring synchronisation between dietary Ca and shell formation spare bone phosphorus resorption and urine excretion. However, the sparing effect of Ca particle on dietary phosphorus has not been experimentally quantified. The majority of papers on phosphorus nutrition used ground calcium and the few using coarse particles confirmed previous observations (Wu et al., 2006; Rao et al. 1999). However, it has been demonstrated that the main nutritional factor alleviating bone osteoporosis in laying hen is the use of particulate Ca (Fleming et al., 2006). It is recommended to introduce two-third of Ca particles in brown egg layers which are laying very early in the morning and 50% in white leghorn layers which are laying later in the day. The extension of the laying period to 90 weeks of age might increase these risks of osteoporosis, which might be prevented by limiting the magnitude of decreased dietary P supply with age in hens. In conclusion, the introduction of particulate calcium and supplementation with phytase of a diet containing at least 0.2% NPP seem a reasonable compromise to avoid mortality and bone osteoporosis of hens for a laying period the duration of which is continuously increasing.

3 Hen calcium nutrition

The strength of the shell depends on the quantity of material and on the fabric of this protective barrier which is controlled by the shell organic matrix (Nys et al., 1999, 2004; Hincke et al., 2012). Nutrients directly provide shell precursors, Ca via diet and bone, carbonate through pulmonary CO_2 and change in acid–base balance, (Mongin, 1978), influence shell formation (e.g. Mn) or the proportion of eggshell through control of egg weight by protein or energy in the diet. Selection for egg production by increasing the duration of the laying period requires better nutritional control of eggshell quality because decrease in shell quality with increasing hen age determines the termination of egg production for a flock. Twenty years ago, limiting the increased egg weight with increasing hen age was carried out by reducing the dietary protein or amino acid levels to minimise the lowering of eggshell percentage. Currently, the change in egg weight with age is far less marked due to genetic selection (Nys et al., 2008). However, control of change in egg weight might again become important with the increased duration of the laying period.

Calcium is the key nutrient influencing shell strength (Classen and Scott, 1982; Hartel, 1990). Calcium requirements have been established at between 0.9% and 1.2% during the growth period of the pullet, and between 2% and 2.5% and between 3.5% and 4.5% just before and during the laying period, respectively. Selection has resulted in earlier sexual maturity and the introduction of a calcium-enriched diet (intermediate diet with 2.5% calcium) should occur two weeks before the appearance of the first eggs (14–16 weeks) to optimise the formation of the medullary bone and shell quality of the first eggs. It is recommended that 1 kg of diet containing 2.2–2.5% Ca be provided for a period of ten days to sexually maturing pullets, but the duration and amount of intermediary diet should be adapted to the precocity of egg production to avoid any delay in providing enough Ca for the eggshell (Pottgueter et al., 2015). Under practical conditions, a conflict of interest between farmers in charge of rearing the pullets (keeping the same diet until hen transfer) and those breeding the hens can delay the introduction of this intermediary diet relative to the initial egg production. In such case, the intermediary period can be shortened and the laying diet introduced earlier or even directly if the level of egg production is relatively high. Introducing a laying hen diet too early can, however, encourage under-consumption, whereas delayed introduction reduces the quality of both the first and subsequent eggs, even though the hen subsequently receives a diet rich in calcium (Roland and Bryant, 2000). In addition, pullets should be trained initially to eat more through the distribution of diets favouring the development of the digestive organs to optimise this transition period. Form of the diet (size of particles, use of crumbles) or the timing of their distribution toward the end of the rearing period will influence feed intake as previously discussed. Hens export 2.2 g calcium per egg and require at least 4 g per day when taking into account mean calcium retention (Nys, 1999). A limited supply of calcium reduces shell strength and egg production, and increases chicken mortality (Hartel, 1990; Keshavarz, 1998a,b; Kebreab et al., 2009). Phase feeding throughout the laying year (initially 3.5% and then 4.5% and finally 5.5% calcium in the feed) might slightly limit the deterioration in shell strength and bone reserve with increased laying hen age, and might improve egg production (Keshavarz and Nakajima, 1993; Safaa et al., 2008; Valkonen et al., 2009).

The selection for increased egg numbers is associated with a daily production of an egg which is laid very early in the morning, especially in brown egg laying hen which lay earlier than white egg laying hen breeds. Consequently, the morning feed intake cannot contribute to the supply of calcium required for eggshell formation, which mainly occurs during the night. All nutritional management (midnight feeding; alternative feeding with various diets; Ca particles) favouring synchronisation of dietary calcium intake with the period of shell formation will reinforce shell strength (Nys, 1999). European legislation relating to bird welfare, however, limits application of lighting programmes, and it has banned nocturnal flashes of light to avoid any sleep disorder in the hens. Hens show, however, remarkable capacity of their calcium metabolism for supplying Ca for the eggshell (Nys, 1993; Nys and Leroy, 2017; Bar, 2008, 2009). First, hens express a specific appetite for calcium (Mongin and Sauveur, 1979; Wilkinson et al., 2011) a few hours before calcification takes place, favouring storage of feed including calcium in the crop and acid secretion induced by dilatation of this organ (Lee et al., 1988). This favours storage and solubilisation of the dietary calcium throughout the night, especially when coarse particles are available to the hens. The provision of coarse calcium particles limits the hen's need to mobilise calcium and phosphorus from the bone reserve, decreases bone phosphorus elimination (Whitehead, 2004; Fleming, 2008) and reinforces bone strength (Guinotte and Nys, 1993; Fleming et al., 1998). Particulate calcium sources reveal different physico-chemical properties and composition in Ca and trace elements (Guinotte and Nys, 1991) and it is important to analyse a new source for trace element composition before its use. The origin of the source of calcium carbonate (from quarries or marine sources) has limited influence on shell quality in contrast to the size of particles (positive effect when larger than 1 mm compared with small particles of 0.2–1 mm, or smaller than 0.2 mm). Particles have to be greater than 0.8 mm in order to be retained in the gizzard (Rao and Roland, 1990). In addition, hens fed mixed particle sizes have greater feed intake than those fed a 100% ground calcium source (Saunders-Blades et al., 2009). Large particles also increase feed intake of calcium when hens are subjected to heat stress. Particle size should be adjusted according to the density and solubility of the limestone source (Saunders-Blades et al., 2009). Low solubility sources can be introduced to supply two-thirds of the calcium as particles of 1–2.5 mm, and a highly soluble marine source as particles of 2–4 mm. The use of Ca particles helps counteract situations unfavourable to shell quality (old hens, heat stress, low dietary Ca intake) and convincingly favours bone strength. It is recommended that the proportion of large particles relative to ground material be increased when the hens are older.

4 Other nutritional factors affecting hen health and eggshell quality

4.1 Fatty acids

A negative interaction of saturated fatty acid with Ca absorption has been demonstrated at high levels of dietary fat (>6%), possibly due to the formation of insoluble soap combining calcium or magnesium with fat (Atteh and Leeson, 1985). The effect seems limited, possibly because calcium absorption in the proximal intestine precedes that of lipids in its distal parts and because of higher secretion of bile salts in hens than in broilers. However, the amount of eggshell has been shown to decrease with the level of energy in the diet (Perez-Bonilla et al., 2012). That might result from the increased fat content. In this experimental work, the egg weight was not modified by the increased energy level and the decrease in eggshell cannot be explained by a change in the ratio of shell to egg weight.

4.2 Vitamins

Shell formation challenges the calcium metabolism in hens. At onset of egg production, the intestinal capacity for absorbing calcium is greatly increased under the control of the active metabolite of vitamin D, the 1,25 dihydroxycholecalciferol ($1,25(OH)_2D_3$). This metabolite is produced at the kidney level, mainly under the influence of the parathyroid hormone which is stimulated by a decreased level of blood ionised Ca. The vitamin D provided by the diet is firstly hydrolysed in the liver into $25(OH)D_3$, its level being mainly related to the dietary supply of vitamin D. The hydrolysis of this precursor is precisely regulated at the kidney level to provide the active metabolite of the vitamin which activates the intestinal absorption of calcium. Mechanism of this regulation is well established in mammals (Khanal and Nemere, 2008; Diaz de Barboza et al., 2015) and birds (Nys, 1993; Bar, 2009; Nys and Leroy, 2017). It is noteworthy that the $1,25(OH)_2D_3$ does not directly stimulate the uterine secretion of calcium in hens. Vitamin D is, however, a key nutrient in controlling eggshell quality by supplying Ca to the shell gland even if the uterine secretion cannot be stimulated by its active metabolite (Bar, 2008, 2009). Vitamin D is therefore essential for maintaining egg production and shell quality, 400 IU being the minimum level according to earlier studies (Whitehead, 1986; 300 IU/kg, NRC 1994; Barroeta et al., 2012); however, the current recommendations are mostly higher in hens producing more than 300 eggs in a laying year (2000 UI, Weber, 2009) and have positive effects on bone strength and egg production as summarised in the review of Barroeta et al. (2012).

High levels of vitamin D_3 (6000–15000 UI) have been shown to improve bone strength (Mattila et al., 2004). Additionally, supplementing an already activated form of vitamin D_3, such as one of the hydroxylated D_3-metabolites, has been proposed to strengthen eggshell quality. Hens with hepatic metabolic disorders, in particular, can be given 25(OH), which has been shown to improve shell quality in the late production stage (Koreleski and Swiatkiewicz, 2005). Replacing vitamin D_3 partly by 25(OH)D_3 might slightly decrease the number of broken eggs and increase egg weight (Weber, 2009; Barroeta et al., 2012) but some studies have shown no effect (Keshavarz, 2003a). The maximum tolerance level was estimated to be 82.5 μg/kg (Barroeta et al., 2012). Calcitriol (1,25$(OH)_2D_3$) has no or a limited influence on eggshell quality up to 5 μg/kg, but might be toxic at 7 μg/kg (Harms et al., 1990; Tsang and Grunder, 1993). The results remain controversial, as some authors pointed out that use of this metabolite in addition to vitamin D did not improve bone strength or hen performance (Barroeta et al., 2012).

Plants usually contain vitamin D_2 but some species and some microalgae, surprisingly, contain vitamin D_3 or its metabolite (Japelt and Jakobsen, 2013) and might be of interest for supplementation in human or animals for stimulating Ca retention. These plants have elicited calcium intoxication in grazing animals due to the presence of 1(OH)D_3 inducing the stimulation of calcium absorption and that of blood ionised Ca. The highest content of 1,25$(OH)_2D_3$ has been found in *S. glaucophyllum* (0.012–0.032 μg/g dry wt) and in *Cestrum diurnum* (1 μg/g dry wt.). The European Safety Authority published an opinion on the use of *S. glaucophyllum* in animal production (EFSA, 2015). In hens, there is evidence of a positive effect on calcium utilisation or eggshell strength in hens fed diets with variable dietary Ca (2.5–3.5%). A trial carried out by the EFSA consortium also showed a positive effect on eggshell quality when dietary calcium was reduced to 2.7%. In a tolerance study (0, 2, 5 and 20 μg of 1,25$(OH)_2D_3$, for 56 days, 2700 hens), they observed no negative effect on performance or mortality, but also no improvement in eggshell quality. This vitamin D metabolite is probably efficient in reducing eggshell defects due to a lower intestinal supply of calcium, but might be inefficient when defects in eggshell are due to the fabric and properties of the material.

In birds, the supply of vitamin E relies on dietary supply. Vitamin E acts as an antioxidant at the cellular level and is crucial for protecting membranes from lipid oxidation (phospholipid, unsaturated fatty acids) which can modify their permeability. It also protects against free radicals. In hens, it might act at the level of ovarian follicles or on the vitelline membrane. Its effect on egg production remains under discussion among authors as outlined by Barroeta et al. (2012). It is, however, noteworthy that a dietary supply of 50–300 UI vitamin E has a positive effect in alleviating some negative effects of heat stress on hen performance or egg quality as shown by numerous authors (Bollengier-Lee et al., 1999; Scheideler and Froning, 1996; Panda et al., 2008; Khan et al., 2011) and can improve the stability of the vitelline membrane of heat stressed hens (Kirunda et al., 2001).

Vitamin K is involved in bone formation and re-modelling as suggested by its effect on osteocalcin, a bone protein reflecting partly osteoblast activity (Whitehead, 2004; Barroeta et al., 2012). Fleming (2008) suggested an association between vitamin K sub-deficiency and hen osteoporosis and supply of 2–3 mg of menadione/kg diet might prevent this defect. The supply of 10 mg menadione increased the cancellous bone volume at 25 weeks of age either by decreasing cancellous bone loss or by prolonging the duration of the cortical bone formation. Further effects were not observed in older hens (Fleming et al., 1998, 2003). Muller et al. (2009) confirmed a positive effect of vitamin K on bone mineralisation when evaluating vitamin K up to 32 mg/kg in old hens and observed no improvement in shell quality.

Vitamin C can be synthesised by the bird kidney but occasionally at levels inferior to the bird's needs. Provision of vitamin C in the laying hen's diet to improve hen performance remains, however, controversial (Barroeta et al., 2012), but might be of interest in situations of stress. In hens exposed to heat stress, the supply of vitamin C (300–400 mg/kg) has been reported to have some beneficial effects on egg production, eggshell quality and bone strength (Kolb and Seehawer, 2001; Chung et al., 2005), but these observations were not confirmed by others as discussed in the review of Barroeta (2012). There are, however, numerous trials showing that vitamin C can improve hen performance, mortality and egg quality especially in old hens, in hens submitted to high or low temperature, during moult or might improve immune function or tolerance to excess salt or vanadium (Barroeta et al., 2012); therefore, these authors concluded on a rather beneficial effect of vitamin C, especially in hens subjected to high temperatures.

A level of at least 2500 IU vitamin A is needed to avoid a drop in egg production, but egg production and tissue level are optimised with 8000 IU in hens (Richter et al., 1996; Barroeta et al., 2012); this vitamin has limited effect on egg quality.

Water-soluble vitamins (B), when supplied above the hen's requirement, have a relatively limited effect on egg quality, but more detail information on the function of these vitamins, their effect of deficiency or high level on egg production and quality are available in the review by Barroeta et al. (2012).

4.3 Electrolyte balance

The electrolyte balance of feed influences birds' acid–base metabolism, which itself is also strongly dependent on shell formation (Mongin, 1978); however, attempts to define the optimum ratio of Na, K and Cl in hens have not been conclusive. Changes in Na + K/Cl ratio as high as 0.4–7.7, obtained by modifying dietary provision of sodium and chlorine, considerably affect blood pH and plasma concentrations of bicarbonate, but they have no effect on shell quality (Mongin, 1978; Kurtoglu and Balev, 2007). However, large excesses of chlorine (>0.7) or sodium (>0.35) diminish shell quality (Hughes, 1988).

4.4 Magnesium

Magnesium (Mg) is involved in the metabolism of amino acid, lipids and also with bone and vitamin D (Morii, 2007). Two thirds of the body Mg is located in the bone. A deficiency in Mg (<0.21%) reduces egg production and the deposition of eggshell (Waddell et al., 1989). However, magnesium has been poorly studied in poultry production because Mg is provided at levels much higher than the chick or hen requirement by plant feedstuffs. A layer diet (corn–soya bean) contains about 0.16% magnesium, that is, fourfold the requirement of laying hens (400 mg/kg, Vogt et al., 1984). The retention of Mg is estimated to be higher than 50%, but its availability may change because of possible interactions with other nutrients as shown in broilers (Shastak and Rodehutscord, 2015). Therefore, there is no chance of observing an Mg deficiency and there is no routine supplementation of Mg in the diet because of these large Mg levels in raw vegetable materials (especially bran, sunflower, rapeseed). In addition, there is no evidence that supplementation improves eggshell quality or that it has detrimental effects on eggshell quality until levels reach 0.77% (Atteh and Leeson, 1985). Nevertheless, an excess of Mg increases water consumption and can increase the occurrence of dirty eggs when the dietary level is higher than 0.6%. This laxative effect of Mg has been confirmed in chicks in which addition of 0.25, 1 or 2 g Mg/kg increases the faeces moisture in a linear manner (Van der Hoeven-Hangor et al., 2013). In broilers, however, a novel interest for Mg resulted from the observation that Mg may influence meat quality and prevent some heat stress induced oxidative damage as recently summarised by Shastak and Rodehutscord (2015); but such an effect has not been explored in laying hens.

4.5 Trace elements

The early literature on trace elements demonstrated that manganese (Mn) and copper (Cu) are essential for eggshell formation. Copper deficiency in hens affects the formation of lysine-derived cross-links and, consequently, the biochemical and mechanical properties of eggshell membranes (Baumgartner et al., 1978; Chowdhury, 1990), resulting in egg shape abnormalities. Eggshells from hens fed on Mn deficient diets (<7 mg) showed lower eggshell thickness, translucent areas due to alterations in eggshell ultra-structure in the mammillary layer and a decreased concentration in polysaccharide precursors of the eggshell matrix (Leach and Gross, 1983). Manganese is a component of Galβ1,3-glucuronosyltransferase (GlcAT-I) involved in proteoglycan synthesis and Mn supplementation increased synthesis of glycosaminoglycan in eggshell and might therefore influence its microstructure and breaking strength (Xiao et al., 2014, 2015). Alternatively, Mn can interfere with the face parallel to the 'c' axis of the $CaCO_3$ crystal and favours an elongation of the calcite crystal along its 'c' axis (Mann et al., 1993). Thus, it is also possible that the presence of trace elements could alter the crystallographic texture of eggshells and, consequently, their mechanical properties. Zinc (Zn) is a component of carbonic anhydrase, a key uterine enzyme in supplying shell carbonate. Inhibition of this enzyme results in lowered bicarbonate ion secretion and, consequently, greatly reduces eggshell weight (Nys et al., 1999). Zinc deficiency (<10 mg/kg) decreased egg production, feed consumption, elicited bone abnormalities and increased mortality in poultry, affecting numerous functions such as the antioxidant system or immune system (Kienholtz et al., 1961; Naz et al., 2016). No negative effect was observed on hen performance and egg quality when the diet contained 28–34 mg Zn/kg (Stahl et al., 1986) or even less when included in a casein type diet (15 mg/kg; Turk et al., 1959).

A hen basal diet (corn–soya bean) supplies 30 mg/kg of Zn, 20 mg/kg of Mn and 6 mg/kg of Cu. These dietary levels are higher than the deficient level but slightly lower than the recommendation for Zn and Mn and, in addition, these elements of plant origin are only partly available. In absence of mineral or organic supply of Mn, Cu and Zn, feed intake and feed conversion are not affected, in contrast to egg weight, egg mass (Bastien et al., 1979; Abdallah et al., 1994; Stefanello et al., 2014). The main effect was, however, observed on eggshell thickness or breaking strength (Mabe et al., 2003; Stefanello et al., 2014). It is, therefore, current practice to supply hen diets with Cu, Zn, Mn from inorganic or organic sources for reinforcing eggshell strength. The mean practical Zn level found in the typical European layer diet is currently in the range of 90–130 mg/kg and fits with the European legislation limit of 120 mg/kg (EU 2016/1095). The effect on eggshell quality is largely controlled by the Mn supplementation, as demonstrated by Sazzad et al. (1994), when supplying 80 mg/kg Mn to a 25 mg/kg basal diet and by Shukla et al. (1993) in quails (80 mg in lieu of 20 mg/kg) or by Xiao et al. (2014, 2015). Manganese provided at levels higher than 70–100 mg/kg had no further effect on eggshell quality (Ochrimento et al., 1992; De Faria et al., 1999; Xiao et al., 2015).

There is little evidence demonstrating a positive effect of Zn when provided at higher levels than the hen requirement (recommendation is 50–60 mg/kg total Zn; EFSA, 2014). When supplied at a high level of 100–200 mg/kg (Stevenson, 1985) or at 90 and 130 mg/kg (Kienholz et al., 1992), Zn did not affect shell thickness. However, diets supplemented with 80 mg/kg Zn as Zn-methionine (Moreng et al., 1992) or 100 mg/kg Zn (Balnave and Zhang, 1993) alleviated decreased eggshell quality and shell weight and reduced percentage of shell defects, in hens exposed to high temperatures and to water containing 2g/l NaCl. In another study involving hens subjected to a hot environment, no corrective effects were observed for eggshell quality in hens fed Zn-methionine supplementation (40 mg/kg Zn; Kita et al., 1997). The source origin influenced the availability of Zn among inorganic (sulphate, oxide) or organic Zn sources. Some organic sources showed slightly higher availability, but the issue remains controversial with presence or absence of difference in the case of Zn (Dale and Strong, 1998; Cyelan and Scheideler, 1999; EFSA, 2014; Swiatkiewicz et al., 2015), but also for Mn sources (Mabe et al., 2003; Lim and Paik, 2003; Xiao et al., 2015).

The eggshell quality of hens fed high dietary Cu (70–140 mg/kg Cu SO4) was not modified by Cu supplementation (Christmas and Harms, 1983) It impaired shell thickness when supplied at 200 mg/kg (Griminger, 1977) or at higher level (250 and 500 mg/kg; Harms and Buresh, 1986), but did not affect egg mass, or even tend to improve feed conversion index (Kim et al., 2016). The negative effect on eggshell quality results partly from the use of sulphate as it disappears when Cu acetate, in lieu of sulphate (250 mg/kg), is used (Al Ankari et al., 1998).

Attempts to improve eggshell quality by supplying boron (100 mg/kg; Qin and Klandorf, 1991), vanadium (20 mg/kg; Ueberschär et al., 1985) or fluoride (6 to 20 mg/l; Coetzee et al., 1997) have been inconclusive. Other metal cations such as nickel (Ni, 100–500 mg/kg) chromium (500–2000 mg/kg) or lead (20–100 mg) decreased eggshell weight (Meluzzi et al., 1996). Chromium picolate improved eggshell thickness at 600 µg/Kg in aged hens, but decreased the Haugh unit at the lower level of 400 µg/kg (Ma et al., 2014). In conclusion, the evidence demonstrating an increased amount of eggshell by feeding additional trace elements is rather limited when mineral deficiencies are avoided. Mn supplementation is, however, of particular interest.

Phytate is identified as an antagonist for the bioavailability of trace minerals in monogastrics, especially for copper and zinc (Schlegel et al., 2016). In cereal-based diets for chickens, without trace mineral supplementation, a higher presence of plant phytate reduces zinc bioavailability from plant origin and possibly that of iron and manganese, but not that of copper. Trace elements are, however, less affected in poultry than in other species due to a lower gastrointestinal pH compared with other species and the inorganic or organic supplementation of zinc, and that of other trace minerals, is poorly affected by plant phytate in poultry (Schlegel et al., 2010, 2013). Microbial phytase releases only an equivalency of 5–10 mg zinc (as sulphate)/500 FTU in chickens. It is expected that, in hens which secrete a high level of acid in the gizzard, the negative interaction between phytate and trace elements should be limited.

4.6 Other additives

There is evidence that some commercial feed additives such as pre and probiotics, organic acids and essential or plant extracts might improve eggshell quality. The literature in this area has been summarised in a recent review by Swiatkiewicz et al. (2015). The results exploring the efficiency of these additives are quite controversial and, in this area of commercial additives, there is the risk of data not being published when showing no positive effect. Fructan, inulin or oligofructose have been reported to improve eggshell breaking strength in aged hens, possibly by increasing the availability of Ca (Swiatkiewicz et al., 2015). Skim milk powder (3% or 4%) has also been reported to improve egg quality (Cesari et al., 2014). A positive effect of inulin and of *Bacillus subtilis* on egg shell quality was also observed by Abdelqader et al. (2013), but other authors did not observe any effect of prebiotics and probiotics as described in the review of Swiatkiewicz et al. (2015).

Short-chain organic acids (0.05% or 0.078%, mixture of formic, butyric, propionic and lactic acid) have also been shown to improve eggshell breaking strength or thickness by 10–17% in aged hens (Sengor et al., 2007; Soltan, 2008), possibly as a consequence of a higher solubility and absorption of calcium. A positive effect was also observed by Swiatkiewicz et al. (2010). Finally, there are also reports demonstrating that a mixture of essential oils might reduce the number of broken eggs (Cabuk et al., 2004) or that herbal products (Sharma et al., 2009) or traditional medicinal herbs (Zhou et al., 2009) can increase eggshell thickness and bone quality; however, others did not observe any effect (Swiatkieswicz et al., 2013).

5 Dietary control of egg yolk colouration

Birds cannot synthesise carotenoids; therefore, yolk colouration depends directly on hen dietary supply. Indeed, the yolk colour depends directly on the level of carotenoids consumed by the hen, its intestinal transfer efficacy and on the chemical composition of the carotenoid source which determines its colouring capacity (Hencken, 1992; Nys, 2000; Hamelin and Altemueller, 2011; Amaya et al., 2014). In poultry, β-carotene does not contribute to yolk or skin colouration and α and β carotene deposition in the yolk is lower than 1% (Hammershoj et al., 2010). Birds mainly accumulate xanthophylls (Hencken, 1992; Nys, 2000; Breithaup, 2007), which correspond to oxycarotenoids containing one or more hydroxyl groups. Deposition of carotenoid in the yolk is rapid (<24 h), corresponding to the duration of the yolk formation. A daily and alternative distribution of different carotenoids with various colours will induce rings of corresponding colour in the yolk when cooked. Eight to ten days are needed for the yolk colour to remain stable and any variation in daily feed intake between hens in a flock can elicit large variation in yolk colour (1–2 DSM colour fan).

The efficiency of oxycarotenoid sources depends on their composition in carotenoids, their digestibility, their rate of transfer, their metabolic modifications, their affinity and yield of deposition in target tissue, and their hue colouration. The effectiveness of deposition of carotenoids varies greatly between sources due to different composition of carotenoids and their efficiency of transfer. The transfer yield of corn gluten xanthophylls to yolk measured by HPLC was the highest for lutein and zeaxanthin (25–30%), but was lower for cryptoxanthin (14–17%) (Looten et al., 2003). The most effective carotenoids are synthetic forms which have been manufactured because of their high transfer and colouring capacity (apo-ester 50–60% and canthaxanthin 30–50%; Breithaupt, 2007; Hamelin and Altemueller, 2011).

The process of extraction of natural sources influences their efficiency; therefore, updated evaluations are required before using a novel product. The digestibility of plant carotenoids, tagetes or paprika can be improved by saponification in poultry, but this observation remains controversial (Breithaupt, 2007). The hydrolysis converts the diester to its free form and was shown to improve the absorption of lutein, zeaxanthin and capsanthin (from red pepper) in poultry, improving their efficiency (×1.5) for yolk colouration (Galobart et al., 2004). Nolan et al. (2016) showed, however, that zeaxanthine and mezzo-zeaxanthin diacetate were more efficiently transferred to yolk than the free form. On the other hand, short-chain saturated fatty acid and long-chain unsaturated fatty acids are known to favour lutein absorption and yolk colouration (Huyghebaert, 1993). By contrast, mycotoxins and coccidiosis will depress the absorption of carotenoids and decrease its tissue accumulation in chicks (Hamilton, 1992; Hernandez-Velasco et al., 2014).

The efficacy of carotenoids depends also on the ability of the source to generate vitamin A by splitting the molecules (Lietz et al., 2010) as the light absorption of carotenoids is due to a chromophore formed by the whole C40 carbon skeleton with its sequential distribution of double and single bonds and its 8 isoprenoid units (Hencken, 1992; Rodriguez-Amaya, 2016). There are also differences in the rate of yolk deposition between different stereoisomers (Hencken, 1992; Hamilton, 1992). The optically active 3R, 3′R carotenoids are predominant in plants and are deposited with a greater efficiency than other isomers, as shown for zeaxanthin. *Trans*-carotenoids are more effective oxycarotenoids than *cis* isomers because of their redder hue and higher stability. The *trans/cis* proportion is variable in commercial synthetic carotenoids (Aureli et al., 2013), depending on the production process. The proportion of *trans*-carotenoids is higher in plants (60–90%) but levels of the *cis*-carotenoids can increase during the manufacturing process or storage under light, heat and oxygen (Leigh Hadden et al., 1999).

Carotenoids are quite sensitive to their physical and chemical environment and continuous efforts are being made by manufacturers to improve stability of the product in optimising the coating of the product (Hamelin and Altemueller, 2011). The synthetic sources are more standardised and stable than plant extract. In premixes, the dominant effect is redox reactions by trace minerals triggered by moisture. Stabilisation and encapsulation are used to protect the carotenoids from degradation and considerable variations can be observed depending on the process developed by the supplier when considering synthetic carotenoids, but also for highly concentrated plant extracts. The recovery rate of different commercial canthaxanthins varied from 66 to 92% was 75.6% for a citranaxanthin product and only 38.5% for capsanthin from a paprika product (Gadient et al., 2010). High variability can also be observed between tagetes or paprika commercial sources (Magnin et al., 2009). Micro emulsification of carotenoids is an alternative process which can improve by 20–30% their efficiency in poultry (Chow et al., 2014).

The main vegetable sources of yellow carotenoids are yellow corn (10–40 mg/kg xanthophyll), corn gluten (180–400 mg/kg in 60% CP), lucerne (350–5040 mg/kg), lucerne concentrates (450–1000 mg/kg) and flowers (marigolds, tagetes; 4500–14000 for meal, –50000 mg/kg for concentrate) and paprika as red carotenoids (Nys, 2000; Breithaupt, 2007; Hamelin and Altemueller, 2011). High carotenoid maize has also been developed by selection for kernels with high zeaxanthin or lutein content, allowing a fourfold increase in yolk colour when supplying a similar amount of maize (Burt et al., 2013). Plants such as corn or lucerne are used as a whole but tagetes, marigold, lucerne and paprika can be extracted to produce highly concentrated and more stable carotenoid sources. Carotenoid concentrations are higher in plants when the protein content of the feedstuff is higher. Wheat and sorghum contain very low levels of carotenoids and do not contribute to egg yolk colouration.

The concentration of total xanthophylls in plant sources varies greatly between strains, with plant maturity at harvest, the dehydration process, duration and conditions of storage, and with the process of extraction when carotenoids are concentrated (Nys, 2000; Breithaupt, 2007; Amaya et al., 2016). The profile of the various xanthophylls in a source is relatively stable and specific to the source (typically 10% carotene and 90% xanthophylls including 45% lutein and 20–30% zeaxanthin and, at lower concentration, cryptoxanthin and zeinoxanthin for corn; 50–75% lutein, 1–7% zeaxanthin and cryptoxanthin, 7–10% violaxanthin and neoxanthin in lucerne) (Nys, 2000; Rodriguez-Amaya, 2016). The deposition of lutein or xanthophyll in maize or lucerne varied from 19 to 27% (Hamelin and Altemueller, 2011). Maize can lose from 25 to 50% of its initial xanthophylls after being dried or stored for one year (Looten et al., 2003). Paprika and tagetes oleoresin are prepared by dehydration, solvent extraction, saponification and stabilisation by spray drying. Paprika effectiveness is lower than that of canthaxanthin. The ratio varies from 3 to 5 for yolk deposition yield and from 1 to 4 for colouration (Breithaupt, 2007). Extracts from algae containing asthaxanthin (3,3′-dihydroxycanthaxanthin) can be used in combination with yellow xanthophylls, but the rate of deposition is lower than for canthaxanthin (Elwinger et al., 1997) and algae remains an expensive source. Alternative sources originating from microalgae (*Chlorella vulgaris*) (Lemahieu et al., 2013; Kotrbacek et al., 2013; Swiakiewicz et al., 2015), photosynthetic bacterium (*Rhodocytus gelatinosus*), fish oil rich in astaxanthin, crab meal or even orange peel can colour egg yolk to a certain degree, but have a low efficiency for commercial use (Nys, 2000).

Synthetic oxycarotenoids which correspond to natural carotenoids (apo-ester, canthaxanthin) have been chosen for their colouring effectiveness, which is two to three times better than carotenoids of vegetable origin, and for their high stability due to encapsulation offering protection against oxidation and degradation. The transfer efficiency of apo-ester was estimated to be from 50 to 55% and that of marigold extract from 16 to 18% (Sirri et al., 2007). These synthetic products are, however, banned in organic production, which relies mainly on more yellow plant sources, explaining the paler yolk colouration of organic eggs. In practice, satisfactory colour of table eggs can be obtained with small amounts of yellow xanthophylls (15–25 mg/kg) combined with 1–2 mg/kg red carotenoids (Nys, 2000; Amaya et al., 2014). The requirement is higher when the egg is cooked or used in prepared foods (noodles, bakery products, mayonnaise, etc.). Egg cooking degrades carotenoids to an extent depending on the source, apo-caroten–ester being more stable than lutein, for which the loss is 19% after egg boiling (Hamelin and Altemueller, 2011).

In the European Union, the supplementation should comply with the European legislation which permits a dietary supply of 80 mg/kg carotenoids with a limit of 8 mg/kg for canthaxantin (Breithaupt, 2007). The European Food Safety Agency assessed and confirmed in 2010 this recommendation for use of canthaxanthin in animal feed (EFSA, 2010). In conclusion, the supply of xanthophylls is crucial to meet the demand of the consumer for a coloured egg yolk and in addition may be of interest for human health as eggs are a good source of zeaxanthin and lutein being more efficiently absorbed in the intestine than other nutrients (Thurnham, 2007). Xanthophylls can help prevent age-related macular disease and possibly other human chronic diseases (cancer, cardiovascular disease) (Rodriguez-Amaya, 2016), but the egg contribution is limited in humans compared with fruit and vegetables. In breeder hens, carotenoids also contribute to reinforcing the antioxidative protection of egg and improve fertility, hatchability and embryonic mortality of fertilised eggs (Surai, 2012).

6 Controlling hen health

Extension of the egg production period is associated with the maintenance of good levels of egg production and quality, but also relies on liveability and mortality of the birds. The exportation of a large amount of egg mass, energy, protein and calcium highly solicited the tissue involved in egg formation. Intense metabolism occurs in the liver for the yolk formation, in the ovary and the oviduct for the formation of other egg compartments. The large exportation of Ca for shell formation challenges the calcium metabolism and bone integrity of hens. The maintenance of these tissues is a prerequisite for extending the laying period. The control of feeding behaviour by management involves adjusting feed to hen requirements to avoid any imbalance. It is also important for hen welfare in term of feather picking which can be at the origin of high mortality in hens or decreased liveability in aged birds.

There is a need for more robust hens, when keeping hens for a longer period, to avoid mortality. Hens need to be more adaptable to variable conditions of production as they will produce throughout the year and face different seasons. The hen's environment can be perfectly controlled in closed building but the development of alternative systems in Europe with free range and floor systems challenges the hens to a greater extent by increasing the variability in their environment. The cost of hen selection limits the number of laying hen lines and the strategy of breeder companies is to select hens adapted to very diverse environments throughout the world, rather than developing lines specific to one particular climate. Nutrition can also alleviate metabolic problems by controlling the development of tissues and, by preventative measures, improving the protection of a particular tissue and maintaining egg quality. Feeding strategies, or use of additives, have been proposed to protect the organs (liver, intestinal tract, bone) involved in the synthesis of egg precursors.

6.1 Preventing hepatic steatosis

The exportation of massive amounts of proteins and lipids in eggs during the laying period challenges hen metabolism and might be at the origin of some metabolic disorders such as fatty liver haemorrhagic syndrome (FLHS). This non-infectious syndrome can be responsible for a large percentage of hen mortality, 3–5%, and can contribute to up to 75% of total mortality (Shini and Bryden, 2009; Trott et al., 2013) . It seems to be favoured by cage production and heat stress. The syndrome corresponds to the excess accumulation of fat in the liver and the appearance of liver haemorrhage which can result in the death of the hen by internal bleeding (Butler, 1976; Shini and Bryden, 2009). These metabolic disorders are mainly observed in commercial production and need particular conditions (e.g. force feeding) in order to be observed in experimental facilities. In some cases, only hepatic steatosis is observed.

In birds, all fatty acid synthesis occurs in the liver, mainly from carbohydrates. That is the case in hens as layer diets contain only a small percentage of lipids. The yolk contains a large amount of lipids and the metabolic activity of the liver is highly stimulated at the bird's sexual maturity to form yolk precursors (Nys and Guyot, 2011). Hens are, therefore, more susceptible to hepatic steatosis as a consequence of genetic, environmental and hormonal influences, but nutritional factors are likely to be crucial as the first clinical sign is a large accumulation of fat in the liver in more than 95% of the cases (Trott et al., 2013). Imbalance in the protein to energy ratio (Hansen and Waltzem, 1993), low protein and high carbohydrate diets (Butler, 1976; Julian, 2005) and low protein high fat diets (Rozemboim et al., 2016) have been shown to favour the appearance of fatty liver haemorrhagic syndrome. Fatty acid synthesis is coupled with glycolysis so a high dietary consumption of carbohydrate markedly increases the fat content of the liver, favouring steatosis (Butler, 1976; Hansen and Waltzem, 1993).

Substitution of carbohydrate by dietary fat has been suggested to reduce hepatic triacylglyceryl synthesis by downregulation of lipogenic enzymes (Butler, 1996; Zhang et al., 2011), and hence might protect against FSLH. However, other authors conclude that a high energy diet, whatever the origin, carbohydrate or fat, induces FLHS (Zhang et al., 2008). That conclusion was confirmed by Rozenboim et al. (2016), who observed a higher incidence of FLSH in hens fed an imbalanced diet with low protein and high energy provided as fat. FLHS was frequent in the 1980s and was observed in hens or males treated with oestrogens which upregulate the gene coding enzyme involved in lipogenesis in the liver (Butler, 1976; Lee et al., 2010; Choi et al., 2013).

This fatty acid disorder can affect the level of egg production but has little effect on egg quality (Lee et al., 2010). The negative effect on egg production is inconsistent, possibly affecting more obese hens with hepatic steatosis than those affected by FLHS. These liver disorders might impact Ca metabolism and bone formation (Jiang et al., 2013), and therefore, it is recommended that dietary 25(OH)D_3 be provided to bypass the liver hydroxylation of vitamin D to improve eggshell quality. Feed restriction can reduce the incidence of fatty liver syndrome, but might also decrease egg production. Supplementation of choline, inositol, vitamin B12, folic acid and vitamin E has been suggested to prevent accumulation of triacylglycerides in the liver and to limit the incidence of hepatic steatosis and FLHS (Butler, 1976) and is frequently used in field conditions.

Choline plays an important role in liver fat metabolism and is part of lecithin, a phospholipid deposited in the yolk. Birds are able to synthesise choline from methionine, but its dietary supply (at least 1050 mg/kg, NRC) is recommended to spare methionine (Barroeta et al., 2012). The requirement for choline, which is mainly provided by dietary feedstuffs, increases in hens supplied with a lower level of methionine. Numerous papers analysed the respective influence of choline and methionine on hen performance and data have been reviewed by Barroeta et al. (2012), who concluded that choline has little influence on hen performance and egg quality when sulphur amino acids meet the hen's requirement. There is evidence, however, that supply of choline prevents the accumulation of fat in the liver (March 1981). A choline deficiency limits exportation of triglycerides by the hepatocytes and reduces lipoprotein synthesis (Zeizel, 1981). The recommendation of choline should be a total of 1200–1500 mg/kg; therefore, 250–500 mg/kg should be supplied in addition to the contribution of dietary feedstuffs (about 1000 mg/kg). An excess of choline slightly decreased egg production . The effect was linear when dietary supplementation varied between 500 and 6800 mg/kg (Zhai et al., 2013).

6.2 Preventing osteoporosis

The incidence of osteoporosis was high in Europe, as indicated by the percentage of hens showing bone fracture at the end of the laying period in early literature (up to 30%, Gregory and Wilkins, 1989). Osteoporosis results from loss of structural bone and progressive skeletal weakness over the laying period and is more severe at end of lay (Fleming et al., 1998; Whitehead and Fleming, 2000). It is associated with the high calcium demands of the laying hen for shell formation which mobilises medullary bone and excretes large amounts of phosphorus because of the desynchronisation between eggshell formation during the night and dietary calcium consumption as discussed in part relative to dietary phosphorus. Therefore, all factors improving the direct supply of dietary calcium will reduce bone mobilisation and phosphorus excretion (Nys, 1999; Fleming, 2008). In particular, the provision of coarse calcium particles limits the hen's needs to mobilise calcium and phosphorus from the bone reserve, decreases bone phosphorus elimination and reinforces bone strength (Guinotte and Nys, 1993; Nys, 1999; Fleming et al., 1998, 2003). The medullary bone is formed at sexual maturity. Therefore, it is important to provide calcium and phosphorus for the proper establishment of this reservoir and to avoid early demineralisation of cortical bone by providing an intermediary diet enriched with calcium (Classen and Scott, 1982). Vitamin K3 and fluoride have also been suggested to affect bone characteristics when provided to immature pullets or mature hens, but the magnitude of the effect is more limited (Fleming et al., 1998, 2003). The protective effect is only partial for the cortical bone, as outlined by Whitehead (2004): the priority which is under the control of oestrogen in laying hens, is to form medullary bone even if detrimental for the cortical bone. There is, in addition, limited structural bone formation in laying hens, It is likely that factors other than nutrition, such as genetics and hen activity also contribute to limit osteoporosis in laying hens (Whitehead and Fleming, 2000; Fleming et al., 2006).

6.3 Preventing keel bone disorders

Keel bone disorders represent one of the most important welfare issues in commercial laying hens, as several studies have found a high prevalence of keel bone fractures in laying hens (Fleming et al., 2004; Sandilands, 2011; Harlander-Matauschek et al., 2015) and because of the development of non-cage systems. Keel bone fractures (KF) correspond to breaks in the bone that will typically manifest as a callus around the fracture site and result from a traumatic bone fracture by collisions with perches or other obstacles or from bad landings (Fleming et al., 2004; Harlander-Matauschek et al., 2015). Keel bone deformation involves sharp, unnatural deviations, or bending of the bone due to long-term pressure on the keel (Wilkins et al., 2004; Pickel et al., 2011).

The incidence is larger in various non-cage systems and ranges from 60% in a non-cage single-tier system (Nicol et al., 2006) to 70% in a multi-aviary system (Freire et al., 2003), or even to 80% in a free range with aerial perches (Wilkins et al., 2011), but seems lower in furnished cage systems (around 50%; Sherwin et al., 2010). A study on numerous flocks in Belgium confirmed the high incidence of keel bone damage (82% fracture and 59% deviation) and identified aviary type, tier flooring material and corridor width as risk factors of keel bone fracture (Heerkens et al., 2016). The incidence and severity of keel bone fractures are generally low at the start of lay, will increase when hens get older and are influenced by the genetic background of the birds (Fleming et al., 2004; Käppeli et al., 2011; Wilkins et al., 2011). Hens free from keel fractures laid more eggs (91.7% vs. 84.9%) of significantly heavier weight (61.9 g vs. 60.2 g), ate less feed (139 g vs. 151 g) and drank less water (212 ml vs. 237 ml) than hens with fractures, probably as a consequence of reduced mobility (Nasr et al., 2013). Lack of bone mass appears to be the underlying cause of keel fracture and deformity, rather than a qualitative modification in bone, as demonstrated by genetic approaches (Fleming et al., 2004; Stratmann et al., 2016.).

Optimising the nutritional factors favouring bone formation or reducing its resorption such as calcium dietary supply and large particle size, dietary phosphorus or vitamin D might reduce the incidence of keel bone damage. A few specific studies have explored nutritional factors. A tentative attempt to supply 25(OH)D_3 has not been successful (Käppeli et al., 2011). Of interest is the observation of Tarlton et al. (2013) demonstrating a lower incidence of keel bone fracture (60% less than a standard diet at 50 weeks of age) in hens receiving a diet supplemented with short-chain omega-3 fatty acid. The same group confirmed this observation in a second experiment, but showed a reverse effect in a second trial when supplying a different level of unsaturated fatty acid, suggesting that excessive quantities of n3, or high levels of C20/22, might increase the incidence (Toscano et al., 2015). The positive effect was associated with better bone strength confirming the influence of bone quality on the incidence of keel bone damage.

In conclusion, more experimental work is needed for exploring if some particular nutritional factors might contribute to reducing the incidence of this welfare issue which has partly emerged from the decision to ban the cage system in favour of alternative systems with high hen density.

6.4 Preventing feather picking

Feather picking and cannibalism are an important welfare issue (Blokhuis et al., 2007), the impact of which might be increased by the increased duration of the laying period as feather loss is frequently observed in flocks of old hens. The risk might increase with the expected ban on beak trimming as it has been shown in some experiments that beak trimming reduced the mortality from 40–51% to 4–8% (Guesdon et al., 2006). However, progress in genetic selection, combined with management on the one hand, and relatively low flock mortality observed in countries where beak trimming has been banned a few years ago on the other hand, support the feasibility of prohibiting beak trimming (Janczak and Riber, 2015).

Feather coverage affects maintenance energy requirement and consequently affects negatively the feed efficiency of layers at normal temperatures. A lower feather covering affected the insulation of hens, favouring heat loss. Consequently, hens with a reduced feathering coverage have a larger energy requirement and loss of feathers induces increased feed consumption, the magnitude of which depends on environmental temperature: +20 g at 12.8°C if the feather coverage corresponded to 50% normal, +7 g at temperature of 23.9°C (Peguri and Coon, 1991).

Feather picking is inversely related to the time spent by the hens in other exploratory activities such as feeding, drinking and foraging. It can be reduced by increasing duration of feeding behaviour, which can be induced by nutrient-diluted, highly insoluble non-starch-polysaccharides-containing diets; insoluble-fibre-rich diets; and diets containing oats or barley instead of wheat (Hetland et al., 2004; Van Krimpen et al., 2005, 2008a,b, 2009; Mateos et al., 2012; Kjaer and Bessei, 2013) or roughage (Steenfeld et al., 2007). In addition, dietary fibre contributes by increasing the feeling of satiety. The appetite for additional fibre and time spent foraging in the litter will depend on the amount of crude fibre in the diet. The presence of fibre litter will therefore be more efficient in reducing feather picking in hens fed a low-fibre diet (Van Krimpen et al., 2009; Mateos et al., 2012). The availability of straw to the hens lowered feather picking (Aerni et al., 2000). Feather picking is increased in cases of protein or amino acid deficiency (Elwinger et al., 2008). Inadequate supply of methionine and lysine stimulates feather picking and, in some cases, cannibalism, as underlined by Kjaer and Bessei (2013). That might result from increased exploratory behaviour which is observed in birds exposed to feed deficiencies. These authors also underlined the potential beneficial effect on feather pecking of dietary tryptophane, a precursor of serotonin. Use of fermented feed, with low attractiveness for hens, was shown to elicit a poorer plumage condition (Engberg et al., 2009). A reduction in incidence of feather picking is associated with a lower mortality as feather picking is associated with cannibalism when severe.

Feather picking during the rearing phase influences development of feather picking and feather conditions in adult hens (Janczak and Riber, 2015). Plumage conditions in hens are better when no feather damage has been observed in pullets at 17–20 weeks of age (Drake et al., 2010). The risk of observing feather pecking during the laying period increased to 90% when it has been observed at an initial stage in pullets (Bestman et al., 2009). Therefore, it might be important to control, by nutrition, feather pecking at an early stage of pullet rearing (Qaisrani et al., 2013). Earlier transfer of bird (<16 weeks), the absence of transfer of pullets to a different farm, or more globally matching the pullet environment to that of hens seems to decrease the severity of feather damage in hens (Drake et al., 2010; Janczak and Riber, 2015). Rearing pullets on straw litter at an early age and during the rearing period has a preventative effect on the development of feather pecking (Aerni et al., 2005; Johnsen et al., 1998; Bestman et al., 2009). It favours hen foraging, which contributes to reducing the incidence of feather picking (Gilani et al., 2013), as underlined in the review of Janczak and Riber (2015). Feed presentation contributed to reducing feather pecking when it increased the duration of feed intake as observed in hens. In pullets, using mashed or crumbled diet instead of pellets, using whole grain or replacing feed hoppers by chain feeders reduced feather pecking (Blokhuis and van der Haar, 1992; Drake et al., 2010; Gilani et al., 2013). In conclusion, management of feed is clearly a key element for reducing feather pecking and should be initiated during the growing phase of the pullets.

7 Conclusions

This chapter has reviewed the role of nutrition in improving eggshell quality and bone health in hens. It has also discussed ways nutrition can help prevent disorders such as hepatic steatosis, osteoporosis and feather picking. As noted in Chapter 1, nutrition remains a crucial factor in optimising egg production, given the genetic similarities between the common breeds used as well as the economic importance of feed as a cost of production.

8 Where to look for further information

Part of the information are issued from reviews published in the *World's Poultry Science Journal* or in proceedings of the European WPSA meetings (European Symposium of Poultry Nutrition, Egg quality meetings, see http://www.wpsa.com/index.php). More details or literature can be found in papers, which are mentioned in the references section.

9 References

Abdallah, A. G., Harms, R. H., Wilson, H. R. and El Husseiny, O. (1994), Effect of removing trace minerals from the diet of hens laying eggs with heavy or light shell weight. *Poult. Sci.* 73, 295–301.

Aerni, V., Brinkhof, M. W. G., Wechsler, B., Oester, H. and Frohlich, E. (2005), Productivity and mortality of laying hens in aviaries: a systematic review. *World Poult. Sci. J.* 61, 130–42.

Aerni, V., El-Lethey, H. and Wechsler, B. (2000), Effect of foraging material and food form on feather pecking in laying hens. *Br. Poult. Sci.* 41, 16–21.

Ahmadi, H. and Rodehutscord, M. (2012), A meta-analysis of responses to dietary non phytate phosphorus and phytase in laying hens. *Poult. Sci.* 91, 2071–8.

Al Ankari, A., Najib, H. and Al Hozab, A. (1998), Yolk and serum cholesterol and production traits as affected by incorporating a supraoptimal amount of copper in the diet of the leghorn hen. *Br. Poult. Sci.* 39, 393–7.

Amaya, E., Becquet, P., Carné, S., Peris, S. and Miralles, P. (2014), Carotenoids in Animal Nutrition. Fefana Publication, p. 96. http://www.fefana.org/sites/default/files/files/booklets/2014-12-04_booklet_carotenoids.pdf

Atteh, J. O. and Leeson, S. (1985), Response of laying hens to dietary saturated and unsaturated fatty acids in the presence of varying dietary calcium levels. *Poult. Sci.* 64, 520–8.

Aureli, R., Schierle, J., Gadient, M. and Murtala, U. F. (2013), Comparaison de deux caroténoïdes jaunes sur l'éfficacité de pigmentation du jaune d'œuf. In *Proceeding 10iéme journée de la Recherche Avicole et Palmipéde à foie gras*, La Rochelle, 26–28 Mars 2013.

Balnave, D. and Zhang, D. (1993), Response of laying hens on saline drinking water to dietary supplementation with various Zn compounds. *Poult. Sci.* 72, 603–9.

Bar, A. (2008), Calcium homeostasis and vitamin D. metabolism and expression in strongly calcifying laying birds. *Comp. Biochem. Phys. Part A.* 151, 477–90.

Bar, A. (2009), Calcium transport in strongly calcifying laying birds: Mechanisms and regulation. *Comp. Biochem. Phys. Part A.* 152, 447–569.

Bar, A., Razaphkovsky, V. and Vax, E. (2002), Re-evaluation of calcium and phosphorus requirements in aged laying hens. *Br. Poult. Sci.* 43(2), 261–9.

Barroeta A. C., Davin, R. and Baucells, M. D. (2012), *Optimum Vitamin Nutrition in Laying Hens, in: Optimum Vitamin Nutrition in the Production of Quality Animal Foods.* 5M. Publishing Benchmark House, Sheffield, UK, pp. 89–135.

Bastien, R. W., Bradley, J. W., Pennington, B. L. and Fergusson, T. M. (1979), Effect of dietary mineral supplements on radius breaking strength and egg characteristic of caged layers. *Poult. Sci.* 58, 90–2.

Baumgartner, S., Brown, D. J., Salevsky Jr., E. and Leach, R. M. (1978), Copper deficiency in the laying hen. *J. Nut.* 108, 804–11.

Bestman, M., Koene, P. and Wagenaar, J. P. (2009), Influence of farm factors on the occurrence of feather pecking in organic reared hens and their predictability for feather pecking in the laying period. *Appl. Anim. Behav. Sci.* 121, 120–5.

Blokhuis, H. J., Fiks van Niekerk, T., Bessei, W., Elson, A., Guémené, D., Kjaer, J., Maria Levrino, G. A., Nicol, C.J., Tauson, R., Weeks, C. A. and Van Der Weerd, H. A. (2007), The LayWel project: Welfare implications of changes in production systems for laying hens. *World Poult. Sci. J.* 63,101–14.

Blokhuis, H. J. and Van Der Haar, J. W. (1992), Effects of pecking incentives during rearing on feather pecking of laying hens. *Br. Poult. Sci.* 33, 17–24.

Bollengier-Lee, S., Williams, P. E. and Whitehead, C. C. (1999), Optimal dietary concentration of vitamin E for alleviating the effect of heat stress on egg production in laying hens. *Br. Poult. Sci.* 40, 102–7.

Breithaupt, D. E. (2007), Modern application of xanthophylls in animal feeding. *Trends Food Sci. Tech.* 18, 501–6.

Butler, E. J. (1976), Fatty liver diseases in the domestic fowl – a review. *Avian Pathol.* 5, 1–14.

Burt, A. J., Caston, L., Leeson, S., Shelp, B. J. and Lee, E. A. (2013), Development and utilization of high carotenoid maize germplasm: proof of concept. *Crop Sci.* 53, 554–63.

Cabuk, M., Bozkurt, M., Kirkpinar, F. and Ozkul, G. (2004), Effect of phytase supplementation of diets with different levels of phosphorus on performance and egg quality of laying hens in hot climatic conditions. *S. Afr. J. Anim. Sci.*, 34, 13–17.

Cesari, V., Mangiagialli, M. G., Giardini, A., Galimberti, B., Carter, S., Gallazzi, D. and Toschi, H. (2014), Egg quality and productive performance of laying hens fed different levels of skimmed milk powder added to a diet containing Lactobacillus acidophilus. *Poult. Sci.* 93, 1197–1201.

Cyelan, N. and Scheideler, S. E. (1999), Effect of eggshell 49, dietary calcium level and hen age on performance and egg shell quality. In *Proceeding of the Alltech's 15th Annual Symposium*, pp. 61–73.

Chow, P. Y., Gue, S. Z., Leow, S. K. and Goh, L. B. (2014), The bioefficacy of micro emulsified natural pigments in egg yolk pigmentation. *Br. Poult. Sci.* 55, 398–402.

Chowdhury, S. D. (1990), Shell membrane system in relation to lathyrogen toxicity and copper deficiency. *World Poult. Sci. J.* 46, 153–69.

Christmas, B. and Harms R. H. (1983), The performance of laying hens as affected by copper sulfate and methionine level. *Poult. Sci.* 62, 389–91.

Chung, M. K., Choi, J. H., Chung, Y. K. and Chee, K. M. (2005), Effects of dietary vitamins C and E on egg shell quality of broiler breeder hens exposed to eat stress. *Asian-Australasian J. Anim. Sci.* 18, 545–51.

Classen, H. L. and Scott, T. A. (1982), Self-selection of calcium during the rearing and early laying periods of White Leghorn pullets. *Poult. Sci.* 61, 2065–74.

Coetzee, C. B., Casey, N. H. and Meyer, J. A. (1997), Fluoride tolerance of laying hens. *Br. Poult. Sci.* 38, 597–602.

Cowieson, A. J., Bedford, M. R., Selle, P. G. and Ravindran, V. (2009), Phytate and microbial phytase: implications for endogenous nitrogen losses and nutrient availability. *World Poult. Sci. J.* 65, 401–18.

Dale, N. and Strong, C. F. (1998), Inability to demonstrate an effect of eggshell 49 on shell quality in older laying hens. *J. Appl. Poult. Res.* 7, 219–24.

De Faria, D. E., Junqueira, O. M. and Sakomura, N. K. (1999), Santana A.E.Efeito de diferentes Niveis de Manganês e fosforo sobre o desempenho e a qualidade da casca dos ovos de podeiras comerciais. *Rev. bras. Zootec.* 28, 105–12.

Diaz de Barboza G., Guizzardi, S., Tolosa D. E. and Talamoni, N. (2015), Molecular aspect of intestinal calcium absorption. *World J. Gastroenterol.* 21, 7142–54.

Drake, K. A., Donnelly, C. A. and Dawkins, M. S. (2010), Influence of rearing and lay risk factors on propensity for feather damage in laying hens. *Br. Poult. Sci.* 51, 725–33.

EFSA (2014), Scientific Opinion on the potential reduction of the currently authorised maximum zinc content in complete feed. *EFSA. J.* 12(5), 3668.

EFSA (2015), Scientific Opinion on the safety of Solanum glaucophyllum standardized leaves as feed material. *EFSA. J.* 13(1), 3967.

Elwinger, K., Lignell, A. and Wilhelmson, M. (1997), Astaxanthin rich algal meal (Haematococcus pluvialis) as carotenoid source in feed for laying hens. In J. Kijowski and J. Pikul (Eds), *Proceedings of the VIH*, Symposium on the Quality Egg and Egg Products, Poznan.

Elwinger, K., Tufvesson, M., Lagerkvist, G. and Tauson, R. (2008), Feeding layers of different genotypes in organic feed environments. *Br. Poult. Sci.* 49, 654–65.

Engberg, R. M., Hammershoslashj, M., Johansen, N. F., Abousekken, M. S., Steenfeldt, S. and Jensen, B. B. (2009), Fermented feed for laying hens: effects on egg production, egg quality, plumage condition and composition and activity of the intestinal microflora. *Br. Poult. Sci.* 50, 228–39.

Fleming, R. H. (2008), Nutritional factors affecting poultry bone health. *Proc. Nut. Soc.* 67, 177–83.

Fleming, R. H., McCormack, H. A., McTeir, L. and Whitehead, C. C. (2003), Effects of dietary particulate limestone, vitamin K3 and fluoride and photostimulation on skeletal morphology and osteoporosis in laying hens. *Br. Poult. Sci.* 44, 683–9.

Fleming, R. H., McCormack, H. A., McTeir, L. and Whitehead, C. C. (2004), Incidence, pathology and prevention of keel bone deformities in the laying hen. *Br. Poult. Sci.* 45, 320–30.

Fleming, R. H., McCormack, H. A., McTeir, L. and Whitehead, C. C. (2006), Relationships between genetic, environmental and nutritional factors influencing osteoporosis in laying hens. *Br. Poult. Sci.* 47, 742–55.

Fleming, R. H., McCormack, H. A. and Whitehead, C. C. (1998), Bone structure and strength at different ages in laying hens and effects of dietary particulate limestone, vitamin J. and ascorbic acid. *Br. Poult. Sci.* 39, 434–40.

Freire, R., Wilkins, L. J., Short, F. and Nicol, C. J. (2003), Behaviour and welfare of individual laying hens in a non-cage system. *Br. Poult. Sci.* 44, 22–9.

Gadient, M., Shierle, J. and Hernandez J. M. (2010), Stability of Red Carotenoids in Stored Poultry Premix. XIIIth European Poultry Conference, August 2010, Tours.

Galobart, J., Sala, R., Rincon-Carruyo, X., Manzanilla, E. G., Vila, B. and Gasa, J. (2004), Egg yolk color as affected by saponification of different natural pigmenting sources. *J. Appl. Poult. Res.* 13, 328–34.

Gilani, A. M., Knowles, T. G. and Nicol, C. J. (2013), The effect of rearing environment on feather pecking in young and adult laying hens. *Appl. Anim. Behav. Sci.* 148, 54–63.

Gregory, N. G. and Wilkins, L. J. (1989), Broken bones in domestic fowl: handling and processing damage in end-of-lay battery hens. *Br. Poult. Sci.* 30, 555–62.

Griminger, P. (1977), Effect of copper sulfate on egg production and shell thickness. *Poult. Sci.* 56, 359–61.

Guesdon, V., Ahmed, A. M. H., Mallet, S., Faure, J. M. and Nys, Y. (2006), Effects of beak trimming and cage design on laying hen performance and egg quality. *Br. Poult. Sci.* 47, 1–12.

Guinotte, F. and Nys, Y. (1991), The effects of particle size and origin of calcium sources on egg shell quality and bone mineralization in egg laying hens. *Poult. Sci.* 70, 583–92.

Guinotte, F. and Nys, Y. (1993), Factors affecting the intestinal calcium availability in laying hens: consequences on shell quality. In Y. Nys (Ed.), *Proceedings of the 5th European Symposium on the Quality of Eggs and Egg Products*, 10 April 2008, pp. 149–62.

Hamelin, C. and Altemueller, U. (2011), Review of the use of carotenoids in the egg yolk and the pigmented broiler. In *Proceedings of the XIV*, Symposium on Egg Quality and Egg Product, 4–8 September 2011, German, WPSA.

Hamilton, P. B. (1992), The use of high-performance liquid chromatography for studying pigmentation. *Br. Poult. Sci.* 71, 718–24.

Hammershoj, M., Kidmoseb, U. and Steenfeld, S. (2010), Deposition of carotenoids in egg yolk by short-term supplement of coloured carrot (Daucus carota) varieties as forage material for egg-laying hens. *J. Sci. Food Agric.* 90, 1163–71

Hansen, R. J. and Walzem, R. L. (1993), Avian fatty liver hemorrhagic syndrome: a comparative review. *Adv. Vet. Sci. Comp. Med.* 37, 451–68.

Harlander-Matauschek, A., Rodenburg, T. B., Sandilands, V. and Tobalske, B. W. (2015), Causes of keel bone damage and their solutions in laying hens. *World Poult. Sci. J.* 71, 461–72.

Harms, R. H., Bootwalla, S. M., Woodward, S. A. and Wilson, H. Q. (1990), Some observations on the influence of vitamin D. metabolites when added to the diet of commercial hens. *Poult. Sci.* 69, 426–32.

Harms, R. G. and Buresh, R. E. (1986), Influence of nutrient density and copper levels of diets upon performance of laying hens. *Nut. Reports Int.* 33, 2617.

Hartel, G. (1990), Evaluation of the dietary interaction of calcium and phosphorus in the high producing laying hens. *Br. Poult. Sci.* 31, 473–94.

Heerkens, J. L. T., Delezie, E., Rodenburg, T. B., Kempen, H., Zoons, J., Ampe, B. and Tuyttens, F. A. M. (2016), Risk factors associated with keel bone and foot pad disorders in laying hens housed in aviary systems. *Poult. Sci.* 95, 482–8.

Hencken G. (1992), Chemical and physiological behavior of feed carotenoids and their effects on pigmentation. *Poult. Sci.* 71, 711–17.

Hincke, M.T., Nys, Y., Gautron, J., Mann, K., Rodriguez-Navarro, A. B. and Mckee, M. D. (2012), The eggshell: structure, composition and mineralization. *Front. Biosci.* 17, 1266–80.

Hughes, A. L., Dahiya, J. P., Wyatt, C. L. and Classen, H. L. (2008), The efficacy of quantum phytase in a forty-week production trial Using white leghorn laying hens fed corn-soybean meal-based diets. *Poult. Sci.* 87, 1156–61.

Hughes, R. J. (1988), Inter-relationships between eggshell quality, blood acid-base balance and dietary electrolytes. *World Poult. Sci. J.* 44, 30–40.

Huyghebaert, G. (1993), The utilisation of oxy-carotenoids for egg yolk pigmentation, Gent Thesis, University of Gent.

Janczak, A. M. and Riber, A. Q. (2015), Review of rearing-related factors affecting the welfare of laying hens. *Poult. Sci.* 94, 1454–69.

Japelt, R. B. and Jakobsen, J. (2013), Vitamin D in plants: a review of occurrence, analysis, and biosynthesis. *Front. Plant Sci.* 4, 136. Published online 13 May 2013.

Jiang, S., Cheng, H. W., Cui, L. H., Zhou, Z. L. and Hou, J. L. (2013), Changes of blood parameters associated with bone remodeling following experimentally induced fatty liver disorder in laying hens. *Poult. Sci.* 92, 1443–53.

Johnsen, P. F., Vestergaard, K. S. and Nørgaard-Nielsen, G. (1998), Influence of early rearing conditions on the development of feather pecking and cannibalism in domestic fowl. *Appl. Anim. Behav. Sci.* 60, 25–41.

Julian, R. J. (2005), Production and growth related disorders and other metabolic diseases of poultry – a review. *Vet. J.* 169, 350–69.

Käppeli, S., Gebhardt-Henrich, S. G., Fröhlich, E., Pfulg, A., Schaublin, G. and Stoffel, M. G. (2011), Effects of housing, perches, genetics, and 25-hydroxycholecalciferol on keel bone deformities in laying hens. *Poult. Sci.*, 90, 1637–44.

Käppeli, S., Gebhardt-Henrich, S. G., Fröhlich, E., Pfulg, A. and Stoffel, M. G. (2011), Prevalence of keel bone deformaties in Swiss laying hens. *Br. Poult. Sci.* 52, 531–6.

Kebreab, E, France, J., Kwakkel, R. P., Leeson, S., Kuhi, H. D. and Dijkstra, J. (2009), Development and evaluation of a dynamic model of calcium and phosphorus flows in layers. *Poult. Sci.* 88, 680–9.

Keshavarz, J. (1998a), Investigation on the possibility of reducing protein, phosphorus, and calcium requirements of laying hens by manipulation of time of access to these nutrients. *Poult. Sci.* 77, 1320–32.

Keshavarz, J. (1998b), Further investigations on the effect of dietary manipulation of protein, phosphorus, and calcium for reducing their daily requirement for laying hens. *Poult. Sci.* 77, 1333–46.

Keshavarz, J. (2003a), A. comparison between cholecalciferol and 25-OH-cholecalciferol on performance and eggshell quality of hens fed different levels of calcium and phosphorus. *Poult. Sci.* 82, 1415–22.

Keshavarz, J. (2003b), Effects of continuous feeding of low-phosphorus diets with and without phytase during the growing and laying periods on performance of two strains of leghorns. *Poult. Sci.* 82, 1444–56.

Keshavarz, J. and Nakajima, S. (1993), Reevaluation of calcium and phosphorus requirements of laying hens for optimum performance and eggshell quality. *Poult. Sci.* 72, 144–53.

Khanal, Q. C. and Nemere, H. (2008), Regulation of intestinal calcium transport. *Ann. Rev. Nutr.* 28, 179–96.

Khan, R. U., Naz, S., Nikousefat, Z., Tufarelli, V., Javdani, M., Rana, N. and Laudadio, V. (2011), Effect of vitamin E in heat-stressed poultry. *World Poult. Sci. J.* 67, 469–77.

Kienholz, W. E., Moren, R. E. and Flinchum, J. D. (1992), Zinc methionine for stressed laying hens. *Poult. Sci.* 71, 829–32.

Kienholz, W. E., Turk, D. E., Sunde, M. L. and Hoekstra, W. G. (1961), Effect of zinc deficiency in the diet of hens. *J. Nutr.* 75, 211–21.

Kim, J. W., Kim, J. H., Shin, J. E. and Kil, D. Y. (2016), Relative bioavailability of copper in tribasic copper chloride to copper in copper sulfate for laying hens based on egg yolk and feather copper concentrations. *Poult. Sci.* 95, 1591–7.

Kirunda, D., Scheideler, S. E. and Mckee, Q. (2001), The efficacy of vitamin E. (DL-alpha-tocopheryl acetate) supplementation in hen diets to alleviate egg quality deterioration associated with temperature exposure. *Poult. Sci.* 80, 1378–83.

Kita, K., Hohmura, H. and Okumura, J. H. (1997), Influence of dietary zinc methionine on eggshell quality in laying hens under hot climate environment. *Jpn Poult. Sci.* 34, 21–6.

Kjaer, J. B. and Bessei, W. (2013), The interrelationships of nutrition and feather pecking in the domestic fowl. *Arch. Geflügelk.* 77, 1–9.

Kolb, E. and Seehawer, J. (2001), Significance and application of ascorbic acid in poultry. *Arch. Geflugelk.* 65, 106–13.

Koreleski, J. and Swiatkiewicz, S. (2005), Efficacy of different levels of a cholecalciferol 25-OH-derivative in diets with two limestone forms in laying hen nutrition. *J. Anim. Feed Sci.* 14, 305–15.

Kotrbacek, V., Skrivan, M., Kopecky, J., Penkava, O., Hudeckova, P., Uhrikova, H. and Doubek, J. (2013), Retention of carotenoids in egg yolks of laying hens supplemented with heterotrophic Chlorella. *Czech J. Anim. Sci.* 58, 193–200.

Kurtoglu, V. and Balev, T. (2007), Effects of sodium bicarbonate, potassium chloride and sodium chloride supplementation on some blood biochemical parameters in laying hens. In *Proceedings of the 16th European Symposium on Poultry Nutrition*, World Poultry Science Association, Strasbourg, France, 26–30 August 2007.

Leach, J. R. M. and Gross, J. (1983), The effect of manganese deficiency upon the ultrastructure of the eggshell. *Poult. Sci.* 32, 499–504.

Lee, B. K., Kim J. S., Ahn H. J., Hwang J. H., Kim J. M., Lee H. T, An B. K. and C. W. Kang. (2010), Changes in hepatic lipid parameters and hepatic messenger ribonucleic acid expression following estradiol administration in laying hens (Gallus domesticus). *Poult. Sci.* 89, 2660–7.

Lee, C. M., Dimaline, R., Moore, S., Vowles, L. and Dockray, G. J. (1988), Reciprocal endocrine interactions between proventriculus and antrum in chickens. *Regul. Peptides* 22, 415–19.

Lei, Q. B., Shi, L. X., Zhang, K. Y., Ding, X. M., Bai, S. P. and Liu, Y. G. (2011), Effect of reduced energy, protein and entire substitution of inorganic phosphorus by phytase on performance and bone mineralization of laying hens. *Br. Poult. Sci.* 52, 202–13.

Leigh, Hadden W., Watkins, R. H., Levy, L. W., Regalado, E., Rivadeneira, D. M., Van Breemen, R. B. and Schwartz, S. (1999), Carotenoid composition of marigold (Tagetes erecta) flower extract used as nutritional supplement. *J. Agric. Food Chem.* 47(10), 4189–94.

Lemahieu, C., Bruneel, C., Termote-Verhalle, R., Muylaert, K., Buyse, J. and Foubert, H. (2013), Impact of feed supplementation with different omega-3 rich microalgae species on enrichment of eggs of laying hens. *Food Chem.* 141, 4051–9.

Lim, H. S. and Paik, I. J. (2003), Effects of supplementary mineral methionine chelates (Zn, Cu, Mn) on the performance and eggshell quality of laying hens. *Asian Australas. J. Anim. Sci.* 16, 1804–8.

Lietz, G., Lange, J. and Rimbach, G. (2010), Mmolecular and dietary regulation of b,b-carotene 15,15'-monooxygenase 1 (BCM01). *Arch. Biochem. Biophys.* 502, 8–16.

Looten, P. H., Wils, D., Saniez, M. G. and Nys, Y. (2003), Xanthophyll profile of corn gluten meal throughout the year and transfer yield into egg yolk. In Y. Nys (Ed.), *Proceedings of 16th European Symposium on the Quality of Poultry Meat* and *10th European Symposium on the Quality of Eggs and Egg Products*, 23–26 September 2003, pp. 186–91; 949–54.

Ma, W., Gu, Y., Lu, J., Yuan, L. and Zhao, R. (2014), Effects of chromium propionate on egg production, egg quality, plasma biochemical parameters, and egg chromium deposition in late-phase laying hens. *Biol. Trace Elem. Res.* 157, 113–19.

Mabe, H., Rapp, C., Bain, M. M. and Nys, Y. (2003), Supplementation of a corn soybean meal diet with manganese, copper and zinc from organic or inorganic sources improves eggshell quality in aged laying hens. *Poult. Sci.* 82, 1903–13.

Magnin, M., Jeanmichel, P. and Mahieu, A. (2009), Stabilité comparée de dix extraits de tagetes et de sept extraits de paprika sources de pigments naturels jaunes et rouges pour les volailles. *Huitièmes Journées de la Recherche Avicole*, St Malo, France, 25–26 Mars 2009, p. 20.

Mann, S., Archibald, D. D., Didymus, J. M., Douglas, T., Heywood, B. R., Meldrum, F. C. and Reeves, N. J. (1993), Crystallization at inorganic-organic interfaces biominerals and biomimetic synthesis. *Science* 261, 1286–92.

March, B. (1981), Choline supplementation of layer diets containing soybean meal as protein supplement. *Poult. Sci.* 60, 823.

Marusich, W. L. and Bauernfeind, J. C. (1981), Oxycarotenoids in poultry feeds. In J. C. Bauerfeind (Ed.), *Carotenoids as Colorants and Vitamin A. Precursors*. Academic Press, New York, pp. 320–462.

Mateos, G. G., Jimenez-Moreno, E., Serrano, M. P. and Lazaro, R. P. (2012), Poultry response to high levels of dietary fiber sources varying in physical and chemical characteristics. *J. Appl. Poult. Res.* 21, 156–74.

Mattila, P., Valaja, J., Rossow, L., Venalainen, E. and Tupasela, T. (2004), Effect of vitamin D2- and D3-enriched diets on egg vitamin D. content, production, and bird condition during an entire production period. *Br. Poult. Sci.* 83, 433–40.

Meluzzi, A., Simoncini, F., Siri, F., Vandi, L. and Giordini, G. (1996), Feeding hens diets supplemented with heavy metals (chromium, nickel and lead). *Arch. Geflugelk* 60, 119–25.

Mongin, P. (1978), Acid-base balance during eggshell formation. In J. E. Piiper (Ed.), *Respiratory Function of Birds, Adult and Embryonic*, Springer Verlag, Berlin.

Mongin, P. and Sauveur, B. (1979), The specific appetite of domestic fowl. In K. N. Boorman and B. M. Freeman (Eds), *Food Intake Regulation in Poutry*, Edinburgh, British Poultry Science, pp. 171–89.

Moreng, R. E., Balnave, D. and Zhang, D., (1992), Dietary zinc methionine effect on eggshell quality of hens drinking saline water. *Poult. Sci.* 71, 1163–7.

Morii, G., shizawa, Y., Morii G. and Durlach J. (Eds), *Perspective in Magnesium Research*, Springer Verlag, London, pp. 263–5.

Müller, J. H., Murakami, A. E., Scapinello, C., Moreira, H. and Varela, E. (2009), Effect of vitamin J. on bone integrity and eggshell quality of white hen at the final phase of the laying cycle. *Revist Brasil Zootecnia* 38, 488–92.

Nahm, J. G. (2007), Efficient phosphorus utilization in poultry feeding to lessen the environmental impact of excreta. *World Poult. Sci. J.* 63, 625–54.

Nasr, M. A. F., Murrell, J. and Nicol, C. J. (2013), The effect of keel fractures on egg production, feed and water consumption in individual laying hens. *Br. Poult. Sci.* 54, 165–70.

Naz, S., Idris, M., Klalique, M. A., Zia-Ur-Rahman, Alhidary, I. A., Abdelrahman, M. M., Khan, R. U., Chand, N., Farooq, U. and Ahmad, S. (2016), The activity and use of zinc in poultry diets. *World Poult. Sci. J.* 72, 159–67.

Nicol, C. J., Brown, S. N., Glen, E., Pope, S. J., Short, F. J., Warriss, P. D., Zimmerman, P. G. and Wilkins, L. J. (2006), Effects of stocking density flock size and management on the welfare of laying hens in single-tier aviaries. *Br. Poult. Sci.* 47, 135–46.

Nolan, J. M., Meagher, K. A., Howard, A. N., Moran, R., Thurnham, D. H. and Beatty, S. (2016), Lutein, zeaxanthin and meso-zeaxanthin content of eggs laid by hens supplemented with free and esterified xanthophylls. *J. Nutr. Sci.* 5, e1–10.

Nys, Y. (1993), Regulation of 1,25(OH)$_2$D$_3$, of osteocalcin and of intestinal and uterine calbindin in hens. In P. J. Sharp (Ed.), *J. Endocrinol.*, pp. 345–58.

Nys, Y. (1999), Nutritional factors affecting eggshell quality. *Czech J. Anim. Sci.* 44, 135–43.

Nys, Y. (2000), Dietary carotenoids and egg yolk coloration. *Archiv. für. Geflugelkunde* 64, 45–54.

Nys, Y. (2014), Phosphorus utilization in laying hens. In *Proceedings of Associazione Scientifica di Avicoltura – Sezione Italiana della*, World's Poultry Science Association, 50° Convegno Nazionale 'Il fosforo nell'alimentazione del pollame: un caso di sostenibilità', Fiera di Forlì, 15 Aprile 2014.

Nys, Y., Burlot, T. and Dunn, I. C. (2008), Internal quality of eggs: any better, any worse? XXIIH. 30 June to 4 July 2008, World's Poultry Science Association (Australian branch), p. 10. CDRom, papers\\wpc08Final00034.

Nys, Y., Frapin, D. and Pointillard, A. (1996), Occurence of phytases in plants, animals and microorganisms. In M. B. Coelho and E. T. Kornegay (Eds), *Phytase in Animal Nutrition and Management: A BASF. Reference Manual*, pp. 213–36.

Nys, Y., Gautron, J., Garcia, Ruiz J. M. and Hincke, M. T. (2004), Avian Eggshell mineralization: biochimical and fonctional characterization of matrix proteins. *Comptes Rendus Palevol* 3, 549–62.

Nys, Y., Hincke, M. T., Arias, J. L., Garcia-Ruiz, J. M. and Solomon, S. E. (1999), Avian eggshell mineralization. *Avian Poult. Biol. Rev.* 10, 143–66.

Nys, Y. and Guyot, N. (2011), Egg formation and chemistry. In Y. Nys, M. Bain and F. Van Immerseel (Eds), *Improving the Safety and Quality of Eggs and Egg Products.* Vol1, *Egg Chemistry, Production and Consumption*, Cambridge, Woodhead Publishing, pp. 83–132.

Nys, Y. and Leroy, N. (2017), Calcium homeostasis and eggshell biomineralization in female chicken. In D. Feldman, W. Pike, R. Bouillon, E. Giovannucci, D. Goltzman and M. Hewison (Eds), *Vitamin D*, Fourth Edition, Elsevier Inc (in press).

Ochrimento, C., Lemser, A., Richter, G., Krause, U. and Bonsak, H. (1992), Effect of manganese content in laying hen feed with different Ca and mineral level on the egg shell quality and bone mineralisation of hens. *Arch Tiernahr.* 42, 25–35.

Panda, A. K., Ramarao, S. V., Raju, M. V. and Chatterjee, R. N. (2008), Effect of dietary supplementation with vitamins E. and C. on production performance, immune responses and antioxidant status of White Leghorn layers under tropical summer conditions. *Br. Poult. Sci.* 49, 592–9.

Peguri, A. and Coon, C. (1993), Effect of feather coverage and temperature on layer performance. *Poult. Sci.* 72, 1318–29.

Pérez-Bonilla, A., Jabbour, C., Frikha, M., Mirzaie, S., Garcia, J. and Mateos, G. G. (2012), Effect of crude protein and fat content of diet on productive performance and egg quality traits of brown egg-laying hens with different initial body weight. *Poult. Sci.* 91, 1400–5.

Pickel, T., Schrader, L. and Scholz, B. (2011), Pressure load on keel bone and foot pads in perching laying hens in relation to perch design. *Poult. Sci.* 90, 715–24.

Pottgueter, R. (2015), Nutrition of hens in extended production cycles – as a practical approach. In *Proceeding of XVH*, European Symposium on the Quality of Eggs and Egg Products, 10–13 May 2015.

Qaisrani, S. N., Van Krimpen, M. M. and Kwakkel, R. P. (2013), Effects of dietary dilution source and dilution level on feather damage, performance, behavior, and litter condition in pullets. *Poult. Sci.* 92, 591–602.

Qin, X. and Klandorf, G. (1991), Effect of dietary born supplementation on egg production, shell quality and calcium metabolism in aged broiler breeder hens. *Poult. Sci.* 70, 2131–8.

Rao, K. S. and Roland, D. (1990), Retention patterns of various sized limestone particles in gizzard of commercial leghorns hens. *Poult. Sci.* 60, 185.

Rao, S. V. R., Ramasubba, R. V. and Reddy, V. Q. (1999), Non-phytin phosphorus requirements of commercial broilers and White Leghorn layers. *Anim. Feed Sci. Technol.* 80, 1–10.

Richter, G., Lemser, A., Ludke, C., Steinbach, G. and Mockel, P. (1996), Studies on the vitamin A. requirements and recommendations for laying hens. *Arch. GeflugelK.* 60, 174–80.

Rodriguez-Amaya, D. B. (2016), *Carotenoids: Chemistry, Biology and Technology*, IFT Press Willey-Blackwell, Oxford, UK, p. 305.

Rogers, P. J. (1995), *The Development of the Brain and Behavior in the Chicken*, CAB International, Wallingford, UK, pp. 95–110.

Roland, D. A. and Bryant, M. (2000), Nutrition and feeding for optimum eggshell quality, 21st World's Poultry Science Congress, 20–24 August, CD. S3.6.03. WPSA. Branche canadienne, Saskatoon (CAN).

Rozenboim, I., Mahato, J., Cohen, N. A. and Tirosh, O. (2016), Low protein and high-energy diet: a possible natural cause of fatty liver hemorrhagic syndrome in caged White Leghorn laying hens. *Poult. Sci.* 95, 612–21.

Safaa, H., Serrano, M. and Valencia, D., Frikha, M., Jiménez-Moreno, E. and Mateos, G. G. (2008), Productive performance and egg quality of Brown laying hens from fifty-eight to seventy-three weeks of age as influenced by level and source of calcium in the diet. *Poult. Sci.* 87, 2043–51.

Sandilands, V. (2011), The laying hen and bone fractures. *Vet. Rec.* 169, 411–12.

Saunders-Blades, J. L., Macisaac, J. L., Korver, D. Q. and Anderson, D. M. (2009), The effect of calcium source and particle size on the production performance and bone quality of laying hens. *Poult. Sci.* 88, 680–9.

Sauveur, B. (1987), Effect of prelaying dietary calcium and phosphorus levels on growth and subsequent laying performance of pullets. In *6th European Symposium on Poultry Nutrition*, Königslutter, WPSA(D), pp. E18–25.

Sazzad, H. M., Bertechini, A. G. and Nobre, P. T. C. (1994), Egg production, tissue composition and mineral metabolism in 2 strains of commercial layers with various levels of manganese in diets. *Anim. Feed Sci. Tech.* 46, 271–5.

Scheideler, S. E. and Froning, G. W. (1996), The combined influence of dietary flaxseed variety, level, form and storage conditions on egg production and composition among vitamin E-supplemented hens. *Poult. Sci.* 75, 1221–6.

Schlegel, P., Jondreville, C. and Nys, Y. (2016), The role of phytate in formulating diets for trace minerals. In *Proceeding of International Phytase Summit*, 7–10 November 2016, Miami, USA.

Schlegel, P., Nys, Y. and Jondreville, C. (2010), Zinc availability and digestive zinc solubility in piglets and broilers fed diets varying in their phytate contents, phytase activity and supplemented zinc source. *Animal* 4, 200–9.

Schlegel, P., Nys, Y. and Jondreville, C. (2013), Bioavailability of zinc sources and their interaction with phytates in broilers and piglets. *Animal* 7, 47–59.

Sengor, E., Yardimci, M., Cetingul, S., Bayram, H., Sahin, G. and Dogan, H. (2007), Effects of short chain fatty acid (SCFA) supplementation on performance and egg characteristics of old breeder hens. *S. Afr. J. Anim. Sci.* 37, 158–63.

Sharma, R. K., Ravikanth, K., Maini, S., Rekhe, D. S. and Rastogi, S. J. (2009), Influence of calcium and phosphorus supplements with synergistic herbs on egg shell quality in late layers. *Vet. World* 2, 231–3.

Shastak, Y. and Rodehutscord, M. (2015), A. review of the role of magnesium in poultry nutrition. *World Poult. Sci. J.* 71, 125–37.

Sherwin, C. M., Richards, G. J. and Nicol, C. J. (2010), Comparison of the welfare of layer hens in 4 housing systems in the UK. *Br. Poult. Sci.* 51, 488–99.

Shini, S. and Bryden, W. L. (2009), Occurrence and control of fatty liver hemorrhagic syndrome (FLHS) in caged hens. A report for the Australian egg Corporation Limited. AECL Publication No UQ-105A. AECL Project No UQ-105. ISSN 1448–1316.

Sirri, F., Laffaldano, N., Minelli, G., Meluzzi, A. and Rosato, M. P. (2007), Comparative pigmentation efficiency of high dietary levels of apo-ester and marigold extract on quality traits of whole liquid egg of two strains of laying hens. *J. Appl. Poult. Res.* 16, 429–37.

Sohail, S. S. and Roland, D. A. (2002), Influence of dietary phosphorus on performance of hy-line W36 hens. *Poult. Sci.* 81, 75–83.
Soltan, M. A. (2008), Effect of organic acid supplementation on egg production, egg quality, and some blood serum parameters in laying hens. *Int. J. Poult. Sci.* 7, 613–21.
Stahl, J. L., Cook, M. E. and Sunde, M. L. (1986), Zinc supplementation – its effects on egg production, feed conversion, fertility and hatachability H. *Poult. Sci.* 65, 2104–9.
Stefanello, C., Santos, T. C., Murakami, A. E., Martins, E. N. and Carneiro, T. C. (2014), Productive performance, eggshell quality, and eggshell ultrastructure of laying hens fed diets supplemented with organic trace minerals. *Poult. Sci.* 93, 104–13.
Stevenson, M. H. (1985), Effect of added cassava root meal, ZnO, and their interaction on the production and quality of eggs from laying hens. *J. Sci. Food Agric.* 36, 909–14.
Stratmann, A., Fröhlich, E. K. F., Gebhardt-Henrich, S. G., Harlander-Matauschek, A., Wurbel, H. and Toscano, M. J. (2016), Genetic selection to increase bone strength affects prevalence of keel bone damage and egg parameters in commercially housed laying hens. *Poult. Sci.* 95, 975–84.
Surai, P. F. (2012), The antioxidant properties of canthaxanthin and its potential effects in the poultry eggs and on embryonic development of the chick. *World's Poult. Sci. J.* 68, 717 –26.
Świątkiewicz, S., Arczewska-Włosek, A., Krawczyk, J., Puchała, M. and Józefiak, D. (2015), Dietary factors improving eggshell quality: an updated review with special emphasis on microelements and feed additives. *World Poult. Sci. J.* 71, 83–94.
Swiatkiewicz, S., Koreleski, J. and Arczewska, A. (2010), Laying performance and eggshell quality in laying hens fed diets supplemented with prebiotics and organic acids. *Czech J. Anim. Sci.* 55, 294–306.
Tarlton, J. F., Wilkins, L. J., Toscano, M. J., Avery, N. C. and Knott, L. (2013), Reduced bone breakage and increased bone strength in free range laying hens fed omega-3 polyunsaturated fatty acid supplemented diets. *Bone* 52, 578–86.
Thurnham, D. H. (2007), Macular zeaxanthin and lutein – a review of dietary sources and bioavailability and some relationship with macular pigment optical density and age related macular disease. *Nut. Res. Rev.* 20, 163–9.
Toscano, M. J., Booth, F., Wilkins, L. J., Avery, N. C., Brown, S. B., Richards, G. and Tarlton, J. F. (2015), The effects of long (C20/22) and short (C18) chain omega-3 fatty acids on keel bone fractures, bone biomechanics, behavior, and egg production in free-range laying hens. *Poult. Sci.* 94, 823–35.
Trott, K. A., Giannitti, F., Rimoldi, G., Hill, A., Woods, L., Barr, B., Anderson, M. and Mete, A. (2013), Fatty liver hemorrhagic syndrome in the backyard chicken: aretrospective histopathologiccase series. *Vet. Pathol.* 51, 787–95.
Tsang, C. P. W. and Grunder, A. A. (1993), Effect of dietary contents of cholecalciferol, 1, 25 dihydroxycholecalciferol and 24, 25 dihydroxycholecalciferol on blood concentration of 25, hydroxycholecalciferol, 1,25 dihydroxycholecalciferol, total calcium and eggshell quality. *Br. Poult. Sci.* 34(5), 1021–1227.
Turk, D. E., Sunde, M. L. and Hoekstra, W. G. (1959), Zinc deficiency experiments with Poultry. *Poult. Sci.* 38, 1256.
Ueberschär, K. H., Vogt, H. and Matthes, S. (1985), The effect of various vanadium levels in broiler and laying rations on the performance of the birds and on residues in tissues and eggs. *Arch. Geflügelk.* 49, 23–30.
Valkonen, E., Rinne, Q. and Valaja, J. (2009), The effects of perches in furnished cages, World Poultry Science Association (WPSA), In *Proceedings of the 8th European Symposium on Poultry Welfare*, Cervia Italy, 18–22 May 2009, p. 16.
Van Der Hoeven-Hangoor, E., Van Der Linde, I. B., Paton, N. D., Verstegen, M. W. A. and Hendriks, W. H. (2013), Effect of different magnesium sources on digesta and excreta moisture contentand production performance in broiler chickens. *Poult. Sci.* 92, 382–91.
Van Krimpen, M. M., Kwakkel, R. P., Reuvekamp, B. F. J. and Verstegen, M. W. A. (2005), Impact of feeding management on feather pecking in laying hens. *World Poult. Sci. J.* 48, 389–98.

Van Krimpen, M. M., Kwakkel, Q. P., Van Der Peet-Schwering, C. M., Den Hartog, L. A. and Verstegen, M. W (2008a), Low dietary energy concentration, high nonstarch polysaccharide concentration, and coarse particle sizes of nonstarch polysaccharides affect the behavior of feather-pecking-prone laying hens. *Poult. Sci.* 87, 485–96.

Van Krimpen, M. M., Kwakkel, R. P., Van Der Peet-Schwering, C. M., Den Hartog, L. A. and Verstegen, M. W. (2008b), Low dietary energy concentration, high nonstarch polysaccharide concentration, and coarse particle sizes of nonstarch polysaccharides affect the behavior of feather-pecking-prone laying hens. *Poult. Sci.* 87, 485–96.

Van Krimpen, M. M., Kwakkel, R. P., Van Der Peet-Schwering, C. M., Den Hartog, L. A. and Verstegen, M. W (2009), Effects of nutrient dilution and nonstarch polysaccharide concentration in rearing and laying diets on eating behavior and feather damage of rearing and laying hens. *Poult. Sci.* 88, 759–73.

Vogt, H., Dewar, W. A., Sauveur, B. and Simmons, P. C. M. (1984), Mineral requirements and recommendations for adult birds. *World Poult. Sci. J.* 40, 183–7.

Waddel A. L., Board, R. G., Scott, V. D. and Tullet, S. G. (1989), Influence of dietary magnesium content in the domestic fowl. *Br. Poult. Sci.* 30, 865–76.

Weber, G. M. (2009), Vitamin requirements and economic responses. 17th European Symposium on poultry nutrition, 23–27 August 2009, WPSA, U.J. Branch, Edinburgh. S3.2.

Whitehead, C. C. (1986), Requirements for vitamins. In *Proceeding Symposium Poultry Science*, London, Butterworth, pp. 173–89.

Whitehead, C. C. (2004), Overview of bone biology in the egg-laying hen. *Poult. Sci.* 83, 193–9.

Whitehead, C. C. and Fleming, Q. (2000), Osteoporosis in cage layers'. *Poult. Sci.* 79, 1033–41.

Whitehead, C. C. and Fleming R. H. (2000), Osteoporosis in cage layers. *Poult. Sci.* 79, 1033–41.

Wilkins, L. J., Brown, S. N., Zimmerman, P. H., Leeb, C. and Nicol, C. J. (2004), Investigation of palpation as a method for determining the prevalence of keel and furculum damage in laying hens. *Vet. Rec.* 155, 547–9.

Wilkins, L. J., McKinstry, J. L., Avery, N. C., Knowles, T. G., Brown, S. N., Tarlton, J. and Nicol, C. J. (2011), Influence of housing system and design on bone strength and keel bone fractures in laying hens. *Vet. Rec.* 169, 414.

Wilkinson, S. J., Selle, P. H., Bedford, M. R. and Cowieson, A. J. (2011), Exploring calcium-specific appetite in poultry nutrition. *World Poult. Sci. J.* 67, 587–95.

Wu, G., Liu, Z. Bryant, M. M. and Roland, D. A. (2006), Comparison of Natuphos and Phyzyme as P\phytase sources for commercial layers fed corn-soy diet. *Poult. Sci.* 85, 64–9.

Xiao, J. F., Wu, S. G., Zhang, H. J., Yue, H. Y., Wang, J., Ji, F. and Qi, G. H. (2015), Bio-efficacy comparison of organic manganese with inorganic manganese foreggshell quality in Hy-Line Brown laying hens. *Poult. Sci.* 94, 1871–78.

Xiao, J. F., Zhang, Y. N., Wu, S. G., Zhang, H. J., Yue, H. Y. and Qi, G. H. (2014), Manganese supplementation enhances the synthesis of glycosaminoglycan in eggshell membrane: a.strategy to improve eggshell quality in laying hens. *Poult. Sci.* 93, 380–8.

Zeizel, S. (1981), Dietary Choline: biochemistry, physiology and pharmacology. *Ann. Rev. Nutr.* 1, 95–121.

Zhai, Q. H., Dong, X. F., Tong, J. M., Guo, Y. M. and Bao, Y. E. (2013), Long-term effects of choline on productive performance and egg quality of brown egg laying hens. *Poult. Sci.* 92, 1824–9.

Zhang, J. D., Chen, D. W. and Yu, B. (2008), Effect of different dietary energy sources on induction of fatty liver-hemorrhagic syndromein laying hens. *Int. J. Poult. Sci.* 7, 1232–6.

Zhang, J. W., Chen, D. W. and Wang, Y. M. (2011), Effect of dietary energy source on deposition and fatty acid synthesis in the liver of the laying hen. *Br. Poult. Sci.* 52, 704–10.

Zhou, Z. L., Deng, Y. F., Tao, Q. S., Hu, Y. F. and Hou J. F. (2009), Effects of Gushukang, a Chinese herbal medicine, on bone characteristics and osteoporosis in laying hens. *Poult. Sci.* 88, 2342–5.

Zit, L., Tumova, E. and Stolc, L. (2009), Effects of genotype, age and their interaction on egg quality in rown-egg laying hens. *Acta Vet. Brno.* 78, 85–91.

Chapter 3

Welfare of laying hens: an overview

Tina Widowski, Teresa Casey-Trott, Michelle Hunniford and Krysta Morrissey, University of Guelph, Canada

1 Introduction

Although the majority of the world's laying hens are still housed in conventional ('battery') cages, public concern for the welfare of laying hens is rapidly changing the methods by which eggs are being produced and marketed in many regions of the world. These changes began in a few individual European countries during the 1980s and culminated in the passing of the EU Council Directive 99/74 for laying hens, which now prohibits the use of non-enriched cage systems in all EU countries since January 2012 (Appleby, 2004). There is currently no federal legislation in the United States, but several individual states have passed laws that essentially ban conventional cages (Mench et al., 2011). In New Zealand, conventional cages are due to be phased out by 2022 under the Code of Welfare for Laying Hens, and in February 2016, Egg Farmers of Canada announced that no new conventional cage systems are to be installed after July 2016. Additionally, numerous food labelling and certification schemes have been developed for the purpose of marketing speciality 'animal welfare-friendly' eggs to ethically conscious consumers around the world. In North America, food processers, supermarkets and restaurant chains are driving massive changes by requiring that egg suppliers adopt housing systems that animal protection groups or consumers consider to be acceptable for animal welfare (Mench et al., 2011). In 2015 and early 2016, an unprecedented number of restaurants, grocers and distributors made pledges to purchase eggs only from non-cage systems, which will require new housing for over 150 million hens in the United States and Canada by 2026.

http://dx.doi.org/10.19103/AS.2016.0012.22

Confinement housing is at the forefront of public concern for the welfare of laying hens. The key issues here are related to lack of space and limited opportunities to perform natural behaviour. However, each type of housing system has costs and benefits, and different welfare problems are associated with alternative housing systems. Additionally, there are welfare concerns for laying hens that are not based solely on how adult laying hens are housed. These include injurious pecking (IP; feather pecking (FP) and cannibalism) and problems related to bone health (osteoporosis and fractures). Moreover, the importance of hens' early life experiences on lifetime health and welfare is getting increasing attention. In this chapter, we will provide an overview of the current state of knowledge on the major welfare issues for laying hens. We will also identify some of the welfare trade-offs across housing systems, as well as new and emerging issues directing future research.

2 Defining and measuring animal welfare

In any discussion of animal welfare, it is important to recognize that stakeholder groups vary in the degree of importance that they place on different aspects of animal welfare. Although the broader community places the most emphasis on freedom of movement and behavioural opportunities for laying hens, the scientific assessment of animal welfare involves multiple measures that capture different viewpoints on what constitutes a good quality of life (QoL) for animals. These include measures of health and biological fitness (e.g. physical condition, mortality, production, indicators of stress), the emotional or subjective experiences of animals (e.g. conditions leading to pain, fear, discomfort, reward, pleasure) as well as the ability to perform natural or species-typical behaviour patterns (Fraser et al., 1997). All three viewpoints are articulated in formal definitions of animal welfare (Table 1). For example, the concept of the Five Freedoms, which originated in the report of the Technical Committee to Enquire into the Welfare of Animals Kept under Intensive Livestock Husbandry Systems (Brambell et al., 1965) and was expanded into its current form by the Farm Animal Welfare Council in the United Kingdom in 1979 (FAWC, 2009), includes aspects of the animal's physical and mental well-being and its ability to express normal behaviour. Recognizing that animal welfare is integral to animal health, the World Organization for Animal Health included a definition for animal welfare in its Terrestrial Animal Health Code in 2007 that included many of the same elements (OIE, 2010).

In contrast with formal definitions, conceptual frameworks for animal welfare are continuously evolving. Previously, they focused mainly on identification and alleviation of negative states, but more contemporary frameworks increasingly emphasize the promotion of positive ones (Mellor, 2016). Current thinking also acknowledges that it is practically impossible to alleviate all negative experiences, but that by providing more opportunities for positive experiences, overall welfare will be improved. The Five Domains Model, for example (Table 1), incorporates four physical/functional domains – nutrition, environment, health and behaviour. Each of the physical domains can give rise to experiences that can range from negative to positive. For example, restricted diet or poor food quality will result in hunger, a sufficient diet will neutralize hunger, whereas a plentiful amount of a good quality diet will result in satiety and gustatory pleasure. The fifth domain comprises the overall mental state (welfare status) of the animal arising from the accumulated experiences from each of the physical domains. The Farm Animal Welfare Council in the United Kingdom has also recently suggested that minimum standards for farm animals

Table 1 Concepts and definitions of animal welfare

The Five Freedoms (from the Farm Animal Welfare Council, 1979)
1 **Freedom from hunger or thirst** by ready access to fresh water and a diet to maintain full health and vigour 2 **Freedom from discomfort** by providing an appropriate environment including shelter and a comfortable resting area 3 **Freedom from pain, injury or disease** by prevention or rapid diagnosis and treatment 4 **Freedom to express (most) normal behaviour** by providing sufficient space, proper facilities and company of the animal's own kind 5 **Freedom from fear and distress** by ensuring conditions and treatment which avoid mental suffering
OIE Definition of Animal Welfare (from the Terrestrial Animal Health Code, 2010)
'Animal welfare means how an animal is coping with the conditions in which it lives. An animal is in a good state of welfare if (as indicated by the scientific evidence) it is healthy, comfortable, well-nourished, safe, able to express innate behaviour, and if it is not suffering from unpleasant states such as pain, fear and distress. Good animal welfare requires disease prevention and appropriate veterinary treatment, shelter, management and nutrition, humane handling and humane slaughter or killing. Animal welfare refers to the state of the animal; the treatment that an animal receives is covered by other terms such as animal care, animal husbandry, and humane treatment.'
Five Domains Model of Animal Welfare (see Mellor 2016)
Physical/Functional Domains Nutrition – quantity and quality of food and water Environment – degree of comfort from physical features (e.g. temperature) Health – degree of disease, injury or physical impairment Behaviour– ability to express rewarding behaviours *Mental State* Includes range of negative experiences (e.g. hunger, discomfort, frustration) and positive experiences (e.g. comfort, pleasure, reward) arising from the physical domains
Quality of Life Scale (QoL)
A good life – the balance of + and – experiences is predominantly + A life worth living – the balance of + and – is generally +; can be achieved by full compliance with a code of practice plus elements that promote positive experiences Point of balance – neutral point; can be achieved by compliance with minimum standards of care A life worth avoiding – balance is negative but can be remedied by veterinary care or change in husbandry A life not worth living – balance is strongly negative and cannot be remedied so that euthanasia is the humane alternative

move beyond the Five Freedoms to provide farm animals with 'a life worth living' (FAWC, 2009). Extending further, the concept of QoL has been developed (Yeates, 2011; Mellor, 2016) as an alternative welfare assessment. It considers the balance between negative (minimizing harms) and positive (providing pleasures) states over the lifetime of the animal (Table 1). While the concept of QoL is still developing and its practical application is currently very limited, this viewpoint may increasingly be considered when weighing the pros and cons of different housing and management practices. The scientific evaluation of negative affective states in animals has advanced considerably using a combination of behavioural and physiological evidence, whereas the scientific evaluation of positive affective states is still in its infancy. However, this approach may receive more attention in the future when developing new standards of care policies for farm animals including laying hens (Edgar et al., 2013).

3 Welfare trade-offs in different housing systems

Comparisons of the welfare of animals in different types of housing systems are problematic since there is a wide range of alternative systems, and welfare often depends on details of system design and management. Several decades of research on the welfare of hens in alternative housing systems have been summarized in various literature reviews and scientific reports (e.g. EFSA-AHAW, 2005; Blokhuis et al., 2007; Laywel project, 2006; Lay et al., 2011; Widowski et al., 2013), including reports comparing various welfare measures from data collected on large-scale commercial farms (e.g. Rodenburg et al., 2008b; Sherwin et al., 2010; Petrik et al., 2015; Blatchford et al., 2016; Weeks et al., 2016). It is clear from these reports that there can be significant trade-offs for different aspects of welfare for hens in different commercial-scale systems.

Conventional cages generally result in good health, hygiene and low mortality, but the environmental barrenness and lack of space impose a high degree of behavioural restriction for hens. It is well established that laying hens are highly motivated to gain access to a nest box to lay their eggs and to perform perching, foraging and dust bathing behaviour (see section below). Even basic activities such as locomotion, stretching and wing flapping are significantly constrained in conventional cages, and the lack of load-bearing exercise reduces bone strength (Widowski et al., 2013).

Housing laying hens in cage-free systems with nests, perches and litter (or free range with access to the outdoors) offers substantial opportunities to perform a greater range of locomotion and engage in natural behaviours, but it can also significantly increase the risk of injury and poor health. Risks are highly dependent on the individual system or individual farm. Laying hens housed in non-cage systems, such as single-tier floor systems and multi-tier aviaries, have a greater risk of infectious diseases, broken bones (see section below), internal and external parasites, and generally have higher rates of mortality than hens housed in cages (Lay et al., 2011; Widowski et al., 2013). A recent report based on data from over 3500 commercial flocks in the European Union indicated that mortality rates were significantly higher and more variable in non-cage and especially in free-range systems compared to cage systems (Weeks et al., 2016). The risks were considerably greater when hens were not beak trimmed and were significantly associated with breed. FP and cannibalism are a major cause of poor welfare and higher mortality in non-cage systems and a significant amount of research is ongoing to identify underlying causes, management risks and means of prevention (see section below). Smothering, which occurs when birds mass together and pile on top of one another, is another cause of mortality in non-cage systems, and this problem is only beginning to be investigated (Barrett et al., 2014). Aspects of system design may be important as hens in group sizes of thousands can crowd together at different times of day to access different resources, for example, to dust bathe on litter (Campbell et al., 2016a).

Enriched colony cages (hereafter referred to as furnished cages) outfitted with nest boxes, perches and scratch mats provide more space and greater opportunities to express motivated behaviour than conventional cages, generally resulting in health, hygiene and physical measures of welfare comparable to conventional cages (Sherwin et al., 2010; Rodenburg et al., 2012; Blatchford et al., 2016). Nests are generally well used by hens, although nest floor material, nest space and degree of enclosure can vary considerably across cage designs. Perches in furnished cages are also generally well used (Regmi et al., 2016) and perching behaviour and/or increased space and locomotion afforded by furnished cages result in greater bone strength (Regmi et al. 2016; Casey-Trott, 2016). From a behavioural standpoint, providing resources to support dust bathing and foraging poses the biggest challenge in furnished cages.

As the designs of alternative housing systems for laying hens rapidly evolve, more research is aimed at identifying specific causes of welfare problems in these systems and developing the means to alleviate those problems.

4 Behavioural requirements

The ability to express 'normal' or innate behaviour is included in definitions of animal welfare; however, it is neither practical nor necessary for hens to be able to perform all types of behaviour (Nicol, 2015). For over 50 years, empirical research has focused on identifying the candidate behaviours that may be important for hens and developing ways to accommodate those specific behaviours in commercial systems. The behaviours that appear to be important to hens are referred to as *behavioural needs* (Hughes and Duncan, 1988) and *behavioural priorities* (Weeks and Nicol, 2006).

One way to identify a behavioural need is to consider whether it is motivated by internal or external stimuli, or both (Widowski, 2015). Logically, behaviours that are mainly internally driven would constitute *needs* (Hughes and Duncan, 1988) because the motivation is independent of the environment. Internally driven behaviour patterns may arise, for example, from a disruption in homeostasis (e.g. dehydration leads to drinking behaviour) or changing hormone levels (e.g. increase in luteinizing hormone leads to nesting behaviour; Appleby et al., 2004). Other behaviours are externally motivated and a hen's motivation to perform them occurs only when environmental stimuli are present. More often, motivated behaviour patterns stem from a combination of internal and external factors.

Determining behavioural needs and priorities can be accomplished in two ways: by measuring preferences and the strength of motivation. Measuring what hens prefer will tell you what they want; measuring how hard a hen will work, or what costs she will pay, to access that preferred resource will tell you how much she wants it (Dawkins, 1983). Another way to determine if a behaviour is a need or priority is to observe whether a hen exhibits signs of a negative mental state (e.g. frustration) if she cannot access a resource or perform the behaviour (Duncan, 1970). There are four commonly agreed-upon behavioural needs of laying hens: nesting, foraging, dust bathing and perching (Weeks and Nicol, 2006). Increasingly, minimum standards for even the most basic designs of sustainable housing for laying hens require that these behaviours be accommodated in some way. More complex systems go even further to promote the full expression of these behaviours in order to increase pleasure or positive mental states rather than just to prevent negative states in their absence.

4.1 Nesting

Based on evidence from a variety of motivation studies, access to a nest site is a high-ranking priority for laying hens. Nesting behaviour is internally motivated by changes in hormone levels that occur during ovulation (Mench, 2009). It is comprised of distinct searching (appetitive phase) and then sitting at the site where the hen lays her egg (consummatory phase). Hens are motivated to nest regardless of the environment, and when hens are housed in the absence of a suitable nest site, they will often extend their searching phase (Cooper and Albentosa, 2003), the sitting phase will be largely absent (Mench, 2009; Duncan, 1970) and they may show behaviour that is consistent with frustration (Cooper and Appleby, 1996). The strength of motivation to access a nest has been demonstrated by the high costs that hens are willing to pay (e.g. hens will pass through narrow gaps, weighted doors and unfamiliar birds to access a nest; Weeks and Nicol, 2006).

The two main factors that influence nest design are enclosure and floor surface (or substrate). Hens prefer to lay in an enclosed space (Appleby and McRae, 1986; Struelens et al., 2008), to nest on friable litter over artificial surfaces (Freire et al., 1996) and on artificial turf above plastic mesh (Guesdon and Faure, 2004), wire (Hughes, 1993) or rubber (Reed and Nicol, 1992). When nests are not attractive or absent altogether, hens may perform pacing behaviour (Zimmerman et al., 2000), displacement behaviour (e.g. preening; Duncan, 1970) or become aggressive (Lundberg and Keeling, 1999). If there are not enough nests, or not enough space in the nests, some hens may extend their searching phase and retain the egg in the shell gland past the point when oviposition would have occurred (potentially causing extra-cuticular calcium deposits, shell deformities; Hughes et al., 1986).

Large communal nests and curtained nest areas dominate current nest designs in non-cage systems and furnished cages. Nest use can be reduced in some models of cages when larger group sizes share a single large nest area, and this can result in more cracked or dirty eggs (Wall, 2011). Recent work by Hunniford et al. (2014) demonstrated that the amount of nest space provided in some furnished cages may be insufficient (Fig. 1). Although hens in large furnished cages (with 55 or 80 hens) were provided a large group nest with adequate floor space (according to current guidelines), on average, 23% of eggs were laid outside of the nest. One reason for this, as confirmed by later studies (Hunniford and Widowski, 2016a,b), may be a fundamental misunderstanding of what hens look for in a nest site; it is not the area per bird that dictates the amount of nest space, but rather the number of enclosed spaces within the nest. This has been demonstrated in other studies that have found that hen's behaviour is affected by the degree of enclosure (Ringgenberg et al., 2015) or by the size of the nest (e.g. hens prefer smaller nests; Ringgenberg et al., 2014).

In cage-free systems, there are further consequences if hens lay outside of the nest – loss of egg revenue, compromised shell quality – but there is also an issue where hens become gregarious and use a disproportionately small number of nests, which in the extreme may lead to smothering (Riber, 2010). Because of these consequences, researchers are investigating nest preferences that are specific to cage-free systems (Ringgenberg et al., 2014; 2015), and ways to decrease gregarious nest use, potentially by spreading out where hens lay their eggs by providing heterogeneous nests (Clausen and Riber, 2012). Additionally, new genetic selection strategies are taking nesting behaviour into account by using nest acceptance as an important selection criterion (Icken et al., 2012) and even by using behaviour in group nests in the selection process (Heinrich et al., 2015).

(a)

(b)

Figure 1 (a) Hens are attracted to the enclosure and flooring surface of nests in most furnished cages. (b) Nests can become crowded when all hens are nesting at the same time and optimal nest space and design needs to be determined. Photo credit Michelle Hunniford.

4.2 Foraging

Foraging is the appetitive stage of feeding behaviour; it involves the active search for food particles by pecking and scratching at the ground (Appleby et al., 2004). Even when food is freely available, modern strains of hens can spend around 40% of the day in foraging activities (Nicol, 2015). In addition, hens will choose to work (e.g. forage through wood shavings) to gain access to more than 30% of their daily food intake even when it is otherwise freely available (Lindqvist et al., 2002). Hens will readily forage in a variety of litter substrates, although they do not seem willing to work very hard for access to it (e.g. by pushing a weighted door; de Jong et al., 2007). However, in conventional cages where substrate is not explicitly provided, hens exhibit foraging behaviour by flicking the food in the feed trough with their beak while simultaneously scratching at the cage floor (Mench, 2009; Nicol, 2015). The importance of foraging to hens is mainly demonstrated by an increased risk for behavioural problems (i.e. FP), when foraging substrates are not provided (see next section).

Hens actively forage in the litter substrates provided in non-cage systems (Rodenburg et al., 2008a; Fig. 2). Currently, furnished cages are designed to support foraging behaviour by depositing small amounts of feed onto some type of scratch mat (e.g. artificial turf or smooth plastic); however, there are mixed results as to how well this actually supports foraging behaviour. Although hens did peck and scratch at smooth plastic mats more than expected, Herwig and Widowski (2013) found that the foraging behaviour did not increase following feed deposition on the mat. In contrast, Guinebretière et al. (2015) showed that both rubber and plastic turf mats were attractive for pecking and scratching, and that foraging was increased when either feed or wheat bran was distributed on them. However, mats are often quite small and feed has been shown to be less attractive than litter for foraging (Guinebretière et al., 2014). Wheat bran appears to be a preferred foraging substrate (Guinebretière et al., 2014), with the added benefits that it can be provided automatically via auger and it is lighter than feed so it stays on the mat longer.

4.3 Dust bathing

Dust bathing involves a complex sequence of motor patterns by which the hen distributes fine substrate under the feathers to remove excess oil, dirt and ectoparasites, and improve feather structure (Mench, 2009). A hen's motivation to dust bathe is affected by an interaction of internal and external factors. The behaviour occurs every 2–3 days regardless of whether litter is provided. When deprived of litter for some time, hens will dust bathe more quickly and for a longer period of time when given access to substrate. In systems without litter substrate, hens will dust bathe on the wire floor (i.e. sham or vacuum dust bathing; Weeks and Nicol, 2006), although it loses most of its functional purpose because abrasion from the cages may lead to poorer feather condition and feather loss. Dust bathing is also motivated by features of the external environment, including the sight of a dusty substrate, bright light, a radiant heat source and the sight or sound of other hens dust bathing (Duncan et al., 1998).

There is little evidence that hens prevented from dust bathing are frustrated, but dust bathing itself may be a pleasurable experience for hens. In a push-door experiment, Widowski and Duncan (2000) compared the strength of motivation of hens that were deprived of substrate versus those that had free access to litter. Although the deprived hens did push more weight, some of the hens with free access showed equal or greater motivation to access the litter site, potentially because it contained peat moss (i.e. a highly preferred litter substrate; de Jong et al., 2007). Providing a highly attractive substrate, in this case, may have stimulated the behaviour.

In non-cage systems, hens are highly motivated to dust bathe on litter areas, which may become extremely crowded during peak periods of synchronous behaviour (Campbell et al., 2016a; Fig. 2). The scratch mats provided to hens in furnished cages, however, generally do not appear to support dust bathing since the majority of dust bathing bouts are still observed on the wire floors (Rodenburg et al., 2008b; Herwig and Widowski,

Figure 2 Hens are attracted to the litter for foraging and dust bathing under the system in aviaries. Sufficient litter space is necessary since dust bathing occurs mainly in the afternoon when piling or smothering is a risk. Photo credit Leanne Cooley.

2013; Guinebretière et al., 2015). Hens in furnished or conventional cages have often been observed tossing feed from the feeder during sham dust bathing on the wire floor (personal observations); however, feed is not an ideal substrate for dust bathing due to its higher lipid content compared to other substrates.

4.4 Perching

Perching is considered to be an anti-predator behaviour (Schrader and Müller, 2009; Appleby et al., 2004). Hens will perch on elevated structures during the day, but have a stronger motivation to perch at night to sleep (Olsson and Keeling, 2002), especially when the perches are elevated (Schrader and Müller, 2009). Perches increase the complexity of the environment, may provide an escape from other hens and improve foot and musculoskeletal health (Hester, 2014). There do not seem to be any behavioural consequences that are a direct result of not providing perches, although hens will use them if they are provided, and may even show signs of frustration if access to a visible perch is thwarted (Olsson and Keeling, 2000). The results of studies like Schrader and Müller's (2009) indicate that the low perches used in some housing designs, like furnished cages, do not provide the height that hens prefer, although they are generally well used (Fig. 3). Perch shape, material and spatial configuration can affect foot health and keel deformations as well as the hens' ability to grip and negotiate complex housing systems (Struelens and Tuyttens, 2009).

Figure 3 (a) Hens standing on a perch in a furnished cage. Photo credit: Egg Farmers of Canada. (b) Hens sitting and standing on perches in an aviary; hens prefer to roost on the highest perches at night. Photo credit: Leanne Cooley.

5 Injurious pecking

The commercial egg industry has historically faced, and is currently facing, challenges in managing behavioural problems within flocks of laying hens. FP and cannibalism, now commonly referred to collectively as 'injurious pecking' (Lambton et al., 2013), are generally associated with high-producing commercial strains, although not exclusively performed by modern breeds (Hocking et al., 2004). IP causes, or has the potential to cause, damage to feathers and/or tissue and different types of bird-to-bird pecking (Table 2). IP has been linked with reduced heat retention due to loss of feathers, decreased egg production, pain and increased mortality (Sedlačková et al., 2004). Aggression can also cause wounds, but it is generally not grouped within the IP umbrella as it is thought to stem from an entirely different aetiology (Savory, 1995; Rodenburg et al., 2008a). Although gentle FP (GFP), severe FP (SFP) and cannibalism (C) have been grouped together within an overarching category, they may also have different aetiological bases. For example, although GFP and SFP may be correlated at a given moment, the performance of GFP during rearing does not predict future performance of SFP (Newberry et al., 2007). Similarly, SFP and C are not always correlated (Hughes and Duncan, 1972; Hocking et al., 2004), which is to say that a highly denuded flock may not have issues with C, yet another quite well-feathered flock could. IP is a multi-factorial problem that occurs in all types of housing systems for laying hens. Environmental factors, nutrition, genetic predisposition, physiological states, and even rearing environments are implicated in playing a role in the development of IP behaviour.

Currently, the two main methods used commercially to prevent and control the development of, or the damage from, IP behaviour are beak treatment (hot blade trimming or infra-red treating) and the use of dim lighting. Numerous studies have shown

Table 2 Types of bird-to-bird pecks

Peck type	Description
Gentle feather peck (GFP)	Gentle pecks directed at the tips of the feathers, usually performed repeatedly in bouts. Recipient usually does not react, and this type of pecking does not result in feather removal.
Severe feather peck (SFP)	Forceful pecks directed at the feathers and involves grasping, pulling, removal and often consumption of removed feathers. Can occur independently or within bouts. Recipient usually reacts or withdraws.
Cannibalistic peck (C)	Pecking directed at the tissue and can cause tissue damage and/or bleeding. Can occur on any part of the body with the recipient reacting (e.g. squawking or withdrawing).
Vent peck	Cannibalistic pecks directed specifically at the cloaca of recipient hen. Can cause tissue damage, intestinal exposure and subsequent death.
Aggressive peck	Overt and forceful peck, usually directed at recipient's head. Aggressive pecks represent behaviour that is part of establishing a dominance hierarchy, or used to aid in resource guarding, and is not synonymous with SFP.

that beak trimming can reduce the damage and mortality caused by and associated with IP (e.g. Morrissey et al., 2016; Weeks et al., 2016); however, beak trimming poses its own welfare issues and the practice has been banned in some European countries. There may be other non-invasive means for shortening beaks. Van der Weerd et al. (2006) showed promising results for reduction of beak length with two types of abrasive materials on the bottom of the feed trough for non-beak trimmed hens. However, Morrissey et al. (2016) recently found no effect of providing abrasive blocks to hens in large furnished cages on beak shape or IP, although the decision to place the abrasive blocks in a fixed position on the sides of the cage may not have attracted pecking behaviour. In a follow-up study, the presence of cuttlebones in the cage resulted in a tendency for shorter beaks over a 10-week period (Morrissey, unpublished data). Reductions in light intensity or use of different colours of light are commonly used on farms as they reduce overall activity levels (including IP) or mask the appearance of feathers that attract pecking. Significantly more SFP and a higher rate of mortality were observed for hens housed with a high light intensity (30 lux) compared to those housed with low light intensity (3 lux) (Kjaer and Vestergaard, 1999). However, the use of dim lighting also raises welfare concerns as it may detrimentally affect eye development in young birds and it may impair vision and a bird's ability to negotiate their environment (Nicol et al., 2013). Additionally, these management tools do not address the underlying motivation for IP. Therefore, a significant research effort is directed at identifying the mechanisms and other means of prevention in commercial conditions, and a number of cross-sectional studies on commercial farms have identified risk factors for IP (Table 3).

Providing hens with litter substrate can reduce the frequency of IP (Huber-Eicher and Weschler, 1997; Aerni et al., 2000). One hypothesis is that FP is redirected foraging behaviour (Blokhuis and Arkes, 1984; Blokhuis, 1986; Huber-Eicher and Weschler, 1997), which was supported by Dixon et al. (2008), who showed that the detailed pecking motor patterns used in SFP were extremely similar to those used for foraging. Modern commercial diets are generally quite energy dense and freely available, which means hens are not required to spend large amounts of time foraging for food. Apart from foraging motivation, there may be nutritional or other aspects of gut function related to IP. Diets high in fibre can reduce IP rates (reviewed in Choct and Hartini, 2003; van Krimpen et al., 2005; Rodenburg et al., 2013), and hens from a high-FP genetic line were observed to eat more of a diet high in fibre than hens from a low-FP line (Kalmendal and Bessei, 2012). High-FP hens have also been shown to consume more feathers than low-FP hens, which affects intestinal rate of passage and suggests that an insoluble component in the diet is important to hens (McKeegan and Savory, 2001; Harlander-Matauschek et al., 2006). The emerging concept of the microbiota–gut–brain axis and its role in brain function, mood and behaviour across species (Cryan and Dinan, 2012) is leading to new research aimed at exploring the relationships between FP, feather eating and intestinal microbiota in laying hens (e.g. Meyer et al., 2012).

Both foraging substrates and high fibre diets have been shown to reduce IP, but some of these effects may be due simply to a shift in time budget, reducing the amount of time that is available to hens to spend pecking at their group mates. Other environmental enrichments may be capable of doing the same, or may in fact be able to fulfil the innate behavioural need for foraging to some degree. For example, hanging white polypropylene ropes in cages has been shown to elicit and maintain attraction (Jones et al., 2002) and has the potential to reduce IP, although these data did not show statistical differences in FP. Extra environmental enrichment in the form of ropes and vertically fixed pecking

Table 3 Selected risk factors affecting the probability or rate of gentle feather pecking (GFP), severe feather pecking (SFP) and feather damage (FD) based on significant results ($P < 0.05$) from a number of published cross-sectional studies (see associated references). Where the feather pecking type was not defined, both SFP and GFP were combined

<table>
<tr><th rowspan="2">Risk factor</th><th colspan="3">Prevalence of FP or FD</th><th rowspan="2">References</th></tr>
<tr><th>GFP</th><th>SFP</th><th>FD</th></tr>
<tr><td>↑ Age</td><td>↓</td><td>↑</td><td>↑</td><td>Drake et al., 2010; Lambton et al., 2010</td></tr>
<tr><td>Beak trimmed</td><td>↑</td><td>↓</td><td>↓</td><td>Lambton et al., 2010; Gilani et al., 2013</td></tr>
<tr><td>↑ Stocking density</td><td colspan="2">↑</td><td></td><td>Huber-Eicher and Audige, 1999</td></tr>
<tr><td>↑ % Flock ranging</td><td>↓</td><td>↓</td><td></td><td>Lambton et al., 2010</td></tr>
<tr><td>Access to perches</td><td>↓</td><td></td><td></td><td>Huber-Eicher and Audige, 1999; Lambton et al., 2010</td></tr>
<tr><td>Delayed access to litter at POL</td><td></td><td></td><td>↑</td><td>Lambton et al., 2010</td></tr>
<tr><td>↑ Temperature</td><td>↓</td><td></td><td></td><td>Lambton et al., 2010; Gilani et al., 2013</td></tr>
<tr><td>Low variation in humidity</td><td>↓</td><td></td><td></td><td>Gilani et al., 2013</td></tr>
<tr><td>Poor air quality</td><td colspan="2">↑</td><td>↑</td><td>Huber-Eicher and Audige, 1999; Drake et al., 2010; Gilani et al., 2013</td></tr>
<tr><td>High light intensity</td><td></td><td></td><td>↑</td><td>Drake et al., 2010</td></tr>
<tr><td>High noise level</td><td>↑</td><td></td><td>↑</td><td>Drake et al., 2010; Gilani et al., 2013</td></tr>
<tr><td>↑ Variation in sound/light levels</td><td>↓</td><td></td><td></td><td>Gilani et al., 2013</td></tr>
<tr><td>Pelleted feed</td><td></td><td>↑</td><td>↑</td><td>Lambton et al., 2010</td></tr>
<tr><td>Chain feeder vs. pan feeder</td><td></td><td>↑</td><td>↑</td><td>Drake et al., 2010; Gilani et al., 2013</td></tr>
<tr><td>↓ Diet changes during rear</td><td>↓</td><td>↓</td><td></td><td>Gilani et al., 2013</td></tr>
<tr><td>Change in farm from rear to lay</td><td></td><td></td><td>↑</td><td>Drake et al., 2010</td></tr>
</table>

mats were well used by hens and significantly reduced IP in large furnished cages on a commercial farm, but had little effect on feather condition or mortality (Morrissey et al., 2016). In a follow-up study with similar enrichments, albeit slightly modified (e.g. more enrichments/hen, larger bunches of ropes, non-cauterized rope tips), hens housed with extra enrichments had better feather cover when the study concluded at 35 weeks of age (Morrissey, unpublished data).

Genetic predisposition plays a large role in the development of IP behaviour.

There are significant breed differences (Hocking et al., 2004; Morrissey et al., 2016), and the performance of IP is a heritable trait (Rodenburg et al., 2003; Buitenhuis et al., 2004). Divergent lines from a single strain have been developed on the basis of levels of IP performance (Kjaer et al., 2001). A lot of research has been devoted to understanding how to incorporate IP behaviour (or lack thereof) into breeding programmes (Muir et al., 2014). Traditionally, breeding companies have focused on health and production parameters for hens housed in individual cages (Craig and Muir, 1996; Jones and Hocking, 1999). This can be useful for commercial hens that are housed in conventional battery cages, where group sizes are small, activity is limited and feather peckers can be easily identified. However, selection based on so few traits has been criticized as one of the reasons for increased behavioural problems in modern farms. In addition, the move towards alternative and socially sustainable housing systems means that breeding companies are looking to adapt their breeding programmes to meet the needs of the modern production system. Genetic selection tools have been identified and breeding companies have begun to house hens in family groups, using both social and behavioural parameters as part of their breeding goals (Ellen et al., 2008; Muir et al., 2014; Sun et al., 2014).

At the individual level, physiological factors and temperament traits have also been implicated in the development of IP behaviour. The onset of egg laying is linked with increased hormone circulation, a spike in cannibalistic pecking and increased mortality (Savory, 1995). The visual stimulus of the exposed cloaca during egg laying can be attractive to young hens and become a pecking target (Savory, 1995). Hormonal fluctuations have been shown to increase IP as large exogenous doses of oestrogen and testosterone increased FP (Wysocki et al., 2010). Increased locomotor behaviour (or hyperactivity) has also been suggested to be linked with increased IP (Kjaer, 2009) and selection for high levels of production have altered this behavioural trait in some strains (Kjaer et al., 2015). In fact, groups of hens that performed more IP and had worse plumage condition also had higher values for a measure of activity (Morrissey et al., 2016).

Although FP is generally considered an undesirable behaviour and an indicator of poor welfare, there is some evidence to suggest that it is used by some hens as a proactive coping mechanism (reviewed in Rodenburg et al., 2004). Jensen et al. (2005) showed that hens that performed more FP were also more active in an open field test, a restraint test, as well as a novel object test, compared to hens that were classified as FP recipients. In addition, high-FP hens showed lower serotonin and dopamine turnover than their low-FP conspecifics (van Hierden et al., 2002). In humans, the serotonergic system is responsible for regulating mood, sleep and appetite, and has even been implicated in disorders such as schizophrenia and obsessive compulsive disorders (Wysocki et al., 2010). It is hypothesized that serotonin may have similar effects in poultry in terms of the role it plays in psychological diseases in humans. After identifying a dose for a serotonin agonist that effectively reduced serotonin turnover, van Hierden et al. (2004) showed that dosing hens with this agonist significantly increased the performance of IP, in particular, SFP behaviour.

Finally, there is evidence to suggest that aspects of the brooding and rearing environment can significantly affect IP behaviour in adulthood (reviewed in Rodenburg et al., 2008a). The presence of a mother hen increases foraging behaviour while reducing FP behaviour (Riber et al., 2007). In the absence of a mother hen, it appears that the light and dark periods during brooding can affect behaviour in similar ways. SFP was greatly reduced for chicks that were reared with a dark brooder from 0 to 5 weeks of age compared to those reared with just a heat lamp (Jensen et al., 2006). These behavioural differences persisted up to 23 weeks of age. Even light periods during the final week of egg incubation resulted in chicks that performed more FP than chicks that had been incubated in complete darkness (Riedstra and Groothuis, 2004).

In summary, IP is a multifactorial behavioural problem with major economic and welfare implications for the egg industry. The consequences of IP are significantly greater in the non-cage and free-range systems that the general public is demanding, while at the same time there are efforts to ban beak trimming due to welfare concerns. To remain sustainable, the industry must strive to find preventative and therapeutic measures that are effective and acceptable to the general public.

6 Skeletal health and bone fractures

Poor skeletal health in commercial laying hens has been well documented as a production issue since the 1950s, a welfare concern since the 1980s, and has been the topic of several extensive scientific reviews over the past few decades (Whitehead and Wilson, 1992; Thorp, 1994; Knowles and Wilkins, 1998; Webster, 2004; Beck and Hansen, 2004; Fleming, 2008). 'Caged layer fatigue' emerged as a problem shortly after the implementation of battery cages with symptoms typically manifesting in the form of spontaneous fractures within the spine or appendicular skeleton resulting in reduced mobility and increased risk of emaciation, starvation or spontaneous mortality if left untreated (Couch, 1955). With advances in nutritional strategies and an improved understanding of the calcium:phosphorus ratio and vitamin D requirements (Taylor and Moore, 1954; Summers and Leeson, 1976), this condition has become relatively infrequent. However, osteoporosis (the development of fragile, brittle bones) remains a significant problem in the layer industry and has been the topic of several reviews (Knowles and Wilkins, 1998; Whitehead and Fleming, 2000; Webster, 2004).

In laying hens, osteoporosis is believed to be a consequence of very effective selection for increased egg production over the last half century. To support egg production, a sufficient layer diet includes high levels of calcium, offering at least 3–4% calcium/day for eggshell production and medullary bone renewal (Summers et al., 1976). However, the rate of calcium deposition into the shell is limited by the gastrointestinal absorption of calcium (Tyler, 1940; Simkiss, 1961), and only 60–75% of the calcium required for the shell is derived from the diet with the remainder drawn from the hen's bone stores (Mueller et al., 1964). The constant demand for calcium puts extreme pressure on a relatively small skeletal system to continuously provide enough calcium to support the roughly 2–2.5 g of calcium required for each egg. The widespread problem of osteoporotic hens brings up the question of whether aspirations to achieve a 500 egg hen are truly sustainable.

Osteoporosis in laying hens is further exacerbated by restrictive housing where load-bearing exercise is limited. Adding perches and dust baths, increasing cage heights, providing larger space allowances and opportunities for flight have all been investigated for their effects on bone strength (Whitehead and Wilson, 1992; Newman and Leeson, 1998; Jendral et al., 2008; Silversides et al., 2012; Enneking et al., 2012; Hester et al., 2013); however, marked improvements to bone health are predominately seen in cases where the opportunities for load-bearing exercise are substantially increased and diverse (Whitehead, 2002). Non-cage housing systems, which allow for flight as well as perching, jumping, walking and dust bathing, provide the highest reward in terms of gaining bone strength. However, actual welfare improvements in terms of skeletal health may be limited as the fracture incidence is also considerably increased in the majority of non-cage systems. Osteoporosis makes hens susceptible to fractures over the course of the laying period (Knowles and Wilkins, 1998; Fleming et al., 1998; Sandilands et al., 2009) and particularly at the end of lay during catching and handling for transport to slaughter (Gregory et al., 1992).

Recent research has highlighted the sternum of the hen, the keel bone, as a site particularly susceptible to fractures and deformations during the laying period (Harlander-Matauschek et al., 2015; Fig. 4). Prevalence rates have been reported to range from 10 to 30% in conventional cages, 20 to 60% in furnished cages and greater than 85%

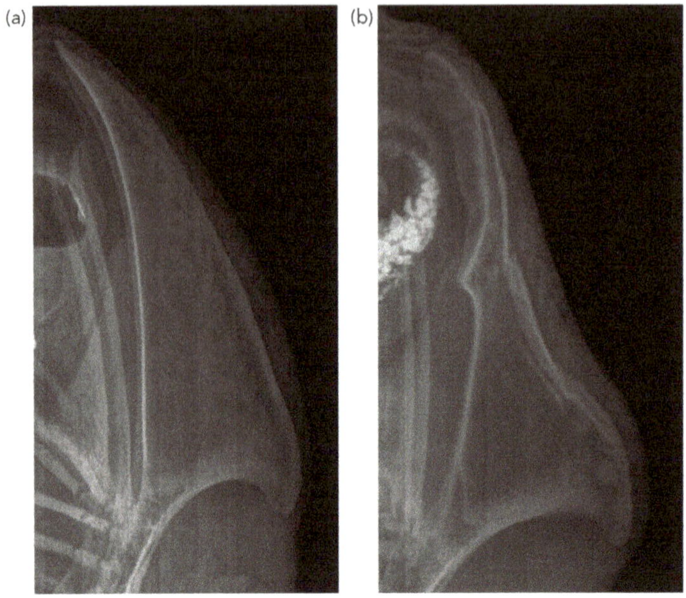

Figure 4 Radiographs showing an intact keel showing (a) an intact keel and (b) a keel with multiple fractures located along the spine and caudal tip of the keel. Photo credit Teresa Casey-Trott.

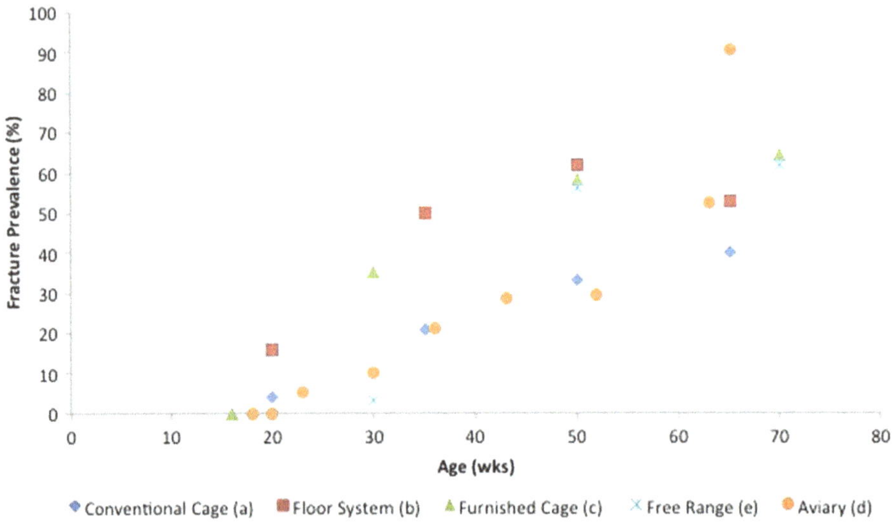

Figure 5 Prevalence of keel fractures over time from hens housed in various adult systems. The data were compiled from several recent longitudinal studies: (a, b) Petrik et al., 2015, (c) Casey-Trott, 2016, (d) Tarlton et al., 2013 and (e) Stratmann et al., 2015.

in non-cage systems (Sandilands et al., 2009; Rodenburg et al., 2008b; Käppeli et al., 2011; Wilkins et al., 2011; Petrik et al. 2015). A sample of longitudinal studies tracking the prevalence of keel bone fractures in various housing systems is described in Fig. 5.

Fractures of the keel have been shown to be painful for hens (Nasr et al., 2012), potentially limiting their mobility in non-cage systems, and are associated with alterations in behaviour in furnished cages (Casey-Trott and Widowski, 2016). The noticeably high rates of keel bone fracture may be a result of a combination of factors such as injuries from collisions in non-cage systems (Stratmann et al., 2015), perch use (Sandilands et al., 2009; Hester et al., 2013; Pickle et al., 2010, 2011), compression fractures due to osteoporosis similar to those seen in humans (Kondo, 2008), early onset of egg production (Gebhardt-Henrich and Frohlich, 2015), nutritional inadequacies (Whitehead, 2004) or genetic factors (Whitehead, 2004; Stratmann et al., 2016) as reviewed by Harlander-Matauschek et al. (2015). The keel is known to continue ossification until 40 weeks of age (Buckner et al., 1949) and is still cartilaginous at the caudal tip in 16-week-old pullets entering the laying period (Casey-Trott, 2016). One area that deserves further research is the role of competition for calcium for eggshell production versus ossification of the keel in young hens and any subsequent links to susceptibility to damage later in life.

Although the aetiology of keel damage is still not well understood, falls and collisions by birds attempting to navigate aviaries are commonly reported (Campbell et al., 2016b). Perch placement (angles and distance between perches), perch material and aviary design can affect the ability of hens to fly or jump between different levels and contribute to the risk of keel damage (Wilkins, et al., 2011; Heerkens et al., 2016). The addition of ramps or ladders between levels has been shown to reduce falls and collisions and decrease the incidence of keel fractures (Stratmann et al., 2015). Other aspects of system design and management such as flooring material and lighting need to be explored in order to design housing systems that pose less of a risk of injury to hens (Harlander-Matauschek et al., 2015).

Osteoporosis has been at the forefront of laying hen research for more than half a century, but the condition is far from being eliminated from commercial flocks around the world. One area with possibly the greatest potential for improvement of skeletal health is genetics (Whitehead, 2002). Bishop et al. (2000) demonstrated that selecting for improved bone strength yielded a moderate to high heritability within only six generations. Fleming et al. (2004) further confirmed this positive finding by demonstrating that the high bone index line had improved keel bone integrity, developing significantly less fresh fractures, and fewer twisted and severely deformed keels compared to the low bone index line. Using birds derived from those same experimental lines, Stratmann et al. (2016) recently demonstrated a significant reduction in keel fractures in the high bone index line compared to the low bone index line and a commercial hybrid in two styles of commercially available aviaries. Prevalence of keel bone fractures was also affected by aviary design with evidence for additive effects between genetic strain and specific housing design.

Unlike improvements made by dietary changes or opportunities for exercise, the studies in genetic selection report both improvements to overall bone characteristics and composition and a reduction in fractures. Lower incidence of old fractures in less selected, heritage lines compared to high-producing commercial lines also suggests a genetic component to osteoporosis (Budgell and Silversides, 2004). Unfortunately, selection for improved bone characteristics has the potential for negative consequences on production, egg traits and efficiency. Bishop et al. (2000) determined an inverse relationship between shell quality and bone strength and Stratmann et al. (2016) reported poorer production and egg quality traits from the high bone index line. However, even within high-producing lines, dramatic differences in fracture rates have been reported, with as high as a 7-fold difference between groups (Clark et al., 2008). Using the relatively short incubation and maturation period of poultry in combination with identifying quantitative trait loci (QTL) to find genetic markers for the bone strength characteristics (Podisi et al., 2012) has the potential to bring about positive changes to skeletal health in a relatively short period of time.

7 Future trends in research

Most research on the welfare of laying hens has focused on housing and management of adult birds. However, the early experiences of chicks and pullets can have lifelong effects on the behaviour, health and welfare of laying hens (Janczak and Riber, 2015), and these effects can even occur prenatally, during incubation or before the egg is laid (Dixon et al., 2015). Research is needed to target these early life stages. This is particularly important when pullets are destined for alternative housing systems. The welfare of hens in non-cage systems requires that the hens housed in them are calm, adapt well to change and novelty,

Figure 6 Example of an open-style rearing system for pullets. (a) Chicks are started in the middle of the system with opportunities to run, jump and perch from one day of age. (b) As the chicks and pullets grow, terraces are raised and opened to provide access to a litter area (not shown). Photo credits: Michelle Hunniford.

Figure 7 (a) Example of a rearing aviary in which chicks are started in furnished cages with perches. (b) The enclosures are opened as the pullets grow to provide access to more vertical space and a litter area. Although chicks do have access to perches from one day of age, opportunities for greater exercise, flying and complex spatial activity come at a later age. Photo credits Leanne Cooley.

and have few behavioural problems such as FP and cannibalism. Fearfulness is particularly problematic in large groups since it is associated with FP, and fearful responses to novel events can cause injury due to piling and smothering. Hens in multi-tier aviaries must also be able to negotiate complex environments where feed, water and nest boxes are located on different levels. Optimal rearing systems for pullets destined for furnished cages also need to be determined.

Practical experience indicates that it is critical that hens housed in complex aviaries also be reared in complex systems. Environmental complexity during rearing affects cognitive as well as physical development. Recent studies have shown that, compared to pullets reared in conventional rearing cages, those reared in aviaries have better working memory and increased ability to perform spatial tasks (Tahamtani et al., 2015), are less fearful and use three-dimensional space more fully (Brantsæter et al., 2016). Rearing pullets with perches (Ehneking et al. 2012) or in aviaries (Regmi et al., 2016) also increases musculoskeletal development and various measures of bone strength (Casey-Trott, 2016).

The designs of rearing aviaries are evolving and currently there are a number of different variations that are commercially available. Some designs are similar to standard rearing cages outfitted with perches where chicks are started in an enclosed cage, but as they grow, the front of the cage is opened to allow access to the barn and litter floors. Other designs offer more vertical space to chicks within the starting cage and include different levels of perches and platforms during the first weeks of life; again, as the birds grow, these cages are opened to allow them access to the barn and litter floors. Additionally, there are designs of pullet aviaries that provide significantly more vertical and horizontal space from one day of age. As chicks grow, platforms within the system are raised and terraces are opened to provide access to the litter floor. These different designs of rearing systems facilitate different types of locomotion (e.g. running, perching, climbing and flying) at different ages. Currently, little is known about the development of locomotory abilities in domestic chickens or how these different designs affect the learning process. New research is aimed at investigating how and when the different types of locomotion develop and at identifying strain differences in the use of vertical space (Kozak et al., 2016).

8 Where to look for further information

Throughout this chapter, a number of classic and recent reviews have been cited for each section. Readers may also find some recently published books helpful.

For a general overview of concepts and experimental approaches used in animal welfare, science readers are referred to *Animal Welfare*, 2nd edition (2011), edited by Michael Appleby, Joy Mench, Anna Olsson and Barry Hughes. Published by CABI.

For descriptions and reviews on different housing systems and welfare issues of laying hens, readers are referred to *Alternative Systems for Poultry – Health, Welfare and Productivity* (2012), edited by Victoria Sandilands and Paul Hocking. Published by CABI as part of their Poultry Science Symposium Series.

For a comprehensive text on behaviour of the laying hens as well as topics important to their welfare, readers are referred to *The Behavioural Biology of Chickens* (2015), edited by Christine Nicol. Published by CABI.

9 References

Aerni, V., El-Lethey, H. and Wechsler, B. (2000), 'Effect of foraging material and food form on feather pecking in laying hens', *Brit. Poult. Sci.*, 41, 16–21.

Appleby, M. C. (2004), 'European ban on conventional cages: History and prospects', *J. Appl. Anim. Welf. Sci.*, 6, 103–21.

Appleby, M. C. and McRae, H. E. (1986), 'The individual nest box as a super-stimulus for domestic hens', *Appl. Anim. Behav. Sci.*, 15, 169–76.

Appleby, M. C., Mench, J. A. and Hughes, B. O. (2004), *Poultry Behaviour and Welfare*, CABI Publishing, Wallingford, UK.

Appleby, M. C., Mench, J. A., Olsson, I. A. S. and Hughes, B. O. (2011), *Animal Welfare* (2nd Edn.), CABI Publishing, Wallingford, UK.

Barrett, J., Rayner, A. C., Gill, R. Willings, T. H. and Bright, A. (2014), 'Smothering in UK range flocks. Part 1: Incidence, location, timing and management', *Vet. Rec.* doi: 10.1136/vr.102327.

Beck, M. M. and Hansen, K. K. (2004), 'Role of estrogen in avian osteoporosis', *Poult. Sci.*, 83, 200–6.

Bishop, S. C., Fleming, R. H., McCormack, H. A., Flock, D. K. and Whitehead, C. C. (2000), 'Inheritance of bone characteristics affecting osteoporosis in laying hens', *Brit. Poult. Sci.*, 41, 33–40.

Blatchford, R. A., Fulton, R. M. and Mench, J. A. (2016), 'The utilization of the Welfare Quality® assessment for determining laying hen condition across three housing systems', *Poult. Sci.*, 95, 154–63.

Blokhuis, H. J. (1986), 'Feather-pecking in poultry: Its relation with ground-pecking', *Appl. Anim. Behav. Sci.*, 16, 63–7.

Blokhuis, H. J. and Arkes, J. G. (1984), 'Some observations on the development of feather-pecking in poultry', *Appl. Anim. Behav. Sci.*, 12, 145–57.

Blokhuis, H. J., Fiks van Niekerk, T., Bessei, W., Elson, A., Guémené, D., Kjaer, J. B., Maria Levrino, G. A., Nicol, C. J., Tauson, R., Weeks, C. A. and Van de Weerd, H. A. (2007), 'The LayWel project: welfare implications of changes in production systems for laying hens', *Worlds Poult. Sci. J.*, 63, 101–14.

Brambell, F. W. R., Barbour, D. S., Barnett, L., Ewer, T. K., Hobson, A., Pitchforth, H., Smith, W. R., Thorpe, W. H. and Winship, F. J. W. (1965), *Report of the Technical Committee to Enquire into the Welfare of Animals Kept Under Intensive Livestock Husbandry Systems*, HM Stationary Office, London, UK.

Brantsæter, M., Nordgreen, J., Rodenburg, T. B., Tahamtani, F. M., Popova, A. and Janczak, A. M. (2016), 'Exposure to increased environmental complexity during rearing reduces fearfulness and increases use of three-dimensional space in lay hens (*Gallus gallus domesticus*)', *Front. Vet. Sci.*, 3, 14.

Buckner, G. D., Insko Jr., W. M., Henry, A. H. and Wachs, E. F. (1949), 'Rate of growth and calcification of the sternum of male and female New Hampshire chickens', *Ky. Agr. Expt. Sta. Bull.*, 430–3.

Budgell, K. L. and Silversides, F. G. (2004), 'Bone breakage in three strains of end-of-lay hens', *Can. J. Anim. Sci.*, 84, 745–7.

Buitenhuis, J., Rodenburg, T. B., Wissink, P. H., Visscher, J., Koene, P., Bovenhuis, H., Ducro, B. J. and van der Poel, J. J. (2004), 'Genetic and phenotypic correlations between feather pecking behavior, stress response, immune response, and egg quality traits in laying hens', *Poult. Sci.*, 83, 1077–82.

Campbell, D. L. M., Makagon, M. M., Swanson, J. C. and Siegford, J. M. (2016a), 'Litter use by laying hens in a commercial aviary: dust bathing and piling', *Poult. Sci.*, 95, 164–75

Campbell, D. L. M., Goodwin, S. L., Makagon, M. M., Swanson, J. C. and Siegford, J. M. (2016b), 'Failed landings after laying hen flight in a commercial aviary over two flock cycles', *Poult. Sci.*, 95,188–97

Casey-Trott. (2016), 'Opportunities for exercise during pullet rearing: effects on bone health and keel bone damage in laying hens' (*Doctoral dissertation*), University of Guelph, Guelph, Ontario, Canada.

https://atrium.lib.uoguelph.ca/xmlui/bitstream/handle/10214/10042/CaseyTrott_Teresa_201609_PhD.pdf?sequence=1&isAllowed=y. (accessed October 2016).

Casey-Trott, T. M. and Widowski, T. M. (2016), 'Behavioral differences of laying hens with fractures keel bones within furnished cages', *Front. Vet. Sci.*, 3, 42.

Choct, M. and Hartini, S. (2003), 'Interaction between nutrition and cannibalism in laying hens', *Recent Adv. Anim. Nutr. Aust.*, 14, 157–62.

Clark, W. D., Cox, W. R. and Silversides, F. G. (2008), 'Bone fracture incidence in end-of-lay high-producing, non-commercial laying hens identified using radiographs', *Poult. Sci.*, 87, 1964–70.

Clausen, T. and Riber, A. B. (2012), 'Effect of heterogeneity of nest boxes on occurrence of gregarious nesting in laying hens', *Appl. Anim. Behav. Sci.*, 142, 168–75.

Cooper, J. J. and Albentosa, M. J. (2003), 'Behavioural priorities of laying hens', *Avian Poult. Biol. Rev.*, 14, 127–49.

Cooper, J. J. and Appleby, M. C. (1996), 'Individual variation in prelaying behaviour and the incidence of floor eggs', *Brit. Poult. Sci.*, 37, 245–53.

Couch, J. R. (1955), 'Cage layer fatigue', *Feed Age*, 5, 55–7.

Craig, J. V and Muir, W. M. (1996), 'Group selection for adaptation to multiple-hen cages: beak-related mortality, feathering, and body weight responses', *Poult. Sci.*, 75, 294–302.

Cryan, J. F. and Dinan, T. G. (2012), 'Mind-altering microorganisms: The impact of the gut microbiota on brain and behaviour', *Nat. Rev. Neurosci.*, 13, 701–12,

Dawkins, M. S. (1983), 'Battery hens name their price: consumer demand theory and the measurement of ethological "needs"', *Anim. Behav.*, 31, 1195–205.

Dixon, L. M., Duncan, I. J. H. and Mason, G. J. (2008), 'What's in a peck? Using fixed action pattern morphology to identify the motivational basis of abnormal feather-pecking behaviour', *Anim. Behav.*, 76, 1035–42.

Dixon, L. M., Sparks, N. H. C. and Rutherford, K. M. D. (2015), 'Early experiences matter: a review of the effects of prenatal environment on offspring characteristics in poultry', *Poult. Sci.*, 95, 489–99.

de Jong, I. C., Wolthuis-Fillerup, M. and van Reenen, C. G. (2007), 'Strength of preference for dustbathing and foraging substrates in laying hens', *Appl. Anim. Behav. Sci.*, 104, 24–36.

Drake, K. A., Donnelly, C. A. and Dawkins, M. S. (2010), 'Influence of rearing and lay risk factors on propensity for feather damage in laying hens', *Brit. Poult. Sci.*, 51, 725–33.

Duncan, I. J. H. (1970), 'Frustration in the fowl', in Freeman, B. M. and Gordon, R. F. (Eds), *Aspects of Poultry Behaviour*, 15–31. British Poultry Science, Edinburgh.

Duncan, I. J. H., Widowski, T. M., Malleau, A. E., Lindberg, A. C. and Petherick, J. C. (1998), 'External factors and causation of dustbathing in domestic hens', *Behav. Process.*, 43, 219–28.

Edgar, J. L., Mullan, S. M., Pritchard, J. C., McFarlane, U. J. C. and Main, D. C. J. (2013), 'Towards a "Good Life" for farm animals: Development of a resource tier framework to achieve positive welfare for laying hens', *Animals*, 3, 584–605.

EFSA-AHAW (2005), 'Scientific report "Welfare aspects of various systems for keeping laying hens"', *EFSA-AHAW*, p. 143.

Ellen, E. D., Visscher, J., van Arendonk, J. A. M. and Bijma, P. (2008), 'Survival of laying hens: Genetic parameters for direct and associative effects in three purebred layer lines', *Poult. Sci.*, 87(2), 233–9.

Enneking, S. A., Cheng, H. W., Jefferson-Moore, K. Y., Einstein, M. E., Rubin, D. A. and Hester, P. Y. (2012), 'Early access to perches in caged white leghorn pullets', *Poult. Sci.*, 91, 2114–20.

Farm Animal Welfare Council (FAWC) (2009), 'Farm animal welfare in Great Britain: Past, present and future' Farm animal Welfare Council, London. https://www.gov.uk/government/uploads/system/uploads/attachment_data/file/319292/Farm_Animal_Welfare_in_Great_Britain_-_Past__Present_and_Future.pdf (accessed 15 June 2016).

Fleming, R. H. (2008), 'Nutritional factors affecting poultry bone health: Symposium on "Diet and bone health"', *Proc. Nutri. Soc.*, 67, 177–83.

Fleming, R. H., McCormack, H. A., McTeir, L. and Whitehead, C. C. (2004), 'Incidence, pathology and prevention of keel bone deformities in the laying hen', *Brit. Poult. Sci.*, 45, 320–30.

Fleming, R. H., McCormack, H. A. and Whitehead, C. C. (1998), 'Bone structure and strength of different ages in laying hens and effects of dietary particulate limestone, vitamin K and ascorbic acid', *Brit. Poult. Sci.*, 39, 434–40.

Fraser, D., Weary, D. M., Pajor, E. A. and Milligan, B. N. (1997), 'A scientific conception of animal welfare that reflects ethical concerns', *Animal Welfare*, 6, 187–2005.

Freire, R., Appleby, M. C. and Hughes, B. O. (1996), 'Effects of nest quality and other cues for exploration on pre-laying behaviour', *Appl. Anim. Behav. Sci.*, 48, 37–46.

Gebhardt-Henrich, S. G. and Frohlich, E. K. (2015), 'Early onset of laying and bumblefoot favor keel bone fractures', *Animals*, 5, 1192–206. doi:10.3390/ani5040406.

Gilani, A.-M., Knowles, T. G. and Nicol, C. J. (2013), 'The effect of rearing environment on feather pecking in young and adult laying hens', *Appl. Anim. Behav. Sci.*, 148, 54–63.

Gregory, N. G., Wilkins, L. J., Austin, S. D., Belyavin, C. G., Alvey, D. M. and Tucker, S. A. (1992), 'Effect of catching method on the prevalence of broken bones in end of lay hens', *Avian Pathol.*, 21, 717–22.

Guesdon, V. and Faure, J. M. (2004), 'Laying performance and egg quality in hens kept in standard or furnished cages', *Anim. Res.*, 53, 45–57.

Guinebretière, M., Beyer, H., Arnould, C. and Michel, V. (2014), 'The choice of litter material to promote pecking, scratching and dustbathing behaviours in laying hens housed in furnished cages', *Appl. Anim. Behav. Sci.*, 155, 56–65.

Guinebretière, M., Michel, V. and Arnould, C. (2015), 'Dust bathing, pecking and scratching behaviour of laying hens in furnished cages are enhanced by the presence of rubber mats an litter distribution', *Appl. Anim. Behav. Sci.*, 171, 128–37.

Harlander-Matauschek, A., Piepho, H. P. and Bessei, W. (2006), 'The effect of feather eating on feed passage in laying hens', *Poult. Sci.*, 85, 21–5.

Harlander-Matauschek, A., Rodenburg, T. B., Sandilands, V., Tobalske, B. W. and Toscano, M. J. (2015), 'Causes of keel bone damage and their solutions in laying hens', *Worlds Poult. Sci. J.*, 71, 461–72.

Heerkens, J. L., Delezie, E., Rodenburg, T. B., Kempen, I., Zoons, J., Ampe, B. and Tuyttens, F. A. (2016), 'Risk factors associated with keel bone and foot pad disorders', *Poult. Sci.*, 94, 482–8.

Heinrich, A., Icken, W., Thurner, S., Wendl, G., Bernhardt, H. and Preisinger, R. (2015), 'Nesting behaviour – a comparison of single nest boxes and family nests', *Europ. Poult. Sci.*, 78. DOI: 10.1399/eps.2014.72.

Herwig, E. and Widowski, T. M. (2013), 'Effect of density and cage size on foraging and dustbathing by laying hens in large furnished cages', *Proceeding of the IX European Symposium on Poultry Welfare*, 39.

Hester, P. Y. (2014), 'The effect of perches installed in cages on laying hens', *Worlds Poult. Sci. J.*, 70, 247–64.

Hester, P. Y., Enneking, S. A., Haley, B. K., Cheng, H. W., Einstein, M. E. and Rubin, D. A. (2013), 'The effect of perch availability during pullet rearing and egg laying on musculoskeletal health of caged white leghorn hens', *Poult. Sci.*, 92, 1972–80.

Hocking, P. M., Channing, C. E., Robertson, G. W., Edmond, A. and Jones, R. B. (2004), 'Between breed genetic variation for welfare-related behavioural traits in domestic fowl', *Appl. Anim. Behav. Sci.*, 89, 85–105.

Huber-Eicher, B. and Audigé, L. (1999), 'Analysis of risk factors for the occurrence of feather pecking in laying hen growers', *Brit. Poult. Sci.*, 40, 599–604.

Huber-Eicher, B. and Weschler, B. (1997), 'Feather pecking in domestic chicks: its relation to dustbathing and foraging', *Anim. Behav.*, 54, 757–68.

Hughes, B. O. (1993), 'Choice between artificial turf and wire floor as nest sites in individually caged laying hens', *Appl. Anim. Behav. Sci.*, 36, 327–35.

Hughes, B. O. and Duncan, I. J. H. (1972), 'The influence of strain and environmental factors upon feather pecking and cannibalism in fowls', *Brit. Poult. Sci.*,13, 525–47.

Hughes, B. O. and Duncan, I. J. H. (1988), 'The notion of ethological "need", models of motivation and animal welfare', *Anim. Behav.*, 36, 1696–707.

Hughes, B. O., Gilbert, A. B. and Brown, M. F. (1986), 'Categorisation and causes of abnormal egg shells: Relationship with stress', *Brit. Poult. Sci.*, 27, 325–37.

Hunniford, M. E., Torrey, S., Bédécarrats, G., Duncan, I. J. H. and Widowski, T. M. (2014), 'Evidence of competition for nest sites by laying hens in large furnished cages', *Appl. Anim. Behav. Sci.*, 161, 95–104.

Hunniford, M. E. and Widowski, T. M. (2016a), 'Rearing environment and laying location affect pre-laying behaviour in enriched cages', *Appl. Anim. Behav. Sci.*, 181, 205–213.

Hunniford, M. E. and Widowski, T. M. (2016b), 'Nest alternatives: Adding a wire partition to the scratch area affects nest use and nesting behaviour of laying hens in furnished cages', *Appl. Anim. Behav. Sci.* (in review).

Icken, W., Cavero, D., Schmutz, M. and Preisinger, R. (2012), 'New phenotypes for new breeding goals in layers', *Worlds Poult. Sci. J.*, 68, 387–400.

Janczak, A. M. and Riber, A. B. (2015), 'Review of rearing-related factors affecting the welfare of laying hens', *Poult. Sci*, 94, 1454–69.

Jendral, M. J., Korver, D. R., Church, J. S. and Feddes, J. J. R. (2008), 'Bone mineral density and breaking strength of white leghorns housed in conventional, modified, and commercially available colony battery cages', *Poult. Sci.*, 87, 828–37.

Jensen, P., Keeling, L., Schütz, K., Andersson, L., Mormède, P., Brändström, H., Forkman, B., Kerje, S., Fredriksson, R., Ohlsson, C., Larsson, S., Mallmin, H. and Kindmark, A. (2005), 'Feather pecking in chickens is genetically related to behavioural and developmental traits', *Physiol. Behav.*, 86, 52–60.

Jensen, A. B., Palme, R. and Forkman, B. (2006), 'Effect of brooders on feather pecking and cannibalism in domestic fowl (*Gallus gallus domesticus*)', *Appl. Anim. Behav. Sci.*, 99, 287–300.

Jones, R. B. and Hocking, P. M. (1999), 'Genetic selection for poultry behaviour: Big bad wolf or friend in need?', *Anim. Welf.*, 8, 343–59.

Jones, R. B., McAdie, T. M., McCorquodale, C. and Keeling, L. J. (2002), 'Pecking at other birds and at string enrichment devices by adult laying hens', *Brit. Poult. Sci.*, 43, 337–43.

Kalmendal, R. and Bessei, W. (2012), 'The preference for high-fiber feed in laying hens divergently selected on feather pecking', *Poult. Sci.*, 91, 1785–9.

Käppeli, S., Gebhardt-Henrich, S. G., Fröhlich, E., Pfulg, A. and Stoffel, M. H. (2011), 'Prevalence of keel bone deformities in Swiss laying hens', *Brit. Poult. Sci.*, 52, 531–6.

Kjaer, J. B. (2009), 'Feather pecking in domestic fowl is genetically related to locomotor activity levels: Implications for a hyperactivity disorder model of feather pecking', *Behav. Genet.*, 39, 564–70.

Kjaer, J. B., Sorensen, P. and Su, G. (2001), 'Divergent selection on feather pecking behaviour in laying hens (*Gallus gallus domesticus*)', *Appl. Anim. Behav. Sci.*, 71, 229–39.

Kjaer, J. B. and Vestergaard, K. S. (1999), 'Development of feather pecking in relation to light intensity', *Appl. Anim. Behav. Sci.*, 62, 243–54.

Kjaer, J. B., Würbel, H. and Schrader, L. (2015), 'Perseveration in a guessing task by laying hens selected for high or low levels of feather pecking does not support classification of feather pecking as a stereotypy', *Appl. Anim. Behav. Sci.*, 168, 56–60.

Knowles, T. G. and Wilkins, L. J. (1998), 'The problem of broken bones during the handling of laying hens – A review', *Poult. Sci.*, 77, 1798–802.

Kondo, K. L. (2008). 'Osteoporotic vertebral compression fractures and vertebral augmentation', *Seminars in Interventional Radiology*, 25, 413–24.

Kozak, M., Tobalske, B., Martins, C., Bowley, S., Würbel, H. and Harlander-Matauschek, A. (2016), 'Use of space by domestic chicks housed in complex aviaries', *Appl. Anim. Behav. Sci.*, 181, 115–21.

Lambton, S. L., Knowles, T. G., Yorke, C. and Nicol, C. J. (2010), 'The risk factors affecting the development of gentle and severe feather pecking in loose housed laying hens', *Appl. Anim. Behav. Sci.*, 123, 32–42.

Lambton, S. L., Nicol, C. J., Friel, M., Main, D. C. J., Mckinstry, J. L., Sherwin, C. M., Walton, J. and Weeks, C. A. (2013), 'A bespoke management package can reduce levels of injurious pecking in loose-housed laying hen flocks', *Vet. Rec.*, 172(16), 423. doi: 10.1136/vr.101067.

Lay, D. C., Fulton, R. M., Hester, P. Y., Karcher, D. M., Kjaer, J. B., Mench, J. A., Mullens, B. A., Newberry, R. C., Nicol, C. J., O'Sullivan, N. P. and Porter, R. E. (2011), 'Hen welfare in different housing systems', *Poult. Sci.*, 90, 278–94.

LayWel. 2006. Welfare implications of changes in production systems for laying hens: Deliverable 7.1.Overall strengths and weaknesses of each defined housing system for laying hens, and detailing the overall welfare impact of each housing system. http://www.laywel.eu/web/pdf/deliverable%2071%20welfare%20assessment.pdf (accessed 15 June 2016).

Lindqvist, C. E. S., Schutz, K. E. and Jensen, P. (2002), 'Red Jungle Fowl have more contra freeloading than White Leghorn layers: Effect of food deprivation and consequences for information gain', *Behaviour*, 139, 1195–209.

Lundberg, A. and Keeling, L. J. (1999), 'The impact of social factors on nesting in laying hens (*Gallus gallus domesticus*)', *Appl. Anim. Behav. Sci.*, 64, 57–69.

McKeegan, D. E. F. and Savory, C. J. (2001), 'Feather eating in individually caged hens which differ in their propensity to feather peck', *Appl. Anim. Behav. Sci.*, 73, 131–40.

Mellor, D. J. (2016), 'Updating animal welfare thinking: Moving beyond the "Five Freedoms" towards "Life Worth Living"', *Animals* 6(3), 21. doi:10.3390/ani6030021.

Mench, J. A. (2009), 'Behaviour of fowl and other domesticated birds', in P. Jensen (Ed.), *The Ethology of Domestic Animals* (2nd ed.), 121–36. CABI Publishing, Wallingford, UK.

Mench, J. A., Sumner, D. A. and Rosen-Molina, J. T. (2011), 'Sustainability of egg production in the United States – The policy and market context', *Poult. Sci.*, 90, 229–40.

Meyer, B., Bessei, W., Zenetek, J. and Harlander-Matauschek, A. (2012), 'Dietary inclusion of feather affects intestinal microbiota and microbial metabolites in growing Leghorn-type chickens', *Poult. Sci.*, 91, 1506–13.

Morrissey, K. L. H., Brocklehurst, S., Baker, L., Widowski, T. M. and Sandilands, V. (2016), 'Can non-beak treated hens be kept in commercial furnished cages? Exploring the effects of strain and extra environmental enrichment on behaviour, feather cover, and mortality', *Animals*, 6, 3. doi: 10.3390/ani6030017.

Mueller, W. J., Schraer, R. and Schraer, H. (1964), 'Calcium metabolism and skeletal dynamics of laying pullets', *J. Nutri.*, 84, 20–6.

Muir, W. M., Cheng, H.-W. and Croney, C. (2014), 'Methods to address poultry robustness and welfare issues through breeding and associated ethical considerations', *Front. Genet.*, 5, 407. doi: 10.3389/fgene.2014.00407.

Nasr, M. A. F., Nicol, C. J. and Murrell, J. C. (2012), 'Do laying hens with keel bone fractures experience pain?', *PLoS ONE*, 7, e42420. doi:10.1371/journal.pone.0042420.

Newberry, R. C., Keeling, L. J., Estevez, I. and Bil ík, B. (2007), 'Behaviour when young as a predictor of severe feather pecking in adult laying hens: The redirected foraging hypothesis revisited', *Appl. Anim. Behav. Sci.*, 107, 262–74.

Newman, S. and S. Leeson. (1998), 'Effect of housing birds in cages or an aviary system on bone characteristics', *Poult. Sci.*, 77, 1492–6.

Nicol, C. J. (2015), *The Behavioural Biology of Chickens*, CABI Publishing, Wallingford, UK.

Nicol, C. J., Bestman, M., Gilani, A-M., De Haas, E. N., de Jong, I. C., Lambton, S., Wagenaar, J. P., Weeks, C. A. and Rodenburg, T. B. (2013), 'The prevention and control of feather pecking: application to commercial systems', *Worlds Poult. Sci. J.*, 69, 775–88.

OIE (World Organization for Animal Health) (2010), Terrestrial Health Code. Chapter 7.1. Introductions to the recommendations for animal welfare. Article 7.1., http://web.oie.int/eng/normes/mcode/en_chapitre_1.7.1.pdf (accessed July 2016).

Olsson, I. A. S. and Keeling, L. J. (2000), 'Night-time roosting in laying hens and the effect of thwarting access to perches', *Appl. Anim. Behav. Sci.*, 68, 243–56.

Olsson, I. A. S. and Keeling, L. J. (2002), 'The push-door for measuring motivation in hens: Laying hens are motivated to perch at night', *Anim. Welfare*, 11, 11–19.

Petrik, M. T., Guerin, M. T. and Widowski, T. M. (2015), 'On-farm comparison of keel fracture prevalence and other welfare indicators in conventional cage and floor-housed laying hens in Ontario, Canada', *Poult. Sci.* 94, 579–85.

Pickle, T., Scholtz, B. and Schrader, L. (2010), 'Perch material and diameter affect particular perching behaviours in laying hens', *Appl. Anim. Behav. Sci.*, 127, 37–42.

Pickle, T., Schrader, L. and Scholz, B. (2011), 'Pressure load on keel bone and foot pads in perching laying hens in relation to perch design', *Poult. Sci.*, 90, 715–24.

Podisi, B. K., Knott, S. A., Dunn, I. C., Burt, D. W. and Hocking, P. M. (2012), 'Bone mineral density QTL at sexual maturity and end of lay', *Brit. Poult. Sci.*, 53, 763–9.

Reed, H. J. and Nicol, C. J. (1992), 'Effects of nest linings, pecking strips and partitioning on nest use and behaviour in modified battery cages', *Brit. Poult. Sci.*, 33, 719–27.

Regmi, P., Smith, N., Nelson, N., Haut, R. C., Orth, M. W. and Karcher, D. M. (2016), 'Housing conditions alter properties of the tibia and humerus during the laying phase in Lohmann white Leghorn hens', *Poult. Sci.*, 95, 198–206.

Riber, A. B. (2010), 'Development with age of nest box use and gregarious nesting in laying hens', *Appl. Anim. Behav. Sci.*, 123, 24–31.

Riber, A. B., Wichman, A., Braastad, B. O. and Forkman, B. (2007), 'Effects of broody hens on perch use, ground pecking, feather pecking and cannibalism in domestic fowl (*Gallus gallus domesticus*)', *Appl. Anim. Behav. Sci.*, 106, 39–51.

Riedstra, B. and Groothuis, T. G. G. (2004), 'Prenatal light exposure affects early feather-pecking behaviour in the domestic chick', *Anim. Behav.*, 67, 1037–42.

Ringgenberg, N., Frölich, E. K. F., Harlander-Matauschek, A, Toscano, M. J., Würbel, H. and Roth, B. A. (2015), 'Nest choice in laying hens: Effects of nest partitions and social status', *Appl. Anim. Behav. Sci.*, 169, 43–50.

Ringgenberg, N., Frölich, E. K. F., Harlander-Matauschek, A, Würbel, H. and Roth, B. A. (2014), 'Does nest size matter to laying hens?', *Appl. Anim. Behav. Sci.*, 155, 66–73.

Rodenburg, T. B., Buitenhuis, J., Ask, B., Uitdehaag, K., Koene, P., van der Poel, J. J. and Bovenhuis, H. (2003), 'Heritability of feather pecking and open-field response of laying hens at two different ages', *Poult. Sci.*, 82, 861–7.

Rodenburg, T. B., Komen, H., Ellen, E. D., Uitdehaag, K. and van Arendonk, J. A. M. (2008a), 'Selection method and early-life history affect behavioural development, feather pecking and cannibalism in laying hens: A review', *Appl. Anim. Behav. Sci.*, 110, 217–28.

Rodenburg, T. B., de Reu, K. and Tuyttens, F. A. M. (2012), 'Performance, welfare, health and hygiene of laying hens in non-cage systems in comparison with cage systems', in Sandilands, V. and Hocking, P. M. (Eds), *Alternative Systems for Poultry*, CABI Publishing, Wallingford, UK.

Rodenburg, T. B., van Hierden, Y. M., Buitenhuis, J., Riedstra, B., Koene, P., Korte, S. M. van der Poel, J. J., Groothuis, T. G. G. and Blokhuis, H. J. (2004), 'Feather pecking in laying hens: New insights and directions for research?', *Appl. Anim. Behav. Sci.*, 86, 291–8.

Rodenburg, T. B., van Krimpen, M. M., de Jong, I. C., de Haas, E. N., Kops, M. S., Riedstra, B. J., Nordquist, R. E., Wagenaar, J. P., Bestman, M. and Nicol, C. J. (2013), 'The prevention and control of feather pecking in laying hens: identifying the underlying principles', *Worlds Poult. Sci. J.*, 69, 361–74.

Rodenburg, T. B., Tuyttens, F. A. M., Reu, K. D., Herman, L., Zoons, J. and Sonck, B. (2008b), 'Welfare assessment of laying hens in furnished cages and non-cage systems: an on-farm comparison', *Anim. Welf.*, 17, 363–73.

Sandilands, C. and Hocking, P. (2012), *Alternative Systems for Poultry: Health, Welfare and Productivity*, CABI Publishing, Wallingford, UK.

Sandilands, V., Moinard, C. and Sparks, N. H. C. (2009), 'Providing laying hens with perches, fulfilling behavioural needs but causing injury?', *Brit. Poult. Sci.*, 50, 395–406.

Savory, C. J. (1995), 'Feather pecking and cannibalism', *Worlds Poult. Sci. J.*, 51, 215–19.

Schrader, L. and Müller, B. (2009), 'Night-time roosting in the domestic fowl: The height matters', *Appl. Anim. Behav. Sci.*, 121, 179–83.

Sedlačková, M., Bilčík, B. and Košťál, L. (2004), 'Feather pecking in laying hens: Environmental and endogenous factors', *Acta Vet. Brno*, 73, 521–31.

Sherwin, C. M., Richards, G. H. and Nicol, C. J. (2010), 'Comparison of the welfare of layer hens in 4 housing systems in the UK', *Brit. Poult. Sci.*, 51, 488–99.

Silversides, F. G., Singh, R., Cheng, K. M. and Korver, D. R. (2012), 'Comparison of bones of 4 strains of laying hens kept in conventional cages and floor pens', *Poult. Sci.*, 91(1), 1–7.
Simkiss, K. (1961), 'Calcium metabolism and avian reproduction', *Biol. Rev.*, 36, 321–67.
Stratmann, A., Fröhlich, E. K. F., Gebhardt-Henrich, S. G., Harlander-Matauschek, A., Würbel, H. and Toscano, M. J. (2015), 'Modification of aviary design reduces incidence of falls, collisions and keel bone damage in laying hens', *Appl. Anim. Behav. Sci.*, 165, 112–23.
Stratmann, A., Fröhlich, E. K. F., Gebhardt-Henrich, S. G., Harlander-Matauschek, A., Würbel, H. and Toscano, M. J. (2016), 'Genetic selection to increase bone strength affects prevalence of keel bone damage and egg parameters in commercially housed laying hens', Poult Sci., 95(5), 975–84.
Struelens, E., Van Nuffel, A., Tuyttens, F. A. M., Audoorn, L., Vranken, E., Zoons, J., Berckmans, D., Odberg, F., Van Dongen, S. and Sonck, B. (2008), 'Influence of nest seclusion and nesting material on pre-laying behaviour of laying hens', *Appl. Anim. Behav. Sci.*, 112, 106–19.
Struelens, E. and Tuyttens, F. A. M. (2009), 'Effects of perch design on behaviour and health of laying hens', *Anim. Welf.*, 18(4), 533–8.
Summers, J. D., Grandhi, R. and Leeson, S. (1976), 'Calcium and phosphorus requirements of the laying hen', *Poult. Sci.*, 55, 402–13.
Summers, J. D. and Leeson, S. (1976), 'Calcium and phosphorus requirements of the laying hen', *Poult. Sci.*, 55, 402–13.
Sun, Y., Ellen, E. D., van der Poel, J. J., Parmentier, H. K. and Bijma, P. (2014), 'Modelling of feather pecking behavior in beak-trimmed and non-beak-trimmed crossbred laying hens: Variance component and trait-based approach', *Poult. Sci.*, 93, 773–83.
Tahamtani, F. M., Nordgreen, J., Nordquist, R. E. and Janczak, A. M. (2015), 'Early life in a barren environment adversely affects spatial cognition in laying hens (*Gallus gallus domesticus*)', *Front. Vet. Sci.*, 2(3). doi: 10.3389/fvets.2015.00003.
Taylor, T. G. and Moore, J. H. (1954), 'Skeletal depletion in hens laying on a low-calcium diet', *Brit. J. Nutr.*, 8, 112–24.
Thorp, B. H. (1994), 'Skeletal disorders in the fowl: A review', *Avian Pathol.*, 23, 203–36.
Tyler, C. (1940), 'Studies of calcium and phosphorus metabolism in relation to the chemical structure of bone: Part 1 Experiments with laying birds', *Biochem. J.*, 34(2), 202–12.
Van der Weerd, H. A., Croxall, R. and Elson, H. A. (2006). 'Beak blunting, an alternative to beak trimming in laying hens?', WPSA UK Branch British Poultry Abstracts, 2–3.
van Hierden, Y. M., de Boer, S. F., Koolhaas, J. M. and Korte, S. M. (2004), 'The control of feather pecking by serotonin', *Behav. Neurosci.*, 118, 575–83.
van Hierden, Y. M., Korte, S. M., Ruesink, E. W., van Reenen, C. G., Engel, B., Korte-Bouws, G. A., Koolhaas, J. M. and Blokhuis, H. J. (2002), 'Adrenocortical reactivity and central serotonin and dopamine turnover in young chicks from a high and low feather-pecking line of laying hens', *Physiol. Behav.* 75, 653–9.
van Krimpen, M. M., Kwakkel, R. P., Reuvekamp, B. F. J., Van Der Peet-Schwering, C. M. C., Den Hartog, L. A. and Verstegen, M. W. A. (2005), 'Impact of feeding management on feather pecking in laying hens', *Worlds Poult. Sci. J.*, 61, 663–86.
Wall, H. (2011), 'Production performance and proportion of nest eggs in layer hybrids housed in different designs of furnished cages', *Poult. Sci.*, 90, 2153–61.
Webster, A. B. (2004), 'Welfare implications of avian osteoporosis', *Poult. Sci.*, 83, 184–92.
Weeks, C. A., Lambton, S. L. and Williams, A. G. (2016), 'Implications for welfare, productivity and sustainability of the variation in reported levels of mortality for laying hen flocks kept in different housing system: a meta-analysis of ten studies', *PLoS ONE* 11(1), e0146394.
Weeks, C. A. and Nicol, C. J. (2006), 'Behavioural needs, priorities and preferences of laying hens', *Worlds Poult. Sci. J.*, 62, 296–307.
Whitehead, C. C. (2002), 'Bone breakage and osteoporosis in laying hens: causes and solutions', *Proc. Aust. Poult. Sci. Sym.*, 14, 61–8.
Whitehead, C. C. (2004), 'Overview of bone biology in the egg-laying hen', *Poult. Sci.*, 83, 193–9.
Whitehead, C. C. and Fleming, R. H. (2000), 'Osteoporosis in cage layers', *Poult. Sci.*, 79, 1033–41.

Whitehead, C. C. and S. Wilson. (1992), 'Characteristics of osteopenia in hens', in C. C. Whitehead (Ed.), *Bone Biology and Skeletal Disorders in Poultry*, 265–80. Carfax Pub, Abingdon, UK.

Widowski, T. M. (2015), 'Why are behavioral needs important?', in T. Grandin (Ed.), *Animal Welfare: A Practical Approach* (2nd ed.), 247–66. CABI Publishing, Wallingford, UK.

Widowski, T. M., Classen H, Newberry, R. C., Petrik, M., Schwean-Lardner, K., Yue Cottee, S. and Cox, B. (2013), 'Code of practice for the care and handling of pullets, layers and spent fowl: Review of scientific research on priority issues', National Farm Animal Council, Canada https://www.nfacc.ca/resources/codes-of-practice/poultry-layers/Layer_SCReport_2013.pdf (accessed May 2016).

Widowski, T. M. and Duncan, I. J. H. (2000), 'Working for a dustbath: Are hens increasing pleasure rather than reducing suffering?', *Appl. Anim. Behav. Sci.*, 68, 39–53.

Wilkins, L. J., McKinstry, J. L., Avery, N. C., Knowles, T. G., Brown, S. N., Tarlton, J. and Nicol, C. J. (2011), 'Influence of housing system and design on bone strength and keel bone fractures in laying hens', *Vet. Rec.*, 169, 414.

Wysocki, M., Bessei, W., Kjaer, J. B. and Bennewitz, J. (2010), 'Genetic and physiological factors influencing feather pecking in chickens', *Worlds Poult. Sci. J.*, 66, 659–72.

Yeates, J. W. (2011), 'Is a "life worth living" a concept worth having?', *Anim. Welf.*, 20(3), 397–406.

Zimmerman, P. H., Koene, P., van Hooff, J. A. R. A. M. (2000), 'Thwarting of behaviour in different contexts and the gakel-call in the laying hen', *Appl. Anim. Behav. Sci.*, 69, 255–64.

Chapter 4

Welfare standards for laying hens

Andy Butterworth, University of Bristol, UK

1 Introduction

Animal welfare is considered by many consumers to be an important attribute of the food that they eat, and there is an implicit assumption that, if retailers sell eggs or poultry, then the food has been produced after taking into account the welfare of the animals. However, these assumptions are not always borne out by the reality of farming systems, and consumers have expressed concern for laying hen welfare at a higher level than their concerns regarding, for example, beef cattle (European Commission, 2005).

'Welfare' has had a number of definitions, ranging from those with a focus on physiological factors (McGlone, 1993), definitions based around subjective feelings (Dawkins, 1990) and definitions based on the capacity for 'coping'. In this case, coping refers to the animals' response to situations in which high levels of well-being are possible – to situations in which the animals' life is questionably worth living (Farm Animal Welfare Council, 2009). Health is often intertwined with welfare in the phrase 'health and welfare', and it is very likely that provision of conditions, which can provide 'good health', would be considered a prerequisite for 'good welfare' (Rushen, 2003).

Laying hens are farmed in many different systems in the world. In the European Union (EU), Council Directive 1999/74/EC categorizes the common laying hen systems into three groups: alternative systems, unenriched cage systems and enriched cage systems. To discuss laying hen standards, it is necessary to discuss standards for particular systems – as is has proved to be virtually impossible to make a 'one standard fits all' scheme for laying hens because of the diversity of systems.

The category 'alternative systems' covers a wide range of systems, ranging from very simple single-level systems to multilevel aviaries with or without free-range facilities, and is used to refer to systems that are not 'conventional cages'. A 'cage system' is operated

http://dx.doi.org/10.19103/AS.2016.0012.23

without the human carers entering the actual enclosure containing the birds and, globally, cages range from single cages containing 3, 4 or 5 birds, arranged in rows or tiers, to the 'enriched cage' or 'furnished cage' now required in the EU.

To give the consumer confidence in laying hen production, many countries have adopted legislative requirements for cage and non-cage system specifications and, in addition to legal requirements, many countries have also adopted industry-led 'quality assurance' (QA) standards, usually implemented through farm assurance schemes – in which the producer is periodically inspected to check compliance of the farm against the 'standard'. This farm assurance overlying legal requirements is now the basis of much of the 'regulation' of laying hen production in the EU, and in some other parts of the world, including Australasia, the United States and Canada. In the United Kingdom and Europe, it was the more integrated animal industries (dairy, eggs, pig meat, poultry) which were the first to adopt quality assurance standards, which included aspects regarding animal welfare, and QA in these sectors has been arguably very successful in terms of industry penetration – partially because these industries are tightly vertically integrated, allowing adoption of a 'standards' approach through and across companies. The EU Council Directive 1999/74/EC on the Welfare of Laying Hens was implemented on 1 January 2012 and required the adoption of enriched or colony cages ('colony' cages in the United Kingdom house between 40 and 80 birds) and the enriched cages have more space and height than the conventional cages they replaced, a nesting area, a scratching area and perches. Picking the United Kingdom as an example, it is perhaps worthwhile to examine how the 'commercial farm assurance standards' system works for laying hens.

2 Commercial welfare standards for laying hens: the case of the United Kingdom

There are a number of commercial standards – these offer the producer the capacity to sell to retailers who require compliance with a 'standard' – and are, therefore, effectively gateway requirements (and this is the financial and structural incentive for producers to be scheme members) to sell eggs and poultry products in line with the expectations and 'standards' of retailers.

2.1 British Egg Industry Council (BEIC) 'Lion Code'

This code has 700 auditable criteria, and is a 'standard' which requires compliance with conditions regarding health, health planning, housing, space, feeder, drinker, health screening – including salmonella testing, humane culling, stockman inspection requirements, record keeping, recording of medicine use. Nearly 90% of UK eggs are now produced within the BEIC scheme and, since the code's original version was implemented in November 1998, 130 billion Lion eggs have been sold, and more than 50 000 independent audits have been undertaken.

2.2 Freedom Food laying hen standard

This standard is an initiative of the Royal Society for the Prevention of Cruelty to Animals (RSPCA), which is owned entirely by the RSPCA, and operates as a commercial, farm assurance body. This operates alongside the charitable animal welfare activities of the RSPCA. The laying hen standards (Freedom Food – RSPCA, 2013) are created by the farm animal and science departments of the RSCPA charity, in association with industry and consumer stakeholders – the RSPCA says that the laying hen standards were created through the combined: *'Knowledge of farming industry representatives, veterinarians, and animal welfare and production research scientists ... gathered, discussed and taken into account ... through standards technical advisory groups and wider consultation groups.'*

The RSPCA standards include many clauses, which are significantly above legal baseline requirements, some 'aspirational' recommendations and, recently, the introduction to the scheme of outcome (animal-based) measures which more directly assess aspects of the birds' welfare.

The RSPCA states that it builds the content of its standards through

- Extrapolation from evidence, experience, knowledge of similar situations and species
- Reference to/comparison with current 'good practice'
- Use of RSCPA primary research/experience
- Application of reasonable/justifiable anthropomorphism (e.g. Qualitative Behavioural assessment-type approach)
- Application of 'common sense'
- Consideration of ethical issues
- Application of the Precautionary Principle ('informed prudence')

The RSCPA standards cover hatcheries, pullet rearing, laying hen farm, transportation and slaughter. Perhaps surprisingly, since these are clearly 'higher level standards', their penetration into the laying hen market is very high.

2.3 Soil Association Certification

This certification has a focus on organic farming, with the stated aims of avoidance of routine use of pharmaceuticals and vaccines, which is seen by many consumers as fitting well with their aims towards 'clean, green and ethical' systems of livestock management. Because organic standards have at their basis EU legislation which defines and separates organic production from conventional production, organic standards tend to have a general international consistency. The Soil Association standards for laying hens incorporate not only all legal requirements with regard to organic production, as required by Commission Regulation (EC) No. 889/2008, but also specific requirements created by the Soil Association to meet the particular needs of the UK consumer for organic product. The standards contain three levels of requirement – what you 'should do', what you 'must do' and what you 'may' do;

- *What you 'should' do: These give the ideal or best organic practice. They say how you should ideally be working.*
- *What you 'must' do: These state the actual requirements, including what you must get our permission for and what you must not do.*
- *What you 'may' do: These state what you can do. We say if you need to get our permission for these or if there are other conditions.*

We have divided these conditions into three categories to be clear about your responsibilities when using them:

- *With justification – you must be able to justify the use of certain products/practices at your inspection with evidence, such as test results, records, forms, a plan etc. For example, you record why you needed to use the product/practice.*
- *With our approval – we must have approved your use of certain products/practices. This may cover more than one use or it may be in your annual plan that we have approved. For example, you produce an annual plan that details the product/practice which your certification officer has approved and you have a copy available for inspection.*
- *With our permission – you must get our permission before each use of certain products/practices. For example, you phone your certification officer for permission every time, they may ask for further documentation.*

3 How are welfare standards implemented?

Standards are inspected against by assessor/auditors/inspectors – trained (and sometimes licenced by government in the case of organic inspectors), personnel who visit the farms at intervals of between 12 and 18 months (dependent on the scheme) and inspect the premises and the records, and examine the animals against the requirements of the written standards.

In the UK and EU, many of the farm assurance schemes operate their assessments to fulfil the requirements of two ISO standards – ISO/IEC 17065 (Conformity assessment, Requirements for bodies certifying products, processes and services) and ISO/IEC 17020 (Conformity assessment; Requirements for the operation of various types of bodies performing inspection). These international standards prescribe the format of agreed processes and operating procedures, including internal audit and quality procedures which the certification or inspection body (the farm assurance companies which make the farm visits) must adopt to demonstrate sufficient control and uniformity in their inspection processes.

If the farm assurance inspection body issues 'certificates' to producers which meet the standards, then they are known as 'certification bodies' – and their performance is checked against ISO/IEC 17065 by the overarching national accreditation bodies (e.g. UKAS in the United Kingdom, RvA in the Netherlands, ENAC in Spain, etc.) (UKAS, 2016; RvA, 2016; ENAC, 2016). This process of international accreditation by the national accreditation bodies helps to ensure that there is a harmonization of 'approach and standard' among countries, and helps to reduce the concern that farm standards and farm assurance are not a 'level playing field' when compared internationally.

The 'quality' of the inspectors and inspection process is critical if the process is considered to be impartial, sensitive to 'non compliances' against the standard, and to be seen by the producers to be applied 'fairly and justly'. It is important to note that the farm assessors do not act as advisors, as conflicts of interest would result if the same people who assessed the farm against standards were also the people offering advice on resolution of assessment problems. The costs of membership of farm assurance schemes are quite variable, with costs of approximately between £120 and £500 per year, depending on the number of animals, the scheme and the complexity of the farming operation inspected and certified.

The role of the supermarkets and consumers in adoption of laying hen standards is important. Consumers indicate that they want animal products that they purchase to have come from animals with good welfare standards. However, this 'wish' is not always translated into rational buying choices and, for much of the market, the key driver is cost. The use of logos or trade marks associated with the use of farm standards has been common to provide visible choice to consumers and, in the EU, there are many labelled egg choices – most of which are backed up by a standard, and an inspection and certification process. International examples include Respectful Eggs (Stonegate, 2016), Clarence Court Eggs (Clarence Court, 2016), Rabbit River Eggs (Rabbit River, 2016), Burnbrae Eggs (BurnBrae Farms, 2016) – and the trend in many countries is towards more choice in eggs as the consumer becomes more aware of choice options.

In many countries, there is effectively a 'quality grading' choice system for eggs displayed on supermarket shelves. The cheapest product (perhaps named 'basic') rarely carries QA logos, but the 'top end' packages will usually be heavily invested in, and linked to, a 'brand' and will often mention the animal welfare benefits that it supports, a clear strategy to target discretionary spending power, indicating perhaps that animal welfare is perceived as an aspect of quality.

To be, and remain, credible and to withstand scrutiny, standards need to be robust, objective and, as far as is possible, transparent. Standards should also not remain static, they should have the capacity to improve or change to suit altered understanding and external pressures to improve, or to adapt. In many areas of standard development, there is currently a move to try to assess the effects on, and the welfare of, the animals, through the use of measures of outputs (what happens to the animal) rather than on resources provided by the system and its management systems (inputs). An example would be – to look at the amount of bedding provided in a barn-laying system (input) or to look at the condition of the birds – feet, feathers, cleanliness (outputs). However, this gradual change from an input-based approach to standards – to an assessment of the measurable effects on the animals – is not happening 'immediately', as practical (time-efficient and accurate) methods take time to develop, test, implement and to get agreement and adoption by industry and standard owners. Around the world, there are some initiatives and research programmes that are spanning the gap between commercial farm standards and animal-based assessment systems – which originally derive from animal welfare science-based assessment methods. I will describe a couple of these 'standards' as I think that they mark a quantum shift in the approach to animal assessment standards in laying hens.

4 Case study 1: the 'AssureWel' scheme

AssureWel (AssureWel, 2106) is a collaborative project in the United Kingdom between the University of Bristol, the Soil Association and the RSCPA, with links to European Farm assurance companies. AssureWel builds on the base of a previous project Lay Wel (LayWel, 2016). The main aim was to integrate animals-based (looking directly at measures that could be made on the animals) assessment into the framework of existing standards and commercial farm assurance schemes. The project has had a planned progression, with laying hens the first to be covered in Years 1–4, dairy cattle starting in Year 2, pigs in Year 3 and broilers, beef cattle and sheep in Years 4–5. AssureWel methods have been 'field-tested' within the Freedom Food and Soil Association Certification schemes by using the assessment of animal welfare outcomes to help determine compliance with the standards within these schemes. AssureWel has identified the following as being important indicators of laying hen welfare, and these criteria have been selected as 'core' measures – consideration including an understanding of how practical they are to assess on-farm:

- *Feather loss*
- *Bird dirtiness*
- *Beak trimming*
- *Antagonistic behaviours*
- *Flightiness*
- *Birds needing further care*
- *Mortality*

For all schemes, this has meant a shift from input-based to outcome-based assessment. Producers are given feedback to help them to monitor and improve welfare on their farms, and benchmark information is available so that egg producers can see how the animals on their farms do in comparison with other farms in the schemes. Advice and discussions on management support are facilitated within the scheme but, as mentioned above, direct advice giving between the assessor and the producer at the time of the assessment is avoided. The measures have been developed for non-cage systems, organic, free-range and barn systems, and AssureWel states that the assessment protocol can be used to 'assess the level of welfare achieved for an individual animal, farming system or farm assurance scheme'. They also promote the idea that the laying hen producers may use the protocols to self-assess flocks, which can help monitor welfare and provide early indicators to identify problems.

AssureWel indicates that the assessment of the outcome measures within an existing visit takes around 10–15 minutes per farm for the sample sizes suggested, and sample sizes are described thus: 'Some measures require an examination of the entire flock. For other measures, a small sample of 50 birds needs to be assessed. However, assessing a larger number of birds will increase the confidence of the sample and be more representative of the whole flock.' AssureWel protocols were designed to be used by 'anyone with relevant livestock experience – and training is required to ensure reliable and accurate results, and this training is available through the AssureWel project'.

The inclusion of a set of AssureWel measures into farm assurance schemes, or for use by companies, producer groups and vets, may allow formal assessment and then feedback and benchmarking (Huxley et al., 2004), which may help to improve farm management and general levels of hen welfare. AssureWel has, by its nature, been exploratory, and many findings on the most effective means to implement and incorporate outcome-based measures into standards have been made as a result of this hybrid between commercial implementation, welfare outcome standards and the inclusion of 'welfare science' measures during practical farm inspection visits.

5 Case study 2: the 'Welfare Quality' scheme

Welfare Quality® set out to develop scientifically based tools to assess animal welfare, brought together information on consumer perceptions and attitudes with existing knowledge on welfare assessment methods, identified four principles (Good Feeding, Good Housing, Good Health, Appropriate Behaviour), and stemming from these, 12 criteria for animal-based measures that address aspects of the welfare state of the animals. The measures used were, in the majority, derived from established research methods described in the literature, and refined or adapted by the Welfare Quality team scientists. An example would be feather cover in relation to layer health. Fölsch et al. (1980) and Grashorn and Flock (1987) used planimetry for estimating de-feathered areas, and a general score for the plumage of the whole body was already in use by Hughes and Duncan (1972) or Hill (1980) and another method for scoring several parts of the body had been used by Meunier-Salaün (1983), Tauson (1984) and Gunnarsson et al. (1995). Tauson (1984) found good observer reliability by using a 1–4 point scale on five body parts of the hen. This scientific background to a single score provides an example of the background history to the measures adopted by Welfare Quality – the measures adopted build in most part on an extensive history of research and publication.

Table 1 shows the 'hierarchical' structure of measures in the Welfare Quality system for laying hens. Where available, a welfare science 'measure' is proposed; however, when

Table 1 Collection of Welfare Quality data for laying hens on-farm

Principle	Welfare criteria		Measures
Good feeding	1	Absence of prolonged hunger	Feeder space
	2	Absence of prolonged thirst	Drinker space
Good housing	3	Comfort around resting	Shape and total length of available perches, evidence of red mites, dust sheet test
	4	Thermal comfort	Panting, huddling
	5	Ease of movement	Stocking density, perforated floors
Good health	6	Absence of injuries	Keel bone deformation, skin lesions, foot pad dermatitis, toe damage
	7	Absence of disease	On-farm mortality, culls on farm, enlarged crops, eye pathologies, respiratory infections, enteritis, parasites, comb abnormalities
	8	Absence of pain induced by management procedures	Beak trimming
Appropriate behaviour	9	Expression of social behaviours	Aggressive behaviours, plumage damage, comb-pecking wounds
	10	Expression of other behaviours	Use of nest boxes, use of litter, enrichment measures, free range, cover on the range, covered veranda
	11	Good human–animal relationship	Avoidance distance test (ADT)
	12	Positive emotional state	Novel object test (NOT), qualitative behaviour assessment (QBA)

no direct animal-based measure is available, or when such a measure is not sensitive or reliable enough, measures of resources are used to check that a welfare criterion is met. This is in agreement with the general consideration that, where appropriate, both resource and animal-based measures can be used to provide as much information for assessment of animal welfare as is possible. The Welfare Quality® assessment measures were evaluated with respect to their validity (measures that assess aspects of the actual welfare of animals), reliability (acceptable inter- or intra-observer repeatability and robustness to external factors, e.g. time of day or weather conditions) and their feasibility.

The measures are made on-farm, or sometimes at the slaughterhouse (e.g. in the case of skin or feather lesions on birds going to slaughter). The data can, if required, then be combined to calculate 'criterion-scores' – a score for the overall welfare under that criterion heading – for example under the heading 'absence of injuries', the measures (a) Keel bone deformation, (b) skin lesions, (c) foot pad dermatitis and (d) toe damage would be made, and then a combined score for 'absence of injuries' can be made by using the aggregation scoring process described by WelfareQuality. It is then theoretically possible to combine criterion scores to calculate principle scores; and finally it is possible, if requested, to combine criteria scores to make a single welfare score for the farm and categorize the farm according to the principle scores.

The scoring formulae required which allows this 'aggregation process' were created by consultation with groups of experts from animal sciences. Weighted sums were calculated, to define weights that produce the same ranking of farms as the one given by experts, and this allowed the possibility for a farm to be given an overall welfare score based on the data collected on the individual measures (measures as described in Table 1). The scores obtained by the farm by using the measures can be used to assign that farm to a welfare category:

Excellent	The welfare of the animals is of the highest level.
Enhanced	The welfare of animals is good.
Acceptable	The welfare of animals is above or meets minimal requirements.
Not classified	The welfare of animals is low and considered unacceptable.

As the Welfare Quality project progressed, information on the preferred order in which the measures could be carried out and the time taken to carry out these measures became available. Tables 2 and 3 show information on how long the assessment measures are described to take within the Welfare Quality protocols, the recommended order of assessment and sample size recommendations.

The process of selecting laying hens for assessment involves the following:

- The same 100 selected hens can be used for the several clinical assessments; these are keel bone deformation, skin lesions, comb abnormalities, comb-pecking wounds, foot pad dermatitis, beak trimming and plumage damage.
- Additionally, for these measures, the following selection method should be used: 100 birds per flock should be selected from various points in the house. Ideally the selection should reflect the various areas in the house (perches, nest boxes, litter, slatted floor, covered veranda, free range). As on commercial farms, this is not feasible, a selection can be made by either penning (corralling) birds or by picking

Table 2 Order for carrying out Welfare Quality measures, sample size and time required for laying hens on-farm

Measure	Sample method or number of birds to sample	Time required (min)
On-farm mortality	Farm records – Establish number of birds lost (not culled) in relation to total number placed	10
Culls on farm	Farm records – Establish number of birds lost (not culled) in relation to total number placed	
Use of nest boxes	Establish the distribution of eggs over rows and nest boxes. If this is not possible, ask animal unit manager and check records Calculate the nest space per bird	5[c]
Huddling	Group observation (3 times, combined to 1 score), is done while doing other work in the house, so only some time required to write outcome down	1
Panting	Group observation (3 times, combined to 1 score), is done while doing other work in the house, so only some time required to write outcome down	1
Qualitative behaviour assessment (QBA)	Observations made at 2–8 points[a]	30[a]
Novel object test (NOT)	Object placed at 4 sites in the house, each site taking 5 minutes waiting, 2 minutes to assess + 1 minute (movement time) + 3 minutes preparation	35
Avoidance distance test (ADT)	21 hens are assessed from 7 different areas, 10 seconds approaching at each site + 20 seconds recording + 30 seconds movement between sites	30
Plumage damage		
Keel bone deformation		
Comb abnormalities		
Comb-pecking wounds	100 birds picked – 10 birds from 10 locations	180–240[-b]
Skin lesions		
Foot pad dermatitis		
Beak trimming		
Shape and total length of available perches	Establish the total perch length and divide it by the number of hens housed.	5[-c]
Stocking density	Establish the total number of birds placed and divide by the available area	15[-c]
Feeder space	Calculate number of feeders × area/length per feeder and divide by number of birds placed	5[-c]
Drinker space	Calculate number of drinkers × area per drinker and divide by number of birds placed	5[-c]

(Continued)

Table 2 (*Continued*)

Measure	Sample method or number of birds to sample	Time required (min)
Perforated floor	Establish the total available slatted floor area, the type and the state of repair	5[c]
Use of litter	Observe dust bathing and/of scratching/ manipulating litter	2
Dust sheet test	Place dust test sheet at the start of observation period and then assess at the end	5
Evidence of red mites Parasites	Check environment of birds. Also check flock in general and more precise 100 birds for parasites. If possible also check dead birds	1
Aggressive behaviour	Check for aggressive behaviour of the birds	
Cover on the range	Check free-range area and make calculation	5
Free range Enrichment measures Covered veranda	Check free-range area, covered veranda and area inside the house	
Toe damage Enlarged crops Eye pathologies Respiratory infections Enteritis	Based on impressions during time spent in the house and when recording 100 birds to 10 birds from different locations	5
	Total	**345–405 minutes** (6–7 hours) [Y]

[a]Qualitative assessment: observation time per spot-5 minutes in case of 4 spots and 10 minutes in case of 2 spots.
[b]Variation mainly due to variation in scoring 100 birds. Not included are the calculations afterwards that need to be done to get the totals.
[c]Not included are the calculations afterwards that need to be done to get the totals.

up individual birds in several areas of the henhouse. The number of places to collect hens is dependent on the housing system and the number of compartments.
- In general, for the various observations and measurements, the person carrying out the protocol can observe birds in various parts of the henhouse.
- The measures which include use of litter, evidence of red mites, parasites and aggressive behaviours are carried out while doing other work in the house, so the time indication in Table 10 reflects only the time required to write outcome down.
- In cage systems, take the birds from different areas of the house and from different tier levels.

A logical sequence for assessment of the measures includes the following:

- general characteristics of the flock
- questionnaire to farmer

- check/copy farm records and official measuring report if available
- assessor bio-security when entering henhouse
- place black paper for dust measurement
- estimate birds panting/huddling
- QBA (this should always be done prior to all other work in the henhouse to prevent any biased judgement)
- NOT-test
- ADT-test
- Clinical scoring
- Space measurements/food water (if necessary)
- Other Questions/Observations
- Observation of interaction during routine activity (this is not always possible and timing will depend on when the farmer can be observed)
- Final remarks, finishing time

6 The use of global standards

The *World Organisation for Animal Health (OIE)* (*Office International des Epizooties*) (OIE, 2106) was formed in 1924 and in May 2003 it changed its name to the World Organisation for Animal Health, but kept its acronym OIE. It has responsibility for improving animal health worldwide and is a reference organization by the World Trade Organization (WTO). In 2016, it has a total of 180 member countries and, in 2001, it identified animal welfare as a priority and started to show international leadership by creating guidelines and standards, which are now known as the OIE Terrestrial Animal Health Code. These codes provide information on expected standards for animal health and welfare, which in many cases mirror regional legislative requirements. With regard to laying hens, some of the terrestrial code statements and clauses may be relevant to production systems across the world, but in general, the level of detail and the thresholds set are already entrenched in law in the EU, and within the law for many developed countries – but application in developing countries has real potential to promote improvements in monitoring and 'standards setting' in laying hen health and welfare in developing countries.

In a similar way, the Food and Agriculture Organisation of the United Nations (FAO) provides national information related to farm animal welfare, and has the objectives to deal with national laws and regulations related to animal welfare and protection, cruelty against animals, standards of humane care and treatment of animals – and to develop codes of practice and recommendations for institutions, stakeholder organizations and policy makers to promote and support animal health and welfare.

6.1 What do governmental bodies and commercial users think about the use of 'outcome measures' in farm standards and in farm assurance?

The European Food Safety Authority (EFSA, 2012), in a report on the use of animal-based measures to assess the welfare of animals, concluded: 'Animal-based measures are the most appropriate indicators of animal welfare and a carefully selected combination of animal-based measures can be used to assess the welfare of a target population in a valid and robust way.'

However, they went on to state that *'in some situations, feasibility and economic aspects may be most important, whereas in others sensitivity and specificity may take priority'* and that *'the systematic collection of standardised field data on animal-based measures and subsequent storage in well-defined databases can in the future assist in better assessing the validity and robustness of animal-based measures and thus the move towards quantitative risk assessment of animal welfare. Such data are also a potentially valuable tool to monitor animal welfare in Members States.'*

Their recommendations included:

- Both animal-based measures and input factors (resource- and management-based measures) should be used in combination when monitoring animal welfare.
- Once a long list of valid and robust animal-based measures has been identified, a short list should be defined in consultation with a diverse group of stakeholders.
- There should be ongoing evaluation of animal-based measures and how they are used to assess the welfare of animals as this is a rapidly expanding area and new measures and new methods of recording established measures are being developed.

For many industry observers, it is the time taken by outcome measures and the potential variability between assessors that raise the greatest concern about a shift in farm assurance standards from resource measures (space, number of drinkers etc.) to assessment of the animals (plumage condition, aggressive behaviours, avoidance distance) – and it is these concerns, rather than objection to the principle of 'measuring things on the animal' and rather than 'things provided for the animal' that have been the major inhibition to the wider adoption of outcome measure-based systems such as AssureWel and Welfare Quality as yet into commercial standards and farm assurance schemes.

7 Where to look for further information

Improving Farm Animal Welfare; Science and Society Working Together: the Welfare Quality Approach (2013). Wageningen Academic Publishers. ISBN 978 90 8686 216 0.

8 References

AssureWel (2016). http://www.assurewel.org/aboutassurewel. Accessed 4 March 2016.
BurnBrae Farms (2016). http://www.burnbraefarms.com/consumer/faqs/naturegg_freerun.htm. Accessed 4 March 2016.
Clarence Court (2016). http://www.clarencecourt.co.uk. Accessed 4 March 2016.
Dawkins, M. S. (1990). From an animal's point of view: motivation, fitness and animal welfare. *Behav. Brain Sci.* 13: 1–61.
ENAC (2016). https://www.enac.es/web/english. Accessed 4 March 2016.
European Commission (2005). Attitudes of consumers towards the welfare of farmed animals Fieldwork: February to March 2005 Publication: June 2005. http://ec.europa.eu/food/animals/docs/aw_arch_euro_barometer25_en.pdf. Accessed 3 March 2016.

European Food Safety Authority (EFSA) (2012). Scientific Opinion Statement on the use of animal-based measures to assess the welfare of animals. EFSA Panel on Animal Health and Welfare (AHAW) 2, 3 European Food Safety Authority (EFSA), Parma, Italy. *EFSA J.* 10(6): 2767.

Farm Animal Welfare Council (FAWC) (2009). Farm Animal Welfare in Great Britain: Past, Present and Future Farm Animal Welfare Council Area 5A, 9 Millbank, c/o Nobel House, 17 Smith Square, London SW1P 3JR. http://www.fawc.org.uk.

Fölsch, D. W., Benelli, A. and Gozzoli, L. (1980). Die Auswirkungen der Boden- und Batteriehaltung mit Unterschiedlicher Besatzdichte auf die Fortbewegung und das Gefieder von Legehennen. In Proc. 6th European Poultry Conference, Hamburg 4: 160–8.

Grashorn, M. and Flock, D. (1987). Genetisch-Statistische Untersuchungen des Befiederungszustandes an weißen (LSL) und braunen (LB) hennen. Lohmann Informationen, 13–19 November/December.

Gunnarsson, S., Odén, K., Algers, B., Svedberg, J. and Keeling, L. (1995). Poultry health and behaviour in a tiered system for loose housed layers, Dept. of Animal Hygiene, SLU, Skara, Sweden. Report, pp. 35–142.

Hill, A. (1980). Feather loss in layers. Poultry Booklet. No. 7. Gleadthorpe Experimental Husbandry Farm, MAFF, Great Britain, pp. 46–52.

Hughes, B. O. and Duncan, I. J. H. (1972). The influence of strain and environment upon feather pecking and cannibalism in fowl. *Br. Poult. Sci.* 13: 525–47.

Huxley, J. N., Burke, J., Roderick, S., Main, D. C. J. and Whay, H. R. (2004). Animal welfare assessment benchmarking as a tool for health and welfare planning in organic dairy herds. *Vet. Rec.* 155(8): 237–9.

LayWel (2016). http://www.laywel.eu. Accessed 4 March 2016.

Meunier-Salaün, M. C. (1983). Relations interindividuelle dans les petits groupe de poules domestiques. Effet des conditions d` elevage Thesis. INRA, France.

McGlone, J. J. (1993). What is animal welfare? *J. Agric. Ethics.* 6: 26–36.

OIE (2016). http://www.oie.int/international-standard-setting/terrestrial-code/access-online/. Accessed 4 March 2016.

Rabbit River Farms (2016). http://www.rabbitriverfarms.com. Accessed 4 March 2016.

RSPCA (2013). Welfare standards for Laying Hens September 2013. http://industry.freedomfood.co.uk/media/30295/RSPCA_welfare_standards_for_laying_hens___September_2013___web.pdf. Accessed 4 March 2016.

Rushen, J. (2003). Changing concepts of farm animal welfare: bridging the gap between applied and basic research. *App. An. Behav. Sci.* 81(3): 199–214.

RvA (2016). https://www.rva.nl/en. Accessed 4 March 2016.

Stonegate (2016). http://www.stonegate.co.uk/productRespec.aspx. Accessed 4 March 2016.

Tauson, R. (1984). Plumage condition in SCWL laying hens kept in conventional cages of different designs. *Acta. Agric. Scand.* 34: 221–30.

UKAS (2016). https://www.ukas.com. Accessed 4 March 2016.

Chapter 5

Welfare issues affecting free-range laying hens

Dana L. M. Campbell, University of New England and CSIRO, Australia; Sarah L. Lambton, University of Bristol, UK; Isabelle Ruhnke, University of New England, Australia; and Claire A. Weeks, University of Bristol, UK

1 Introduction

Free-range laying hen production systems take a step down a different path in the progression of domestic chicken farming. They take hen groups away from intensive, environmentally controlled, indoor housing and put them back outdoors to provide flocks with more natural, minimally controlled surroundings. Consumers prefer the concept of free-range production, particularly in countries where weather allows outdoor access for most of the year (e.g. Australia, Denmark, the United Kingdom), as these systems are viewed as more natural and ethical (Schröder and McEachern, 2004). Free-range systems are thus rapidly increasing in popularity to meet consumer demand. However, highly productive modern genotypes, selected for cage production, are harder to manage in large flocks and outdoor environments with limited control of environmental conditions. Research into the behaviour, health and welfare of birds in non-cage environments has increased in recent years. Unfortunately, in some cases, a high risk of poor outcomes for hens kept in non-cage and especially free-range conditions has been identified (Fossum et al., 2009; Sherwin et al., 2010; Elson, 2015; Weeks et al., 2016). To sustainably improve free-range housing and management, we must understand how hens behave in

http://dx.doi.org/10.19103/AS.2016.0012.25

these systems, the welfare challenges they face and what modifications will, in practice, adequately address hens' needs.

There is no standard definition of 'free-range'. Producers may use systems that comprise static sheds with an outdoor area or mobile sheds that rotate on accessible pasture (Singh et al., 2015; Sherwin et al., 2013). Free-range systems may or may not include 'winter gardens' – covered outdoor areas or verandas (e.g. Steenfeldt and Nielsen, 2015; Bestman and Wagenaar, 2014) – and hens' age of access to different parts of the entire system can be staggered (Gebhardt-Henrich et al., 2014). Regulations or guidelines for parameters of free-range systems vary between countries (e.g. indoor/outdoor stocking density, flock size, age of first access and daily hours of available access), and in some cases, independent auditing bodies will have further stipulations (e.g. shelter on range) for certification labels (e.g. RSPCA Assured). Organic laying hen systems are typically also free-range. However, the general international consensus is that free-range hens need daily access to outdoor, open-air runs, with the assumption that most hens will be accessing the outdoors on most days. It is this assumption that makes the free-range system unique. Hens are provided with access to an open-air environment, but on a daily basis and often multiple times per day, hens make a choice whether or not to use this outdoor space. This choice results in individual hens potentially being exposed to very contrasting conditions within the same housing environment, which may impact their welfare in different ways. The vast majority of welfare issues that are present, or are more prevalent, within free-range systems ultimately hinge on the hens' use of the outdoor range.

There has recently been an excellent review of factors affecting range use including weather, stocking density, flock size, rearing environment and availability of outdoor shelter (Pettersson et al., 2016), and thus we will not consider these in depth here. Briefly, across multiple studies, range use is affected by weather, with hens preferring calm, mild days for ranging rather than weather extremes, although tolerance of weather by different flocks may depend on the typical weather to which they are accustomed (Pettersson et al., 2016). There is an inverse relationship between stocking density (both indoor and outdoor), flock size and range use (Pettersson et al., 2016; Campbell et al., 2017), with the presence of outdoor shelter encouraging birds to venture onto the range (Pettersson et al., 2016).

The areas of major welfare concern for free-range laying hens reviewed in this chapter include individual differences in range use, behavioural expression on the range, correlations between ranging and health variables, parasite loads, adequate nurition, grass impaction, feather pecking and cannibalism.

2 Use of the outdoor range

2.1 Variation in use of the range

Across a collection of international group-level studies, the number of hens on the range simultaneously in commercial and experimental systems is typically less than 50% but sometimes less than 10% of the flock (Pettersson et al., 2016), although proportions can be higher in smaller experimental flocks (Campbell et al., 2017). More information on range use by individual hens is gradually becoming available with the use of technologies such as radio-frequency identification (RFID). These RFID systems come in many different configurations, but essentially all are designed to track the movement of individually-tagged hens in and out of the entrances (pop holes) to the outdoor range, providing detailed data on how often hens go outside and how long they spend on the range during any single visit (Richards et al., 2011; Gebhardt-Henrich et al., 2014; Campbell et al., 2017). With the increasing use of this individual tracking technology, we see emerging patterns indicating range usage by specific flocks is actually much higher than previously thought. For example, 47–90% of tagged birds (5–10% of flock; 2000 to ≥9000 hens/flock) visited the free-range area at least once during the course of the trial (Gebhardt-Henrich et al., 2014), 80% of tagged birds (10% of flock; 1500 hens/flock) registered on the pop holes frequently (Richards et al., 2011) and 66.5–80.5% of tagged birds (50% of flock; 150 hens/flock) accessed the range on a daily basis (Campbell et al., 2017). However, we also find that, within the same flock, there is considerable individual variation in how long birds spend outside each day, which may range from less than an hour to over 6 hours (Campbell et al., 2017); how many separate visits to the range are made each day, ranging from fewer than 5 visits to over 40 visits per day (Campbell et al., 2017); whether range usage is daily or irregular; and in a small percentage of birds, whether the range is even used at all (Richards et al., 2011; Campbell et al., 2017). These findings are interesting, given the perceived value of providing outdoor access for hens. In this regard, it has been shown that pop hole usage decreases in birds with keel fractures (Richards et al., 2012), suggesting that painful injuries inhibit range access. However, to date, there is no other evidence regarding the health status of birds that remain indoors. A few recent studies have used individual behavioural tests, including tonic immobility, open field test and manual restraint, to indicate that birds staying indoors show higher levels of fear (Hartcher et al., 2016; Campbell et al., 2016a; Larsen et al., 2016) and greater stress reactions (Hernandez et al., 2014), suggesting that different 'personality traits' exist between indoor- and outdoor-preferring hens. These tests were all performed following a period of range access and, currently, no preceding tests to predict range use have been conducted. Given the scope of individual variation within separate hen flocks, the choice itself of different environments provided within the free-range system may improve welfare (Nicol et al., 2009), in addition to any benefits of the range resource itself. Comparisons of individual behavioural expression between outdoor- and indoor-preferring hens warrant further investigation, which should be undertaken concurrently with an evaluation of house design and resource provision. For example, Whay et al. (2007) associated consistent differences in the hens' attitude, behaviour and performance with the type of floor and the presence or absence of perches, indicating that the welfare of the hens was inferior when they were housed on plastic floors with no perches. They postulated that the greater use of the range in some flocks was down to a dislike of the indoor environment.

2.2 Behavioural expression on the range

An enriched range has the potential to enable hens to fully express their behavioural repertoire in a more 'natural' environment. Some specific behaviours which hens have a high motivation to express include dust bathing, perching and foraging (Weeks and Nicol, 2006), where the latter is reported to comprise a majority of the daily time budget in the Red Junglefowl (*Gallus gallus*) ancestors (Dawkins, 1989). As previously discussed, hens venturing onto the range are influenced by a number of different variables (Pettersson et al., 2016), as is also the case with where they go once outside on the range. Several studies have observed that hens do not always use all available areas of the range, with birds staying closer to the pop holes rather than venturing to the outer portions of the range (Harlander-Matauschek et al., 2001; Hegelund et al., 2005; Steenfeldt and Nielsen, 2015; Chielo et al., 2016). In smaller, experimental flocks, range area usage may be more uniform (Campbell et al., 2017). A wide variety of shelters, shade and enrichment will encourage more hens both onto the range and to use more of the available range area (Zeltner and Hirt, 2008a; Nagle and Glatz, 2012; Rault et al., 2013). These ensure a better distribution of birds and prevent overuse of one area, that which is typically close to the pop holes. This, in turn, influences the behaviours exhibited on the range, although currently only a handful of studies have documented range behavioural expression and the relationships with both ground and aerial range coverage.

A recent study by Chielo and colleagues (2016) looked at six commercial free-range flocks on four farms in Lincolnshire, United Kingdom, with flock sizes ranging from 3900 to 23 548 birds aged between 27 and 55 weeks. Hens were provided activity kits approximately 15–35 m away from the shed that comprised a mini shelter, dust bath and perches. Saplings were also planted between 10 and 50 m from the shed. For observational purposes, the range was divided into three distinct zones consisting of (1) the 'apron zone' (0–10 m from shed, 4.1% of available outdoor area), with no additional enrichments or tree cover and sparse ground vegetation; (2) the 'enriched zone' (10–50 m from shed, 21.2% of available outdoor area), with manmade enrichments, natural tree cover and low grass pasture with patches of taller plants and patches of bare earth, particularly where hens had scratched out dust bathing pits; and (3) the 'outer zone' (50 m and beyond from shed, 74.7% of available outdoor area), consisting of mostly grass pasture. Scanning observations revealed that most hens were present in the apron area, resulting in a much higher stocking density for this area when compared to the enriched belt and outer zones. There were also corresponding linear differences in the nearest-neighbour distances, with hens dispersed most in the outer zone and least in the apron zone. Observations of a sample of birds in each zone showed varied proportions of different activities in the three range zones with four behaviours (standing, pecking, walking and foraging) representing over 90% of the overall activity of the hens (Table 1). Dust bathing was observed in only 0.2% of samples.

A similar study by Larsen and colleagues (2014) was undertaken on a single commercial free-range farm located in Victoria, Australia, housing 8000 hens at 62 weeks of age. The range was divided into four distinct 'patches'. These patches consisted of (1) a 30-m-high eucalyptus tree providing a large canopy cover, (2) *Acacia* wattle trees providing 1–2 m high dense canopy cover, (3) bare sand and gravel ground, and (4) bamboo-like dense

Table 1 The mean (±SEM) percentage of time spent performing the four most recorded behaviours from each daily sample of 20 hens per range zone (apron 0–10 m from shed, enriched 10–50 m from shed and outer 50 m+ from shed). Means within a row with different superscripts differed significantly.

Behavioural states	Apron	Enriched	Outer	SEM	*F*-value	*P*-value
Standing	16.6[a]	8.9[b]	4.2[c]	0.61	52.87	<0.001
Pecking	9.5[a]	10.9[a]	3.4[b]	0.57	24.27	<0.001
Walking	9.2[a]	8.3[a]	13.2[b]	0.49	14.40	<0.001
Foraging	1.7[a]	7.1[b]	15.8[c]	0.68	53.63	<0.001

[a–c] $P < 0.001$

Source: Reproduced from Chielo et al. (2016).

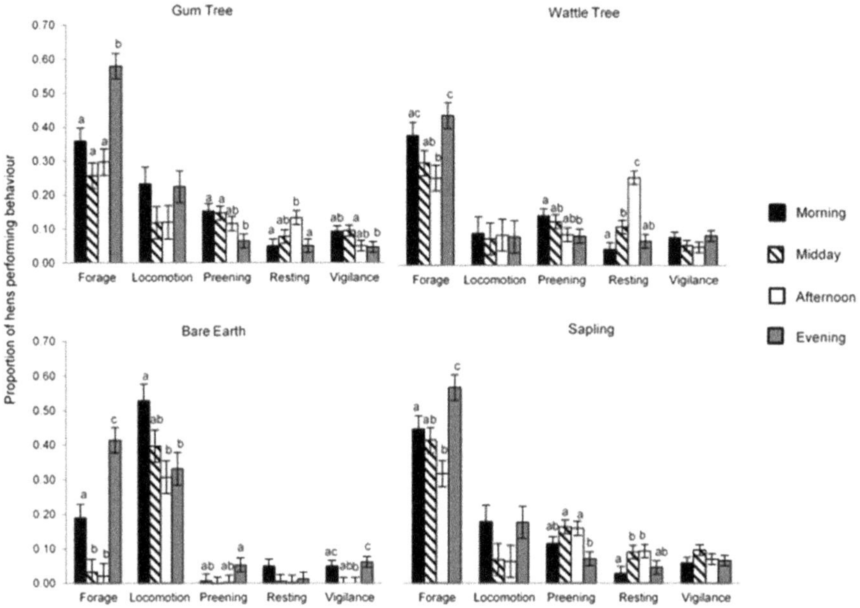

Figure 1 The proportions of hens performing foraging, locomotion, preening, resting and vigilance during the morning (0701–1030 h), midday (1031–1400 h), afternoon (1401–1730 h) and evening (1731–2100 h) in four distinct locations within the range (gum tree, wattle tree, bare earth and sapling) of a commercial farm in Victoria, Australia. Behaviours at each location with different connecting letters differed significantly over time ($P < 0.05$). Reproduced from Larsen et al. (2014).

leafy vegetation. Scanning observations of the numbers of hens in each location and the behaviours they performed showed patches 1 and 2, with the dense canopy cover having higher numbers of hens and with less diurnal variation in hen presence. The five most common behaviours observed were foraging, preening, locomotion, resting and vigilance with more behaviours performed under canopy cover and mostly walking and foraging observed out on the bare earth (Larsen et al., 2014; Fig. 1). These studies indicate that the behavioural repertoire and time budgets of hens on the range depend on both the features provided on the range and how far from the house these are situated.

Finally, in a recent study of six small experimental flocks (of 150 birds each) of 25–28-week-old hens, located in New South Wales, Australia, the mean proportions of birds performing dust bathing, foraging and resting were simultaneously compared between indoors and outdoors (indoor dust bathing 1.69%, outdoor dust bathing 3.53%; indoor foraging 1.51%, outdoor foraging 35.59%; indoor resting 6.94%, outdoor resting 1.37%). The experiment was carried out primarily for investigating impacts of outdoor stocking density and, thus, bird groups had access to different-sized range areas with no shelter provided and with varying levels of grass and weed coverage between the ranges (Campbell et al., 2017). However, across all groups, birds preferred to dust bathe and forage on the range and to rest indoors (Campbell et al., 2016b). The proportions of hens dust bathing both indoors, and overall, were considerably less than those observed in hens on friable litter areas in commercial aviaries (~15–20% of birds on the litter observed dust bathing, Odén et al., 2002, some observations over 90% of birds on the litter observed dust bathing, Campbell et al., 2016c). In contrast, in studies described here (Larsen et al., 2014; Campbell et al., 2016b), proportions of birds foraging (~36–60%) on the range were much greater than previously reported in indoor percheries (5.8–9.1% of birds observed foraging: Carmichael et al., 1999; 5.9% of birds observed foraging: Channing et al., 2001), suggesting that outdoor ranges may allow hens to fulfil their behavioural need to forage better than indoor housing systems, although variation in hen strains and experimental methodology between studies must also be taken into account.

Further behavioural observations across more flocks, specifically in relation to changing ground conditions with different weather patterns (e.g. mud vs dry dirt) or as hens denude range areas in fixed housing, and hens' adaptation to changed surroundings in mobile flocks would aid in the understanding of range use and its value to hens. Finally, hens' use of the range area has implications beyond freedom of behavioural expression. For example, hens staying close to the pop holes can lead to a build-up in parasitic faecal egg counts (Heckendorn et al., 2009), muddied areas with high traffic can track dirt inside and sparsely distributed hens on the outer range may be more vulnerable to predation (Moberly et al., 2004).

3 Mortality, health and correlations with range use

The advent of RFID technology that provides individual tracking data of range use also enables specific individual hen health and welfare variables to be correlated with the use of the range. Individual outdoor-preferring hens had higher body weights than indoor-preferring hens in one study (Singh et al., 2016), but body weight was unrelated or minimally correlated with range use in other studies (Hartcher et al., 2016; Campbell et al., 2016b). Although the same strain of birds was used in all studies, variation in flock size, range size and age of birds may have impacted the results. Outdoor-preferring birds also had improved digestive function possibly through ingestion of stones and grit, but differences in specific characteristics of gut function were dependent on the type of range usage; more and shorter-duration visits resulted in higher gizzard weight than fewer and longer-duration visits (Singh et al., 2016). Hours spent outside have been negatively correlated with toenail length (Campbell et al., 2016b) and positively correlated with plumage condition (Mahboub et al., 2004). However, longer periods out on the grass have also been associated with physiological indicators of increased stress as measured by heterophil:lymphocyte ratios and basophil leucocytes of individual birds (Mahboub et al., 2004). Observations of visually-tagged hens revealed negative correlations between use of the outdoor area and incidence of footpad dermatitis (Rodriguez-Aurrekoetxea and Estevez, 2016). Group-level observations have shown flocks that were observed or reported to have higher range usage had general indicators of improved health (e.g. darker comb colour and better body condition), reduced mortality and better plumage (Whay et al., 2007; Bestman and Wagenaar, 2014). These measures show the time outside ranging may have positive impacts on bird health but free-range flocks also experience many health issues which compromise their welfare.

Free-range/organic laying hens are prone to health problems that are also prevalent in hens housed in other systems including keel damage (Sherwin et al., 2010; Donaldson et al., 2012; Richards et al., 2012; Bestman and Wagenaar, 2014), foot lesions, injurious pecking and ecto/endoparasites (Sherwin et al., 2010; Bestman and Wagenaar, 2014; Hinrichsen et al., 2016a). Comparisons made between housing systems by Sherwin and colleagues (2010) showed free-range hens fared better in some aspects such as plumage damage but worse in others such as the prevalence of vent pecking. However, recent review studies comparing the mortality of laying hens in free-range systems with that of hen mortality in conventional and other non-cage housing do show, on an international scale, that mortality in free-range production is higher (Stokholm et al., 2010; Elson, 2015; Weeks et al., 2016). The causes of high mortality can vary between farms and between regions. For example, one survey of free-range farmers in Australia reported the leading cause of mortality to be predation (42%), followed by cannibalism (39%), heat stress (35%) and grass impaction (27%), where more than one cause of mortality could be listed on the survey (Singh et al., 2015). In studies of 18 organic flocks across Denmark, mortality ranged between 8.6 and 62.3%, with an average of 22.5%. The very high mortality in some of the flocks was partly attributed to an outbreak of *Pasteurella*, but predation and piling (leading to smothering) also caused mortality (Hegelund et al., 2006). Similarly, disease outbreak caused up to 91% mortality in one organic flock in Denmark with major outbreaks in 7 of the total 15 flocks studied (Stokholm et al., 2010). In a further survey of 25 free-range farms across the United Kingdom, farmer-reported mortality up to 70 weeks of age was 1.8–21.4% indicating wide variation in factors affecting separate flocks, although specific causes of mortality were not stated (Whay et al., 2007). Finally, Fossum and colleagues (2009) necropsied hens from various Swedish farms, thus providing more accurate diagnoses than what may be reported in survey questionnaires, and found that free-range hens showed a greater risk of bacterial and parasitic diseases and cannibalism in comparison with hens from cages (Table 2; Fossum et al., 2009). Additionally, the presence of pecking wounds was a predisposing factor to *Escherichia coli* infections (Fossum et al., 2009). Thus, if mortality is an indicator of welfare, free-range laying hens do not fare very well in this respect. The infectious causes may be associated with reduced biosecurity in free-range environments, rapid spread of infections/diseases in large flocks of birds, reduced ability to disinfect between flocks and less environmental control with build-up of wet litter inside or wet mud outside that harbours pathogens. Further research is needed to understand whether free-range hens are potentially more susceptible to pathogens if, for example, they are stressed by changing environmental conditions (e.g. heat and cold stress), stressed by the physiological demands of high production and high ranging

Table 2 Risk analysis regarding main disease categories between laying hen flocks in conventional and enriched cages and flocks in free-range housing with access to outdoor areas, including organic flocks.

Disease category	Observed # of flocks from free-range housing	Expected # of flocks if same risk as in cages	Excess # of flocks in free-range housing compared to cages
Bacterial diseases	17	1	16***
Viral diseases	1	1	0
Parasitic diseases	5	0	5***
Cannibalism	6	0	6***

*** indicates significance at $P < 0.001$. Reproduced from Fossum et al. (2009).

activity, or compromised by high parasite loads. Research into early rearing management strategies that may enhance hen immunity (Moe et al., 2010; Arbona et al., 2011) or preventative strategies in layer flocks is needed to reduce mortality of hens housed in free-range systems.

4 Predators and pathogens

4.1 Predation

Predation is a risk that is specific to free-range housing systems and is a relatively frequently reported cause of mortality (Moberly et al., 2004; Hegelund et al., 2006; Singh et al., 2015; Bestman and Wagenaar, 2014; Bestman and Ouwejan, 2016), although actual losses from predation can vary (e.g. average of 2% flock loss: Moberly et al., 2004; range of 0–14.2% flock loss: Hegelund et al., 2006; up to 487 birds 'missing' from a single flock: Bestman and Ouwejan, 2016). Birds of prey and/or foxes are the most commonly documented threats (Moberly et al., 2004; Bestman and Wagenaar, 2014; Bestman and Ouwejan, 2016), where hens in low densities in the outer parts of the range may be at higher risks of being caught. Recent observations on commercial farms in the Netherlands showed chickens attacked by raptors were typically healthy and attempted to fight back, with bystander chickens also attacking the predator or fleeing (Bestman and Ouwejan, 2016). Individual farms may be at greater risk than others where, within a study across the United Kingdom, up to 1000 birds (11.3%) were reported to be lost to fox predation on one farm, possibly related to management factors (Moberly et al., 2004). However, another study found no effect of flock, weather or farm-based factors in predicting the presence of raptors and attacks (Bestman and Ouwejan, 2016). Observations on-farm showed that free-range hens were unafraid of raptors in close proximity, including raptors eating their prey (Bestman and Ouwejan, 2016) and in an experimental study, birds rapidly returned to the range after recent predator threats (Zeltner and Hirt, 2008b). Predation losses may be reduced by good fencing and provision of shelter on the range, although trees and bushes can also enable raptor attacks by providing perching locations to predators (Bestman and Ouwejan, 2016). In Australia, for example, guard animals such as alpacas and Maremma dogs are permitted as live-in companions with free-range flocks for hen protection. Overall, laying hens are prey animals and outdoor access does make them targets for local predators; thus, more research is needed on possible preventative measures.

4.2 Avian influenza

Among the infectious diseases to which domestic fowl are susceptible when allowed access to an outdoor area is avian influenza (AI). Infection with AI represents not only a health and welfare concern for the affected birds, but also a public health concern, as some strains can infect humans. Once a virus is identified within domestic laying hens, euthanasia is often required to prevent the spread of the infection. However, vaccinations may be applied before flock contamination (Bestman and Wagenaar, 2014). The major outbreak within the United States during 2015 (not limited to free-range farms) highlighted the catastrophic repercussions of infections with a highly pathogenic strain where spread is rapid and mortality typically 100% (Alexander and Capua, 2008; Swayne, 2008). AI viruses, found in wild birds, particularly waterfowl, are spread through contact with infected birds, their respiratory secretions, or excreta. A recent survey of Australian free-range farms reported 52% of producers surveyed had seen wild birds (sparrows and finches) inside the sheds, and on average each farm had at least one water body on which the majority of producers reported seeing wild waterfowl (Scott et al., 2016). Furthermore, data suggest the presence of endemic low pathogenic AI H5 and H7 in Australian wild birds, which have the potential to be introduced to commercial poultry and mutate into highly pathogenic AI (Grillo, 2015). This risk from wild birds is further exacerbated by on-range feeding, which can attract wild birds, with approximately half of Australian producers surveyed reporting this practice (Ruhnke et al., 2015a). Management and environmental conditions on the range, such as feed and open water sources, are important risk factors in the transmission of AI (Feare, 2010). Although hens in all housing systems can be affected by AI and strict biosecurity procedures can minimise risk and spread of infection, it remains that free-range systems are typically less environmentally controlled and permit contact with wild birds.

4.3 Endoparasites

Parasite infection in free-range housing is inevitable when birds have access to ranges or soil surfaces within sheds that have been contaminated by previous batches of birds, by fellow flock mates (Höglund and Jansson, 2011) or by wild bird species. Thus, endoparasites that are an infrequent problem in caged flocks are widespread welfare issues on free-range farms (Jansson et al., 2010). For example, in one report, *Ascaridia galli* was present in 63.8% of sampled birds from free-range/organic systems compared with only 5% of birds sampled from caged systems (Permin et al., 1999). Prevalence in free-range systems is, however, similar to hens housed in enclosed deep-litter barns (*A. galli* 41.9%, Permin et al., 1999), with some studies finding no differences in the prevalence of ascarid infections between free-range/organic and litter-housed hens (Jansson et al., 2010). Other studies of free-range, including organic systems, report a high prevalence of infected flocks or infected birds within flocks (Pennycott and Steel, 2001; Kaufmann et al., 2011; Sherwin et al., 2013; Bestman and Wagenaar, 2014; Hinrichsen et al., 2016b). Necropsies of 749 laying hens from 18 separate organic free-range farms in Germany showed almost all hens (99.5%) were infected with at least one type of helminth parasite with an average of 218.4 worms per hen (Kaufmann et al., 2011). Thapa and colleagues (2015) studied 892 hens across 55 flocks from eight European countries to find that the prevalence of *A. galli* parasites was 69.5%, with an average worm burden of 10 worms per hen. Levels of infection are often underestimated because early larval stages are barely visible and can remain within the intestinal tissues for long periods, whereas generally fewer adult worms are present in the lumen. Interestingly, although endoparasite infection is a major concern when providing hens access to an outdoor range, a negative association has been reported between pasture access time (based only on opening and closing of pop holes) and worm burden (Thapa et al., 2015) and range use (as measured by direct counts of hens on the range) and *Heterakis* faecal egg counts (Sherwin et al., 2013). This relationship may eventuate from reduced levels of excreta indoors and thus a reduced chance of reinfection (Sherwin et al., 2013).

Parasite burden in hens can result in visible behavioural differences such as reduced movement (e.g. walking, running, fluttering and flying) and increased feed intake (Gauly et al., 2007) as well as reduced feed conversion efficiency (Daş et al., 2010) and reduced growth (Gauly et al., 2002). The stimulation of the immune system caused by nematode infections also frequently results in compromised long-term immunocompetence of the affected bird (Hørning et al., 2003). Concurrent infections can also become more severe. For example, *A. galli* is reported to act as a vector for *Salmonella enterica* (Chadfield et al., 2001) and can alter the effects of a concurrent infection with *Pasteurella multocida*, the causative agent of fowl cholera (Dahl et al., 2002). A recent study conducted on 11 free-range layer farms in Denmark demonstrated an association between mortality and level of helminth infection. The mortality was twice as high on farms with a high level of infection (average >200 eggs per gram faeces (EPG)) in the summer compared with farms showing a low level of infection (average ≤200 EPG) (Hinrichsen et al., 2016b). Regular anthelmintic treatments can be effective (Bestman and Wagenaar, 2014), but in some cases they may reduce the hens' own immunity to the parasites (Höglund and Jansson, 2011) and anthelmintic resistance is an increasing concern worldwide (Craig, 1993; Sangster, 1999; Shalaby, 2013). More research is needed on whether birds can be vaccinated for protection against parasitic nematodes (Andersen et al., 2013), or selected for resistance (Gauly et al., 2002).

5 Ranging, nutrition and grass impaction

Globally, commercial brown laying hens genetically selected for intensive indoor cage production are commonly used in free-range systems. Genetic selection criteria for these birds in the past has focused predominantly on productivity and improved feed conversion ratio rather than on other traits, such as sociability, feeding and foraging, or ranging behaviour. Furthermore, traditional feed recommendations were developed based on housing in climate-controlled environments with limited hen movement. Thus, free-range hens can exhibit reduced performance compared to hens housed in conventional cages and poorer body condition. Undernourished hens or hens with an imbalanced nutrient supply are more likely to develop infectious diseases (Gross, 1992; Klasing, 1998). Diet and efficient feed utilisation can also influence hen behaviour where low-efficiency hens showed more walking, pacing, escape and aggressive behaviours than high-efficiency hens that slept and rested more frequently; low-efficiency hens also showed poorer plumage and more agitation (Braastad and Katle, 1989). The nutritional management of free-range birds is therefore critical to their welfare but remains challenging in several ways.

As discussed in previous sections, the freedom of choice that free-range hens experience results in highly diverse range usage between individuals. The energy requirements of birds that range frequently and still maintain daily egg production are thus higher compared to hens that prefer to stay in the barn. Not only is additional energy for increased locomotion required, but exposure to the variable climate between indoors and outdoors has to be taken into account. Active cooling or warming of the body is energy-consuming (Arad and Marder, 1982; Gonyou and Morrison, 1983; Roland et al., 1996). The thermoneutral temperature of a commercial brown laying hen is confined by the lower and upper critical temperatures, 10°C and 25°C (Es et al., 1973), with the ideal humidity at least 40% (Hy-line International, 2014; Institut de Selection de Animale BV, 2014). Consequently, hens on the range are in need of more energy. However, reduced feed intake through overconsumption of pasture, for example, increases the likelihood of reduced body condition and death due to energy deficiency. Nutritional intervention such as adding vitamin E is highly warranted when specific challenges, such as heat stress – a significant cause of mortality in some free-range environments (Singh et al., 2015) – have an impact on the health status of laying hens

(Bollengier-Lee et al., 1998; Lin et al., 2006). Greater-sized ranges of large commercial flocks or very low stocking densities in smaller mobile flocks can increase the distances between the hen and the feed resources available in the shed, which may limit frequent feed access. Subsequently, several feeding strategies have been implemented in commercial free-range enterprises. In some countries, an overall need of an additional 15% metabolisable energy is frequently taken into account when feed formulation is performed (GfE, 1999). However, as this one diet is then fed to the entire flock, the hens that prefer to stay in the barn are oversupplied, leading to animal fattening, which is not ideal from an economic point of view. Choice feeding is another alternative feed strategy that can be implemented. Research has shown that birds are able to balance between their nutrient requirements and nutrient intake if they are given an appropriate choice (Emmans, 1977; Summers and Leeson, 1978). However, in order to ensure the uptake of a balanced diet, certain criteria have to be taken into account such as considering the colour, taste and position of the feed, as well as the number of feed choices (Pousga et al., 2005). Furthermore, the concept of choice feeding has been proven under standardised research conditions, but has not been validated for large commercial flocks. On-range feeding is another alternative feeding strategy applied in some countries (e.g. Australia, Ruhnke et al., 2015a). While this strategy may be beneficial for the hens on the range, the biosecurity risk associated with this practice cannot be overestimated. Feed and water on the range frequently attracts wild birds, which are potential vectors for diseases and may also attract rodents.

Another nutritional welfare issue associated with free-range systems is the impact of access to pasture. In general, the consumption of large amounts of fibrous fodder can reduce the intake of a balanced feed, leading to undernourishment in energy and essential nutrients. Specifically, monogastrics such as chickens have a very limited ability to access fibre as a nutrient source (Fengler and Marquardt, 1988; Choct et al., 1996). The curiosity of hens in combination with the ability to choice feed may result in overconsumption of pasture, severely affecting commercially formulated feed intake (Singh and Cowieson, 2013). Additionally, the quality and quantity of pasture intake can vary significantly among individual hens due to flock behaviour and range design (Walker and Gordon, 2003; Hegelund et al., 2005) affecting overall flock uniformity. Although pasture intake may not be an issue in ranges comprised mostly of bare earth and gravel, in a survey of Australian free-range farms, for example, 82% of layer farmers reported that their range is never stripped of vegetation (Ruhnke et al., 2015c). This high level of pasture intake and the inability to digest large quantities of fibrous material can additionally lead to gut impaction, frequently associated with reduced body condition and production and, in severe cases, death (Ruhnke et al., 2015b,c). Studies looking at the use of multi-enzymes for better nutrient utilisation (Iqbal et al., 2016) or addition of apple cider vinegar to reduce grass impaction (Ruhnke et al., 2015b) are potential strategies to improve the nutrition of free-range laying hens. Additionally, specific nutritional management can reduce the risk of injurious pecking (van Krimpen et al., 2005).

6 Feather pecking and cannibalism

6.1 Injurious pecking

Injurious pecking (gentle and severe feather pecking, vent pecking and cannibalism) are prevalent welfare concerns internationally. These problems are not specific to free-range systems (EFSA, 2005), but they are more difficult to manage in loose housing, where pecking birds are difficult to identify and have access to larger numbers of victims. Injurious pecking can be the leading cause of death in some flocks (Fossum et al., 2009), with mortality up to 36.1% in one study (Stokholm et al., 2010). Interestingly, although feather pecking is highly prevalent in free-range flocks, the more the use of the outdoor range, the better the plumage condition. This is a consistent finding in large commercial flocks irrespective of methodology: group-level farmer reports of percentages of hens using the outdoor areas (Green et al., 2000; Bestman and Wagenaar, 2003, 2014), group-level live observations between flocks of birds (Nicol et al., 2003; Whay et al., 2007; Lambton et al., 2010), individual observations of hen location on commercial farms (Chielo et al., 2016; Rodriguez-Aurrekoetxea and Estevez, 2016) and individual tracking of range use via electronic transponders (Mahboub et al., 2004). However, the relationship between range use and plumage condition has not been found in all studies (Hegelund et al., 2006; Hartcher et al., 2016). Controlled experimental studies have also shown that, although frequencies of general pecking behaviour (i.e. any behaviour that uses the beak) are similar between birds housed in single-tier aviaries with access to an outdoor range and birds in single-tier aviaries with no outdoor access, the frequencies of feather pecking decrease in birds with outdoor access (Shimmura et al., 2008).

The underlying causes, risk factors for and possible methods for prevention and control of injurious pecking are widely studied (Rodenburg et al., 2008; Nicol et al., 2013; Rodenburg et al., 2013), with evidence indicating that pecking is redirected foraging behaviour (Blokhuis, 1986; Huber-Eicher and Wechsler, 1998; Newberry et al., 2007). This is supported by commercial farm observations showing that higher levels of foraging behaviour correlate with reductions in feather pecking (Lambton et al., 2010; Gilani et al., 2013). Varying interrelationships are found between the injurious pecking behaviours with birds dead from cannibalism more likely to show poorer plumage quality (Kjaer and Sørensen, 2002), and the likelihood of both vent pecking and cannibalism increasing with the rate of severe feather pecking (Lambton et al., 2015). However, in some cases the correlation between gentle feather pecking and feather loss is low with feather loss attributed instead to aggression and general feather wear (Whay et al., 2007). Gunnarsson and colleagues (1999) also found that vent pecking could occur independently of feather pecking.

The extent of feather pecking damage is also related to beak trimming, which is a common method for limiting the extent of pecking injury (e.g. Lambton et al., 2010), but trimming creates separate welfare issues for the birds (see Chapter 26). Beak trimming is either banned in several countries (e.g. Sweden; Fossum et al., 2009) or not permitted by certain labels such as 'organic' (e.g. Bestman et al., 2009) or the egg labelling system within Austria, for example (Niebuhr et al., 2005). Therefore, other management strategies for minimising feather pecking, including those that encourage range use, must be implemented. In a survey of 642 Austrian free-range and organic farms with intact-beak hens, the risk of high levels of injurious pecking was reduced by increased farmer-experience (numbers of years of having layers in the hen house) and even distribution of daylight entering the indoor houses, and unsurprisingly was associated with reduced feather cover (Niebuhr et al., 2005). An exploratory study of organic rearing farms within Austria and Germany concluded that feather pecking may be minimised by providing sufficient elevated perch space, adequate drinking places, regular scattering of grain to encourage foraging behaviour and maintaining good, friable litter quality during the laying period (Knierim et al., 2008). Bestman and colleagues (2009) found 90% of flocks showing feather pecking during rearing continued to show feather pecking as layers and suggested that lower stocking density and availability of litter during rearing may reduce the development of feather pecking behaviour. Further work on an experimental farm in Turkey, showed less feather pecking in hens that were permitted 'early' range access at 18 and 20 weeks compared to birds permitted range access at 22 weeks of age (Petek et al., 2015). In a different experimental study, plumage condition at 35 weeks was unaffected by range access at 4 or 16 weeks, although direct measurements of range use were not made (Kjaer and Sørensen, 2002). Furthermore, rates of vent pecking across commercial farms in the United Kingdom decreased with range use at 40 weeks, but increased with range use at 25 weeks (Lambton et al., 2009). In a study of 63 flocks from 26 organic farms within the Netherlands, a higher number of cockerels and a higher percentage of cover outside (thus increasing range usage) were variables found to be associated with lower levels of feather pecking (Bestman and Wagenaar, 2003). Similarly, a summary of data on plumage damage from over 1000 flocks within the United Kingdom also found a higher percentage of canopy cover on the range resulted in less plumage damage at end of lay (Bright et al., 2011). Finally, evidence-based management factors have been identified that can reduce levels of feather pecking and these have been applied successfully on commercial farms within the United Kingdom as will be presented in the following case study.

6.2 Case study: feather pecking management

Injurious pecking in free-range systems has a number of implications for the industry. As well as being an important cause of mortality, as discussed above, it is more widely symptomatic of reduced welfare in the flock, causing pain to victims and indicating shortcomings in the environment of the pecking bird (Gentle and Hunter, 1991; Weeks and Nicol, 2006). Furthermore, the associated plumage damage is correlated with higher feed consumption and lower feed conversion rates (Herremans et al., 1989; Leeson and Morrison, 1978). Thus, the overall effect of injurious pecking both undermines the ethos of free-range housing systems, that of improved welfare for their inhabitants, and could also significantly reduce the productivity of those systems (El-Lethey et al., 2000; Huber-Eicher and Sebo, 2001; Nicol et al., 2013). Although beak trimming may help to manage the problem, the procedure presents welfare problems of its own, as stated above (and discussed in Chapter 26). Consequently, in the last 20–30 years, much research has focused on the causes of, and risk factors for, these behaviours, reviewed by Rodenburg et al. (2013), in the hope of providing the industry with the information necessary to tackle the problem. Many factors have been associated with increased rate or risk of injurious pecking, including high stocking densities (Bilčík and Keeling, 2000; Hansen and Braastad, 1994; Savory, et al., 1999); diets low in insoluble fibre (Hetland et al., 2003; Steenfeldt et al., 2007); lack of access to litter at an early age (Gunnarsson et al., 1999), poor range use, frequent diet changes, use of bell drinkers, and litter restriction (Green et al., 2000; Nicol et al., 2003); feeding mash rather than pelleted feed, spreading feed on the floor, high temperatures inside the laying house (Lambton et al., 2010) and use of lighting in nest boxes (Pötzsch et al., 2001); and counter-intuitively, providing more aerial perches, or perches >50 cm in height, and more and/or longer pop holes (Lambton et al., 2015 but see Lambton et al., 2010; Green et al., 2000).

Despite the volume of research, the problem remains prevalent (e.g. Lambton et al., 2010, 2015). Its persistence may, in part, be due to the fact that it is multifactorial, so where generic advice is provided to farmers it is difficult for them to identify which are the most important risk factors to tackle for their flock. With this in mind, a team at the University of Bristol in the United Kingdom set out to test a package of practicable management strategies. Researchers performed a systematic review of over 300 publications to identify risk factors for injurious pecking. Risk factors supported by robust scientific evidence were identified and discussed with stakeholders from the industry, government, welfare charities, retailers, veterinarians and academics in order to identify practical management strategies, which could be used on-farm to mitigate the risks of injurious pecking. Forty-six management strategies were identified. The research team then recruited 53 treatment farms in the United Kingdom; these farms were visited by a member of the team and a tailored list of management strategies was drawn up to tackle risk factors they identified on-farm. Treatment farms were followed throughout the laying period with visits at 20, 30 and 40 weeks of age (±3 weeks), when data on injurious pecking and mortality were collected. Simultaneously, 47 control farms were also visited, but not provided with a bespoke management package. Although they were free to employ any management strategies they wished, they were not given any guidance or information by the research team.

By the end of the study (reported by Lambton et al., 2013), the treatment flocks on average, employed more than three more management strategies (from the list created by the research team) than the control flocks. Treatment flocks showed less plumage damage, with lower plumage scores and 10% fewer birds affected by plumage damage at 20 weeks of age. Treatment flocks also had lower rates of severe feather pecking, with a tendency towards lower rates of gentle feather pecking and vent pecking. Perhaps more interestingly, the analyses showed that, irrespective of treatment or control status, as the number of management strategies employed increased: plumage damage decreased, rates of gentle and severe feather pecking decreased, likelihood of vent pecking decreased and cumulative mortality at 40 weeks decreased. Thus, it seems that, when presented in a form tailored to meet the specific needs of the farm, the outcomes of more than 20 years of research can be translated into real improvements in free-range laying hen welfare, with the potential to improve productivity. Such findings are gradually being disseminated; for example, the FeatherWel website, which arose from the project, has been viewed internationally, and many of the strategies are being incorporated in assurance schemes and guidelines worldwide.

7 Summary

The poultry industry is an example of what can be achieved through application of science. Modern laying hens kept in houses of several thousand birds can lay almost an egg a day, producing some 20 kg of eggs from around 44 kg of carefully formulated compound feed. Compare this with the seasonal production of a few dozen eggs from small flocks in mobile housing and fed farmyard scraps that was the norm until the 1950s. Advances in scientific understanding of genetics, nutrition, light and thermal environment have all been applied, leading to an incredible increase in productivity. In terms of feed conversion efficiency, egg production is among the most efficient of animal-derived foods for humans. In early laying hen production, a greater understanding of disease-causing agents and methods of transmission led to caged housing systems that separated birds from their faeces, alongside the development of vaccines. Yet now, with the recognition of the hen as a sentient being with behavioural needs, cages are increasingly viewed as an unacceptable way of keeping hens by society as a whole. Moreover, the expanding human population not only increases the demand for eggs but also drives a need for sustainable production in the environmental sense. It becomes imperative to reduce wastage, including from high levels of mortality, and, thus the uptake of science, such as measures to reduce injurious pecking (case study), is of increasing importance. Although farmers may adopt practices that clearly lead to greater profitability, there tends to be resistance to changing practices where the benefit is not clear-cut and the techniques are time-consuming (Palczynski et al., 2016). Thus, some of the current research is towards incorporating social science to better understand the expectations of society and also towards improving knowledge sharing and stimulating the uptake of better practices among farmers. For example, the EU Hennovation project is examining whether grassroots-led initiatives lead to improved uptake of science-based knowledge and a greater development and sharing of innovative approaches. Evidence is emerging of the value of including outcome measures in assurance schemes: Mullan and colleagues (2016) have reported improved plumage in free-range flocks after incorporating feather cover as an outcome measure in two United Kingdom assurance schemes and farmers have been encouraged to monitor their own flocks.

Well-managed free-range units can achieve good hen health and welfare with a reduced carbon footprint, particularly if alternative sources of protein from insects or home-grown crops can be utilised. Current research is directed at achieving even more sustainable and welfare-friendly production by facilitating the uptake of evidence-based techniques and knowledge as well as advancing the underpinning science. The contribution and scale of village flocks to food security and the local economy of many developing countries should not be overlooked (Alders and Pym, 2009). Here, some light-touch science has led to significant increases in production by, for example, crossing modern hybrids with native breeds, introducing extended lighting, and developing efficient incubation techniques, cost-effective dietary supplements and vaccines against diseases such as Newcastle disease.

8 Future trends in research

With increasing consumer demand for free-range eggs in the context of an expanding human population, production needs to be sustainable. In particular, future research needs to address the principal causes of high mortality associated with the system. Studies may also develop new approaches to knowledge sharing in order to improve productivity, health and welfare outcomes. In this chapter, we have discussed current welfare issues that are either specific to or highly prevalent in current commercial free-range systems. We highlight here for future attention, a few research areas that are trending.

RFID technology is gradually being applied to the laying hen sector, and these techniques, along with other types of electronic sensors (Bosco et al., 2010; Quwaider et al., 2010; Nakarmi et al., 2014), are able to provide detailed information of individual hen space use and behaviour while living in large flocks (Siegford et al., 2016). This opens up a realm of possibilities in correlating health variables with known behavioural patterns and having a more accurate understanding of daily patterns and the variability of individual hens. The challenge will then be towards developing systems that can provide good welfare on an individual as well as a flock level. The application of electronic technology in commercial flocks can provide automated, remote monitoring, long-term data collection and hen tracking that otherwise would not be feasible through direct observation. The data provided through electronic monitoring may help unravel the relationship between range use and bird health to determine whether exposure to fresh air, variable weather, long distances to traverse and a supplemented diet improve or hinder the welfare of current domesticated laying hen strains.

Also poorly understood is the influence of social dynamics within the often very large numbers of birds that are housed together in free-range systems. Laying hens are thought to not form a hierarchy in groups of 120 individuals or more (compared to groups of 10 individuals: D'Eath and Keeling, 2003), and within large indoor flocks, this may reduce aggression between birds (Nicol et al., 1999). But currently there are few data available on how social interactions may influence range use. Birds may be socially motivated to go outside, dominant hens may occupy the pop holes, subgroups may form between individuals staying indoors versus individuals preferring daily outdoor access, or subgroups may form in different areas of the range as previously found within large flocks housed in indoor tiered aviary systems (Odén et al., 2000). Chielo and colleagues (2016) did find an inverse relationship between hen numbers and nearest-neighbour distances, indicating that hens prefer to spread out when more space is available, but groups may also cluster in highly preferred areas of the range (Larsen et al., 2014). Social dynamics can play a role in the transmission of feather pecking (Zeltner et al., 2000; McAdie and Keeling, 2002), the transmission of fear and stress within hen groups (de Haas et al., 2012) and social facilitation of dust bathing behaviour (Duncan et al., 1998). Thus, range access patterns including using available range space may also be influenced by group-level dynamics and this warrants further research.

Management factors continue to play a vital role in free-range systems. Further research is warranted to determine whether there is a certain amount of time a bird 'should' spend outdoors each day for optimal health and welfare, what is the ideal time for first opening of the pop holes (Lambton et al., 2009; Petek et al., 2015) and how birds are impacted when range access is removed for extended periods of time during disease outbreak or inclement weather (Bestman and Wagenaar, 2003, 2014). Additionally, management interventions within the rearing environment, such as providing enrichment, may better prepare birds for successfully using an outdoor environment, through for example, reducing fear levels (Reed et al., 1993; Grigor et al., 1995) or improving spatial cognition (Krause et al., 2006). The effects of highly variable light levels between indoor and outdoor environments on range use and injurious pecking are also poorly understood. Furthermore, breeding selection for strains of birds showing improved adaptation to environmental change may be better suited to outdoor housing, thus improving hen use of the range.

Finally, stricter environmental control may become a necessity for protection of and maintaining good welfare for large numbers of birds, and thus other loose-housed systems that provide both a choice of environments, natural light and ventilation may be viable alternative options. These systems could be in the form of enclosed winter gardens or new designs that meet consumer preferences while providing controlled housing such as the Rondeel™ system (van Niekerk and Reuvekamp, 2011; Elson, 2015). However, with any system, research would still be necessary to determine best management practices for ensuring sustainable production and optimal hen welfare.

9 Where to look for further information

In addition to the literature cited in this chapter, a good review article for non-specialist poultry readers is by Weeks, et al. (2011), *Managing Health Risks in Free Range Laying Hens*, a Morrisons Farming Programme publication, and the Poultry Science symposium book edited by Perry (2004), *Welfare of the Laying Hen*. Recommendations of key societies, organisations and websites for the latest information would include the following:

- World's Poultry Science Association (www.wpsa.com) organises regular national and international poultry conferences and publishes scientific reviews in the *World's Poultry Science Journal*.
- LayWel (www.laywel.eu) was an EU project that examined the welfare and productivity of various housing systems for laying hens with particular emphasis on welfare.
- AssureWel (www.assurewel.org) incorporates animal welfare outcomes in assurance schemes and has a useful page on feather scoring.
- FeatherWel (www.featherwel.org) offers practical strategies for managing injurious pecking that have been tested in practice but are derived from scientific evidence.
- Australian Poultry Science Symposium hosts annual conferences presenting a wide range of free-range and other poultry research.
- PoultryHub in Australia (www.poultryhub.org) based at the University of New England, Armidale, NSW, provides up-to-date scientific data on multiple aspects of free-range and other poultry systems.

Additional international research projects include Hennovation (www.hennovation.eu), which is active in five European countries investigating the effectiveness of developing and sharing innovative ideas using a new model of grassroots networking for knowledge exchange. Feather pecking is a main topic also included in the project wiki at www.henhub.eu.

Top research centres for future research and potential collaboration include the *Animal Welfare and Behaviour Group*, University of Bristol, United Kingdom; *Wageningen UR Livestock Research* and *Louis Bolk Institute*, the Netherlands; and *Animal Welfare Science Centre*, University of Melbourne, Australia.

10 References

Alders, R. G. and Pym, R. A. E. 2009. Village poultry: Still important to millions eight thousand years after domestication. *World's Poultry Science Journal*, 65, 181–90.

Alexander, D. J. and Capua, I. 2008. Avian influenza in poultry. *World's Poultry Science Journal*, 64, 513–32.

Andersen, J. P., Norup, L. R., Dalgaard, T. S., Rothwell, L., Kaiser, P., Permin, A., Schou, T. W., Fink, D. R., Jungersen, G., Sørensen, P. and Juul-Madsen, H. R. 2013. No protection in chickens immunized by the oral or intra-muscular immunization route with *Ascaridi galli* soluble antigen. *Avian Pathology*, 42, 276–82.

Arad, Z. and Marder, J. 1982. Comparison of the productive performances of the Sinai Bedouin fowl, the White Leghorn and their crossbreds: Study under natural desert conditions. *British Poultry Science*, 23, 333–8.

Arbona, D. V., Anderson, K. E. and Hoffman, J. B. 2011. A comparison of humoral immune function in response to a killed Newcastle's vaccine challenge in caged vs. free-range Hy-line brown layers. *International Journal of Poultry Science*, 10, 315–19.

Bestman, M. W. P. and Wagenaar, J-P. 2003. Farm level factors associated with feather pecking in organic laying hens. *Livestock Production Science*, 80, 133–40.

Bestman, M. W. P., Koene, P. and Wagenaar, J-P. 2009. Influence of farm factors on the occurrence of feather pecking in organic reared hens and their predictability for feather pecking in the laying period. *Applied Animal Behaviour Science*, 121, 12–125.

Bestman, M. W. P. and Wagenaar, J-P. 2014. Health and welfare in Dutch organic laying hens. *Animals*, 4, 374–90.

Bestman, M. W. P. and Ouwejan, J. 2016. Predation of free-range laying hens. *Proceedings of the 50th Congress of the International Society for Applied Ethology*, 12–15 July, Edinburgh, UK, p. 283.

Bilčík, B. and Keeling, L. J. 2000. Relationship between feather pecking and ground pecking in laying hens and the effect of group size. *Applied Animal Behaviour Science*, 68, 55–66.

Blokhuis, H. J. 1986. Feather-pecking in poultry: its relation with ground-pecking. *Applied Animal Behaviour Science*, 16, 63–7.

Bollengier-Lee, S., Mitchell, M. A., Utomo, D. B., Williams, P. E. V. and Whitehead, C. C. 1998. Influence of high dietary vitamin E supplementation on egg production and plasma characteristics in hens subjected to heat stress. *British Poultry Science*, 39, 106–12.

Bosco, A. D., Mugnai, C., Sirri, F., Zamparini, C. and Castellini, C. 2010. Assessment of a global positioning system to evaluate activities of organic chickens at pasture. *Journal of Applied Poultry Research*, 19, 213–18.

Braastad, B. O. and Katle, J. 1989. Behavioural differences between laying hen populations selected for high and low efficiency of food utilisation. *British Poultry Science*, 30, 533–44.

Bright, A., Brass, D., Clachan, J., Drake, K. A. and Joret, A. D. 2011. Canopy cover is correlated with reduced injurious feather pecking in commercial flocks of free-range laying hens. *Animal Welfare*, 20, 329–38.

Campbell, D. L. M., Hinch, G. N., Dyall, T. R., Warin, L., Little, B. A. and Lee, C. 2017. Outdoor stocking density in free-range laying hens: radio-frequency identification of impacts on range use. *Animal*, 11, 121–30.

Campbell, D. L. M., Hinch, G. N., Downing, J. A. and Lee, C. 2016a. Fear and coping styles of outdoor-preferring, moderate-outdoor and indoor-preferring free-range laying hens. *Applied Animal Behaviour Science*, 185, 73–7.

Campbell, D. L. M., Hinch, G. N., Downing, J. A. and Lee, C. 2016b. Outdoor stocking density in free-range laying hens: effects on behaviour and welfare. *Animal*, in press, doi:10.1017/S1751731116002342.

Campbell, D. L. M., Makagon, M. M., Swanson, J. C. and Siegford, J. M. 2016c. Litter use by laying hens in a commercial aviary: dust bathing and piling. *Poultry Science*, 95, 164–75.

Carmichael, N. J., Walker, A. W. and Hughes, B. O. 1999. Laying hens in large flocks in a perchery system: influence of stocking density on location, use of resources and behaviour. *British Poultry Science*, 40, 165–76.

Chadfield, M., Permin, A., Nansen, P. and Bisgaard, M. 2001. Investigation of the parasitic nematode *Ascaridia galli* (Shrank 1788) as a potential vector for *Salmonella enterica* dissemination in poultry. *Parasitology Research*, 87, 317–25.

Channing, C. E., Hughes, B. O. and Walker, A. W. 2001. Spatial distribution and behaviour of laying hens housed in an alternative system. *Applied Animal Behaviour Science*, 72, 335–45.

Chielo, L. I., Pike, T. and Cooper, J. 2016. Ranging behaviour of commercial free-range laying hens. *Animals*, 6, 28, doi:10.3390/ani6050028.

Choct, M., Hughes, R. J., Wang, J., Bedford, M. R., Morgan, A. J. and Annison, G. 1996. Increased small intestinal fermentation is partly responsible for the anti-nutritive activity of non-starch polysaccharides in chickens. *British Poultry Science*, 37, 609–21.

Craig, T. M. 1993. Anthelmintic resistance. *Veterinary Parasitology*, 46, 121–31.

Dahl, C., Permin, A., Christensen, J. P., Bisgaard, M., Muhairwa, A. P., Petersen, K. M. D., Poulsen, J. S. D. and Jensen, A. L. 2002. The effect of concurrent infections with *Pasteurella multocida* and *Ascaridia galli* on free range chickens. *Veterinary Microbiology*, 86, 313–24.

Daş, G., Kaufmann, F., Abel, H. and Gauly, M. 2010. Effect of extra dietary lysine in *Ascaridia galli*-infected grower layers. *Veterinary Parasitology*, 170, 238–43.

Dawkins, M. S. 1989. Time budgets in red junglefowl as a baseline for the assessment of welfare in domestic fowl. *Applied Animal Behaviour Science*, 24, 77–80.

de Haas, E. N., Kops, M. S., Bolhuis, J. E., Groothuis, T. G. G., Ellen, E. D. and Rodenburg, T. B. 2012. The relation between fearfulness in young and stress-response in adult laying hens, on individual and group level. *Physiology & Behavior*, 107, 433–9.

D'Eath, R. B. and Keeling, L. J. 2003. Social discrimination and aggression by laying hens in large groups: from peck orders to social tolerance. *Applied Animal Behaviour Science*, 84, 197–212.

Donaldson, C. J., Ball, M. E. E. and O'Connell, N. E. 2012. Aerial perches and free-range laying hens: The effect of access to aerial perches and of individual bird parameters on keel bone injuries in commercial free-range laying hens. *Poultry Science*, 91, 304–15.

Duncan, I. J. H., Widowski, T. M., Malleau, A. E., Lindberg, A. C. and Petherick, J. C. 1998. External factors and causation of dustbathing in domestic hens. *Behavioural Processes*, 43, 219–28.

EFSA, 2005. Opinion of the Scientific Panel on Animal Health and Welfare on a request from the Commission related to the welfare aspects of various systems of keeping laying hens. *The EFSA Journal*, 127, 1027.

El-Lethey, H., Aerni, V., Jungi, T. W. and Wechsler, B. 2000. Stress and feather pecking in laying hens in relation to housing conditions. *British Poultry Science*, 41, 22–8.

Elson, H. A. 2015. Poultry welfare in intensive and extensive production systems. *World's Poultry Science Journal*, 71, 449–59.

Emmans, G. C. 1977. The nutrient intake of laying hens given a choice of diets, in relation to their production requirements. *British Poultry Science*, 18, 227–36.

Es, A. V., Aggelen, D. V., Nijkamp, H. J., Vogt, J. E. and Scheele, C. W. 1973. Thermoneutral zone of laying hens kept in batteries. *Journal of Animal Physiology and Animal Nutrition*, 32, 121–9.

Feare, C. J. 2010. Role of wild birds in the spread of highly pathogenic avian influenza virus H5N1 and implications for global surveillance. *Avian Diseases*, 54, 201–12.

Fengler, A. I. and Marquardt, R. R. 1988. Water-soluble pentosans from rye: II. Effects on rate of dialysis and on the retention of nutrients by the chick. *Cereal Chemistry*, 65, 298–302.

Fossum, O., Jansson, D. S., Etterlin, P. E. and Vågsholm, I. 2009. Causes of mortality in laying hens in different housing systems in 2001 to 2004. *Acta Veterinaria Scandinavica*, 51, doi: 10.1186/1751-0147-51-3.

Gauly, M., Bauer, C., Preisinger, R. and Erhardt, G. 2002. Genetic differences of *Ascaridia galli* egg output in laying hens following a single dose infection. *Veterinary Parasitology*, 99–107.

Gauly, M., Duss, C. and Erhardt, G. 2007. Influence of *Ascaridia galli* infections and anthelmintic treatments on the behaviour and social ranks of laying hens (*Gallus gallus domesticus*). *Veterinary Parasitology*, 146, 271–80.

Gebhardt-Henrich, S. G., Toscano, M. J. and Fröhlich, E. K. F. 2014. Use of outdoor ranges by laying hens in different sized flocks. *Applied Animal Behaviour Science*, 155, 74–81.

Gentle, M. J. and Hunter, L. N. 1991. Physiological and behavioral-responses associated with feather removal in *Gallus-Gallus Var Domesticus*. *Research in Veterinary Science* 50, 95–101.

GfE, 1999. Empfehlungen zur Energie- und Nährstoffversorgung der Legehennen und Masthühner (Broiler) [German]; Recommendations for the energy and nutrient supply of laying hens and broilers. DLG-Verlagsgesellschaft GmBh, Frankfurt am Main, Germany.

Gilani, A-M., Knowles, T. G. and Nicol, C. J. 2013. The effect of rearing environment on feather pecking in young and adult laying hens. *Applied Animal Behaviour Science*, 148, 54–63.

Gonyou, H. W. and Morrison, W. D. 1983. Effects of defeathering and insulative jackets on production by laying hens at low temperatures. *British Poultry Science*, 24, 311–17.

Green, L. E., Lewis, K., Kimpton, A. and Nicol, C. J. 2000. Cross-sectional study of the prevalence of feather pecking in laying hens in alternative systems and its association with management and disease. *Veterinary Record*, 147, 233–8.

Grigor, P. N., Hughes, B. O. and Appleby, M. C. 1995. Effects of regular handling and exposure to an outside area on subsequent fearfulness and dispersal in domestic hens. *Applied Animal Behaviour Science*, 44, 47–55.

Grillo, T. 2015. Contribution to the 2012 Avian Influenza in Wild Birds Surveillance Program. RIRDC Publication No 15/016 RIRDC Project No PRJ-008337, https://rirdc.infoservices.com.au/items/15-016. Accessed 22 July 2016.

Gross, W. B. 1992. Effects of ascorbic acid on stress and disease in chickens. *Avian Diseases*, 688–92.

Gunnarsson, S., Keeling, L. J. and Svedberg, J. 1999. Effect of rearing factors on the prevalence of floor eggs, cloacal cannibalism and feather pecking in commercial flocks of loose housed laying hens. *British Poultry Science* 40, 12–18.

Hansen, I. and Braastad, B. O. 1994. Effect of rearing density on pecking behavior and plumage condition of laying hens in 2 types of aviary. *Applied Animal Behaviour Science*, 40, 263–72.

Harlander-Matauschek, A., Felsenstein, K., Niebuhr, K. and Troxier, J. 2001. The effect of group size on locomotion of laying hens in outdoor runs. *Proceedings of the 6th European Symposium on Poultry Welfare*, 1–4 September, Zollikofen, Switzerland.

Hartcher, K. M., Hickey, K. A., Hemsworth, P. H., Cronin, G. M., Wilkinson, S. J. and Singh, M. 2016. Relationships between range access as monitored by radio frequency identification technology, fearfulness, and plumage damage in free-range laying hens. *Animal*, 10, 847–53.

Heckendorn, F., Häring, D. A., Amsler, Z. and Maurer, V. 2009. Do stocking rate and a simple run management practice influence the infection of laying hens with gastrointestinal helminths? *Veterinary Parasitology*, 159, 60–8.

Hegelund, L., Sørensen, J. T., Kjaer, J. B. and Kristensen, I. S. 2005. Use of the range area in organic egg production systems: effect of climatic factors, flock size, age and artificial cover. *British Poultry Science*, 46, 1–8.

Hegelund, L., Sørensen, J. T. and Hermansen, J. E. 2006. Welfare and productivity of laying hens in commercial organic egg production systems in Denmark. *NJAS – Wageningen Journal of Life Science*, 54, 147–55.

Hernandez, C. E., Lee, C., Ferguson, D., Dyall, T., Belson, S., Lea, J. and Hinch, G. 2014. Personality traits of high, low and non-users of a free range area in laying hens. *Proceedings of the 48th Congress of the International Society for Applied Ethology*, 29 July to 2 August, Vitoria-Gasteiz, Spain, p. 89.

Herremans, M., Decuypere, E. and Siau, O. 1989. Effects of feather wear and temperature on predition of food intake and residual food consumption. *British Poultry Science*, 30, 15–22.

Hetland, H., Svihus, B. and Moe, R. 2003. Effect of feed structure on performance and welfare in laying hens housed in conventional and furnished cages. *Acta Agriculturae Scandinavica*, 53, 92–100.

Hinrichsen, L. K., Riber, A. B. and Labouriau, R. 2016a. Associations between and development of welfare indicators in organic layers. *Animal*, 10, 953–60.

Hinrichsen, L. K., Labouriau, R., Engberg, R. M., Knierim, U. and Sørensen, J. T. 2016b. Helminth infection is associated with hen mortality in Danish organic egg production. *Veterinary Record*, doi:10.1136/vr.103614.

Höglund, J. and Jansson, D. S. 2011. Infection dynamics of *Ascaridia galli* in non-caged laying hens. *Veterinary Parasitology*, 180, 267–73.

Hørning, G., Rasmussen, S., Permin, A. and Bisgaard, M. 2003. Investigations on the influence of Helminth parasites on vaccination of chickens against Newcastle Disease virus under village conditions. *Tropical Animal Health and Production*, 35, 415–24.

Huber-Eicher, B. and Wechsler, B. 1998. The effect of quality and availability of foraging materials on feather pecking in laying hen chicks. *Applied Animal Behaviour Science*, 55, 861–73.

Huber-Eicher, B. and Sebo, F. 2001. The prevalence of feather pecking and development in commercial flocks of laying hens. *Applied Animal Behaviour Science*, 74, 223–31.

Hy-line International 2014. Hy-line Red Book – an online management guide. Hy-line International, Iowa, USA, www.hylineredbook.com.

Iqbal, Z., Sharma, N., Sharma, N. K., M'Sadeq, S., Perez-Marinaldo, R. A., Ramirez-Cuevas, S., Roberts, J., Hilliar, M., Singh, M., Wu, S., Swick, R. A. and Ruhnke, I. 2016. Effect of pasture and feed additives on performance and egg quality in ranging laying hens. *Proceedings of the 27th Australian Poultry Science Symposium*, 14–17 February, Sydney, NSW, Australia, p. 145.

Jansson, D. S., Nyman, A., Vågsholm, I., Christensson, D., Göransson, M., Fossum, O. and Höglund, J. 2010. Ascarid infections in laying hens kept in different housing systems. *Avian Pathology*, 39, 525–32, doi:10.1080/03079457.2010.527923.

Kaufman, F., Daş, G., Sohnrey, B. and Gauly, M. 2011. Helminth infections in laying hens kept in organic free range systems in Germany. *Livestock Science*, 141, 182–7.

Kjaer, J. B. and Sørensen, P. 2002. Feather pecking and cannibalism in free-range laying hens as affected by genotype, dietary level of methionine + cystine, light intensity during rearing and age at first access to the range area. *Applied Animal Behaviour Science*, 76, 21–39.

Klasing, K. C. 1998. Nutritional modulation of resistance to infectious diseases. *Poultry Science*, 77, 1119–25.

Knierim, U., Staack, M., Gruber, B., Keppler, C., Zaludik, K. and Niebuhr, K. 2008. Risk factors for feather pecking in organic laying hens – starting points for prevention in the housing environment. *16th IFOAM Organic World Congress*, Modena, Italy, 16–20 June 2008, available at http://orgprints.org/11775 (accessed June 2016).

Krause, E. T., Naguib, M., Trillmich, F. and Schrader, L. 2006. The effects of short term enrichment on learning in chickens from a laying strain (*Gallus gallus domesticus*). *Applied Animal Behaviour Science*, 101, 318–27.

Lambton, S. L., Knowles, T. G., Yorke, C. and Nicol, C. J. 2009. Risk factors affecting the development of vent pecking and cannibalism in loose housed laying hens flocks. *Proceedings of the 8th European Symposium on Poultry Welfare*, 18–22 May, Cervia, Italy, p. 13.

Lambton, S. L., Knowles, T. G., Yorke, C., Nicol, C. J. 2010. The risk factors affecting the development of gentle and severe feather pecking in loose housed laying hens. *Applied Animal Behaviour Science*, 123, 32–42.

Lambton, S. L., Nicol, C. J., Friel, M., Main, D. C. J., McKinstry, J. L., Sherwin, C. M., Walton, J. and Weeks, C. A. 2013. A bespoke management package can reduce levels of injurious pecking in loose-housed laying hen flocks. *Veterinary Record*, 172, 423. doi:10.1136/vr.101067.

Lambton, S. L., Knowles, T. G., Yorke, C. and Nicol, C. J. 2015. The risk factors affecting the development of vent pecking and cannibalism in free-range and organic laying hens. *Animal Welfare*, 24, 101–11.

Larsen, H., Cronin, G., Hemsworth, P., Smith, C. and Rault, J-L. 2014. Behaviour of free-range laying hens in distinct outdoor environments. *Proceedings of the 48th Congress of the International Society for Applied Ethology*, 29 July to 2 August, Vitoria-Gasteiz, Spain, p. 186.

Larsen, H., Cronin, G., Hemsworth, P., Smith, C., Rault, J-L. 2016. Fearfulness and access to the outdoor range in commercial free-range laying hens. *Proceedings of the 50th Congress of the International Society for Applied Ethology*, 12–15 July, Edinburgh, UK, p. 143.

Leeson, S. and Morrison, W. D. 1978. Effect of feather cover on feed-efficiency in laying birds. *Poultry Science*, 57, 1094–6.

Lin, H., Jiao, H. C., Buyse, J. and Decuypere, E. 2006. Strategies for preventing heat stress in poultry. *World's Poultry Science Journal*, 62, 71–86.

Mahboub, H. D. H., Müller, J. and van Borell, E. 2004. Outdoor use, tonic immobility, heterophil/lymphocyte ratio and feather condition in free-range laying hens of different genotype. *British Poultry Science*, 45, 738–44.

McAdie, T. M. and Keeling, L. J. 2002. The social transmission of feather pecking in laying hens: effects of environment and age. *Applied Animal Behaviour Science*, 75, 147–59.

Moberly, R. L., White, P. C. L. and Harris, S. 2004. Mortality due to fox predation in free-range poultry flocks in Britain. *Veterinary Record*, 155, 48–52.

Moe, R. O., Guémené, D., Bakken, M., Larsen, H. J. S., Shini, S., Lervik, S., Skjerve, E., Michel, V. and Tauson, R. 2010. Effects of housing condition during the rearing and laying period on adrenal reactivity, immune response and heterophil to lymphocyte (H/L) ratios in laying hens. *Animal*, 4, 1709–15.

Mullan, S., Szmaragd, C., Cooper, M. D., Wrathall, J. H. M., Jamieson, J., Bond, A., Atkinson, C. and Main, D. C. J. 2016. Animal welfare initiatives improve feather cover of cage-free laying hens in the UK. *Animal Welfare*, 25, 243–53.

Nagle, T. A. D. and Glatz, P. C. 2012. Free range hens use the range more when the outdoor environment is enriched. *Asian-Australasian Journal of Animal Sciences*, 25, 584–91.

Nakarmi, A. D., Tang, L. and Xin, H. 2014. Automated tracking and behaviour quantification of laying hens using 3D computer vision and radio frequency identification technologies. *Transactions of the ASABE*, 57, 1455.

Newberry, R. C., Keeling, L. J., Estevez, I. and Bilčík, B. 2007. Behaviour when young as a predictor of severe feather pecking in adult laying hens: The redirected foraging hypothesis revisited. *Applied Animal Behaviour Science*, 107, 262–74.

Nicol, C. J., Gregory, N. G., Knowles, T. G., Parkman, I. D. and Wilkins, L. J. 1999. Differential effects of increased stocking density mediated by increased flock size, on feather pecking and aggression in laying hens. *Applied Animal Behaviour Science*, 65, 137–52.

Nicol, C. J., Pötzsch, C., Lewis, K. and Green, L. E. 2003. Matched concurrent case-control study of risk factors for feather pecking in hens on free-range commercial farms in the UK. *British Poultry Science*, 44, 515–23.

Nicol, C. J., Caplen, G., Edgar, J. and Browne, W. J. 2009. Associations between welfare indicators and environmental choice in laying hens. *Animal Behaviour*, 78, 413–24.

Nicol, C. J., Bestman, M., Gilani, A-M., de Haas, E. N., de Jong, I. C., Lambton, S., Wagenaar, J. P., Weeks, C. A. and Rodenburg, T. B. 2013. The prevention and control of feather pecking: application to commercial systems. *World's Poultry Science Journal*, 69, 775–8.

Niebuhr, K., Zaludik, K., Baumung, R., Lugmair, A. and Troxler, J. 2005. Injurious pecking in free-range and organic laying hen flocks in Austria. Animal Science Papers and Reports 23, Supplement 1, 195–201. *Presented at the 7th European Symposium on Poultry Welfare*, 15–19 June, Lublin, Poland.

Odén, K., Vestergaard, K. S. and Algers, B. 2000. Space use and agonistic behaviour in relation to sex composition in large flocks of laying hens. *Applied Animal Behaviour Science*, 67, 307–20.

Odén, K., Keeling, L. J. and Algers, B. 2002. Behaviour of laying hens in two types of aviary systems on 25 commercial farms in Sweden. *British Poultry Science*, 43, 169–81.

Palczynski, L. J., Buller, H., Lambton, S. L. and Weeks, C. A. 2016. Farmer attitudes to injurious pecking in laying hens and to potential control strategies. *Animal Welfare*, 25, 29–38.

Pennycott, T. W. and Steel, F. 2001. Parasitic worms in commercial free-range poultry flocks in England and Wales. *Veterinary Record*, 149, 428.

Permin, A., Bisgaard, M., Frandsen, F., Pearman, M., Kold, J. and Nansen, P. 1999. Prevalence of gastrointestinal helminths in different poultry production systems. *British Poultry Science*, 40, 439–43.

Perry, G. C., ed. 2004. Welfare of the laying hen. Poultry Science Symposium Series, Vol 27. CABI Publishing, Oxfordshire, UK.

Petek, M., Topal, E. and Cavusoglu, E. 2015. Effects of age at first access to range area on pecking behaviour and plumage quality of free-range layer chickens. *Archives Animal Breeding*, 58, 85–91.

Pettersson, I. C., Freire, R. and Nicol, C. J. 2016. Factors affecting ranging behaviour in commercial free-range hens. *World's Poultry Science Journal*, 72, 137–49.

Pötzsch, C. J., Lewis, K., Nicol, C. J. and Green, L. E. 2001. A cross-sectional study of the prevalence of vent pecking in laying hens in alternative systems and its associations with feather pecking, management and disease. *Applied Animal Behaviour Science*, 74, 259–72.

Pousga, S., Boly, H. and Ogle, B. 2005. Choice feeding of poultry: a review. *Livestock Research for Rural Development*, 17(4).

Quwaider, M., Daigle, C., Biswas, S., Siegford, J. and Swanson, J. 2010. Development of a wireless body-mounted sensor to monitor location and activity of laying hens in a non-cage housing system. *Transactions of the ASABE*, 53, 1705–13.

Rault, J-L., van de Wouw, A. and Hemsworth, P. 2013. Fly the coop! Vertical structures influence the distribution and behaviour of laying hens in an outdoor range. *Australian Veterinary Journal*, 91, 423–6.

Reed, H. J., Wilkins, L. J., Austin, S. D. and Gregory, N. G. 1993. The effect of environmental enrichment during rearing on fear reactions and depopulation trauma in adult caged hens. *Applied Animal Behaviour Science*, 36, 39–46.

Richards, G. J., Wilkins, L. J., Knowles, T. G., Booth, F., Toscano, M. J., Nicol, C. J. and Brown, S. N. 2011. Continuous monitoring of pop hole usage by commercially housed free-range hens throughout the production cycle. *Veterinary Record*, 169, 338.

Richards, G. J., Wilkins, L. J., Knowles, T. G., Booth, F., Toscano, M. J., Nicol, C. J. and Brown, S. N. 2012. Pop hole use by hens with different keel fracture status monitored throughout the laying period. *Veterinary Record*, 170, 494. doi:10.1136/vr.100489.

Rodenburg, T. B., Komen, H., Ellen, E. D., Uitdehaag, K. A. and van Arendonk, J. A. M. 2008. Selection method and early-life history affect behavioural development, feather pecking and cannibalism in laying hens: A review. *Applied Animal Behaviour Science*, 110, 217–28.

Rodenburg, T. B., Van Krimpen, M. M., De Jong, I. C., de Haas, E. N., Kops, M. S., Riedstra, B. J., Nordquist, R. E., Wagenaar, J. P., Bestman, M. and Nicol, C. J. 2013. The prevention and control of feather pecking in laying hens: identifying the underlying principles. *World's Poultry Science Journal*, 69, 361–74.

Rodriguez-Aurrekoetxea, A. and Estevez, I. 2016. Use of space and its impact on the welfare of laying hens in a commercial free-range system. *Poultry Science*, 95, 2503–13.

Roland, D. A., Bryant, M. M., Rabon, H. W. and Self, J. 1996. Influence of calcium and environmental temperature on performance of first-cycle (Phase 1) commercial Leghorns. *Poultry Science*, 75, 62–8.

Ruhnke, I., DeKoning, C., Drake, K., Choct, M. and Singh, M. 2015a. Feeding practices in Australian free-range egg production. *Proceedings of the 20th European Symposium for Poultry Nutrition*, 24–27 August, Prague, Czech Republic.

Ruhnke, I., DeKoning, C., Drake, K., Glatz, P., Walker, T., Skerman, A., Hunt, P., Sommerlad, M., Choct, M. and Singh, M. 2015b. Free range farm demographics and practices in Australia. *Proceedings of the 26th Australian Poultry Science Symposium*, 9–11 February, Sydney, NSW, Australia, p. 260.

Ruhnke, I., Cowling, G., Sommerlad, M., Swick, R. and Choct, M. 2015c. Gut impaction in free-range hens. *Proceedings of the 26th Australian Poultry Science Symposium*, 9–11 February, Sydney, NSW, Australia, pp. 242–4.

Sangster, N. C. 1999. Pharmacology of anthelmintic resistance in cyathostomes: will it occur with the avermectin/milbemycins? *Veterinary Parasitology*, 85, 189–204.

Savory, C. J., Mann, J. S. and Macleod, M. G. 1999. Incidence of pecking damage in growing bantams in relation to food form, group size, stocking density, dietary tryptophan concentration and dietary protein source. *British Poultry Science*, 40, 579–84.

Schröder, M. J. A. and McEachern, M. G. 2004. Consumer value conflicts surrounding ethical food purchase decisions: a focus on animal welfare. *International Journal of Consumer Studies*, 28, 168–77.

Scott, A. B., Singh, M., Hernandez-Jover, M., Barnes, B., Glass, K., Moloney, B., Lee, A., Groves, P. and Toribio, J-A. 2016. On-farm surveys to inform avian influenza risk assessment model. *Proceedings of the 27th Australian Poultry Science Symposium*, 14–17 February, Sydney, NSW, pp. 196–9.

Shalaby, H. A. 2013. Anthelminthics resistance; how to overcome it? *Iranian Journal of Parasitology*, 8, 18–32.

Sherwin, C. M., Richards, G. J. and Nicol, C. J. 2010. Comparison of the welfare of layer hens in 4 housing systems in the UK. *British Poultry Science*, 51, 488–99.

Sherwin, C. M., Nasr, M. A. F., Gale, E., Petek, M., Stafford, K., Turp, M. and Coles, G. C. 2013. Prevalence of nematode infection and faecal egg counts in free-range laying hens: relations to housing and husbandry. *British Poultry Science*, 54, 12–23.

Shimmura, T., Suzuki, T., Hirahara, S., Eguchi, Y., Uetake, K. and Tanaka, T. 2008. Pecking behaviour of laying hens in single-tiered aviaries with and without outdoor area. *British Poultry Science*, 49, 396–401.

Siegford, J. M., Berezowski, J., Biswas, S. K., Daigle, C. L., Gebhardt-Henrich, S. G., Hernandez, C. E., Thurner, S. and Toscano, M. J. 2016. Assessing activity and location of individual laying hens in large groups using modern technology. *Animals*, 6, doi:10.3390/ani6020010.

Singh, M. and Cowieson, A. J. 2013. Range use and pasture consumption in free-range poultry production. *Animal Production Science*, 53, 1202–8.

Singh, M., Ruhnke, I., DeKoning, C., Drake, K., Glatz, P., Walker, T., Skerman, A., Hunt, P., Sommerlad, M. and Choct, M. 2015. Free range poultry survey 2014 – farm demographics and practices. Poultry CRC final report – unpublished data.

Singh, M., Hernandez, C. E., Lee, C., Hinch, G. and Cowieson, A. J. 2016. Wanderers versus stay at home: who has the better guts? *Proceedings of the 27th Australian Poultry Science Symposium*, 14–17 February, Sydney, NSW, Australia, pp. 78–81.

Steenfeldt, S., Kjaer, J. B. and Engberg, R. M. 2007. Effect of feeding silages or carrots as supplements to laying hens on production performance, nutrient digestibility, gut structure, gut microflora and feather pecking behaviour. *British Poultry Science*, 48, 454–68.

Steenfeldt, S. and Nielsen, B. L. 2015. Welfare of organic laying hens kept at different indoor stocking densities in a mulit-tier aviary system. I: egg laying and use of the veranda and outdoor area. *Animal*, 9, 1509–17.

Stokholm, N. M., Permin, A., Bisgaard, M. and Christensen, J. P. 2010. Causes of mortality in commercial organic layers in Denmark. *Avian Diseases*, 54, 1241–50.

Summers, J. D. and Leeson, S. 1978. Dietary selection of protein and energy by pullets and broilers. *British Poultry Science*, 19, 425–30.

Swayne, D. E., ed. 2008. *Avian Influenza*. Blackwell Publishing, Iowa, USA.

Thapa, S., Hinrichsen, L. K., Brenninkmeyer, C., Gunnarsson, S., Heerkens, J. L. T., Verwer, C., Niebuhr, K., Willett, A., Grilli, G., Thamsborg, S. M., Sørensen, J. T. and Mejer, H. 2015. Prevalence and magnitude of helminth infections in organic laying hens (*Gallus gallus domesticus*) across Europe. *Veterinary Parasitology*, 214, 118–24.

van Krimpen, M. M., Kwakkel, R. P., Reuvekamp, B. F. J., van der Peet-Schwering, C. M. C., Den Hartog, L. A. and Verstegen, M. W. A. 2005. Impact of feeding management on feather pecking in laying hens. *World's Poultry Science Journal*, 61, 663–86.

van Niekerk, T. and Reuvekamp, B. 2011. Rondeel™, a new housing design for laying hens. *Lohmann Information*, 46, 25.

Walker, A. and Gordon, S. 2003. Intake of nutrients from pasture by poultry. *Proceedings of the Nutrition Society*, 62, 253–6.

Weeks, C. A., Lambton, S. L. and Williams, A. G. 2016. Implications for welfare, productivity and sustainability of the variation in reported levels of mortality for laying hen flocks kept in different housing systems: A meta-analysis of ten studies. *PLoS ONE*, 11, e0146394.doi:10.1371/journal.pone.0146394.

Weeks, C. A., Coles, G. C., Stafford, K. A. and Parsons, D. G. 2011. Managing health risks in free range laying hens. *Morrisons Farming Programme publication.* Available at http://www.

morrisons-corporate.com/Global/0_FarmingPage/Managing%20health%20risks%20in%20free-range%20laying%20hens.pdf. Accessed 2 August 2016.

Weeks, C. A. and Nicol, C. J. 2006. Behavioural needs, priorities and preferences of laying hens. *World's Poultry Science Journal*, 62, 296–307.

Whay, H. R., Main, D. C. J., Green, L. E., Heaven, G., Howell, H., Morgan, M., Pearson, A. and Webster, A. J. F. 2007. Assessment of the behaviour and welfare of laying hens on free-range units. *Veterinary Record*, 161, 119–28.

Zeltner, E., Klein, T. and Huber-Eicher, B. 2000. Is there social transmission of feather pecking in groups of laying hen chicks? *Animal Behaviour*, 60, 211–16.

Zeltner, E. and Hirt, H. 2008a. Factors involved in the improvement of the use of hen runs. *Applied Animal Behaviour Science*, 114, 395–408.

Zeltner, E. and Hirt, H. 2008b. A note on fear reaction of three different genetic strains of laying hens to a simulated hawk attack in the hen run of a free-range system. *Applied Animal Behaviour Science*, 113, 69–73.

Chapter 6

Beak trimming of laying hens: welfare costs and benefits

Dorothy McKeegan, University of Glasgow, UK

1 Introduction

Beak trimming (also known as debeaking, beak tipping or beak treatment) refers to the partial amputation of the beak, a procedure routinely carried out soon after hatching to prevent injurious pecking in laying hens and broiler breeders. For many years, the most common method of beak trimming was the application of a hot blade (HB) which simultaneously cuts and cauterises the upper and lower mandible, but in the last decade, this approach has been generally superseded by infrared (IR) beak trimming. Here, the beak is subject to a high-intensity IR energy source which causes subsequent loss of the beak tip. In the past, HB beak trimming in early life was sometimes followed by a second trim in adulthood (Hester and Shea-Moore, 2003), but IR trimming is applied only to chicks.

Beak trimming gives rise to numerous welfare concerns including loss of normal beak function (reduced ability to feed, drink and preen), acute pain, debilitation, reduced sensation and the possibility of chronic pain. As a result, some countries have banned (or are in the process of banning) beak trimming as a routine mutilation in commercial poultry production. While such bans are laudable in their aims, their implementation is problematic because controlling feather pecking and cannibalism in laying hens remains a major challenge. Despite many years of extensive study, our understanding of the causation of injurious pecking is still deficient and outbreaks of damaging pecking remain unpredictable.

For many, the justifiability of beak trimming rests on a cost/benefit analysis based on assessing the welfare harms caused by beak trimming and weighing these against the

http://dx.doi.org/10.19103/AS.2016.0012.26

potential welfare benefits of reducing the prevalence of feather pecking and cannibalism. This ethical analysis requires careful examination of the evidence for negative welfare outcomes resulting from beak trimming, and of the evidence that the practice is the best way to reduce or prevent feather pecking. The availability of other approaches to reduce injurious pecking is also relevant and research is ongoing to determine whether alternative control methods are sufficiently reliable to dispense with beak trimming. A growing body of evidence suggests that IR trimming has welfare advantages over HB trimming, potentially increasing its acceptability and this has led to it being the only legally allowed method in some countries and, in Europe, only via derogation in EU Council Directive 1999/74/EC. In this chapter, the welfare costs and benefits of beak trimming will be reviewed and relevant research findings will be critically evaluated. To provide historical context and as a basis for comparison, research pertaining to HB beak trimming will be included, but the focus will be on IR beak trimming as the primary method being utilised worldwide.

2 Feather pecking and cannibalism

Bird-to-bird pecking was categorised by Savory (1995) into five types: aggressive pecking, gentle feather pecking, severe feather pecking, tissue pecking and vent pecking. This categorisation highlights that feather pecking and cannibalism are considered to be unrelated to aggression; aggressive pecking is related to dominance rank where pecks may be aimed at subordinate birds or delivered in relation to an agonistic encounter. Aggressive pecks normally consist of a single vigorous peck aimed at the comb, head, neck or back, resulting in the immediate withdrawal (and often accompanying vocalisation) of the recipient (Savory, 1995). Although repeated aggressive pecks can cause feather and tissue damage, various studies have demonstrated that feather pecking and aggressive pecking have different underlying mechanisms (Rodenburg et al., 2013). Recently, however, it has been suggested that there may be some genetic correlation between them (Bennewitz et al., 2014; Grams et al., 2015).

Gentle feather pecking is often aimed at the feather tips or litter particles on the plumage of a dustbathing conspecific. It tends to consist of bouts of several pecks, is usually non-damaging and is ignored by the recipient. Severe feather pecking results in beak-inflicted plumage damage by the pecking, pulling out and eating of feathers. Such severe feather pecks and feather pulls are repeated persistently and can be aimed at any part of the body. These more forceful pecks and pulls can cause feather damage and may eventually cause the recipient to withdraw. There is little evidence that gentle feather pecking evolves into severe feather pecking. Indeed, various studies have failed to find a relationship between these behaviours (Savory and Griffiths, 1997; Kjaer and Vestergaard, 1999; Newberry et al., 2007) and they have distinct gene expression patterns (Hughes and Buitenhuis, 2010). Removed feathers are often eaten (McKeegan and Savory, 1999), resulting in bare patches (often on the back, tail or near the cloaca). Denuded areas of skin exposed by feather removal are novel and may attract tissue pecking (cannibalism), causing rapid wound development. Open wounds and blood attract further attention, and the severity of injuries can accelerate in a very short time. Vent pecking is a particular type of cannibalism in which pecks are aimed specifically at the cloaca, sometimes with no other pecking damage having taken place (Newberry, 2004). Death of the recipient occurs rapidly through tissue damage and loss of blood.

Various theories exist on the causation of feather pecking, but the weight of evidence supports the idea that damaging feather pecking is related to foraging, where laying hens are more likely to develop the behaviour when a suitable foraging substrate is not available, especially in early life (Blokhuis, 1989; Huber-Eicher and Wechsler, 1997; Johnsen et al., 1998). Additionally, a wide range of factors have been shown to influence the incidence and severity of feather pecking including light (type and intensity, e.g. Kjaer and Vestergaard, 1999), current type and availability of litter substrate (e.g. Nicol et al., 2001), stocking density and group size (e.g. Simonsen et al., 1980), diet (e.g. Van Krimpen et al., 2005), environmental enrichment (e.g. Daigle et al., 2014), production system (e.g. Keeling, 1994), genetic variation (e.g. Rodenburg and Koene, 2003), age (e.g. Huber-Eicher and Wechsler, 1997), hormonal state (Hughes, 1973) and fear and stress (e.g. Rodenburg et al., 2004). This extensive list illustrates the highly multifactorial nature of severe feather pecking and its complex aetiology helps at least to explain our ongoing inability to predict this behaviour in commercial flocks.

Injurious pecking remains a significant welfare and economic problem in the egg industry and it is generally accepted that pecking problems are more difficult to control in large group and non-cage housing systems than in conventional cages (Rodenburg et al., 2013). This is because birds predisposed to severe feather pecking are not contained and have wide access to more potential victims, and because it is more difficult to identify and remove feather peckers in these systems (Rodenburg et al., 2013). Since 2012, with the advent of the EU ban on conventional cages in Europe, there has been increased use of loose housing systems and thus the issue of damaging pecking has been in sharp focus. However, severe feather pecking occurs in all production systems including cage, barn, organic and free-range housing (Nicol et al., 2013), and the threat of serious pecking outbreaks (associated with significant welfare and economic costs) has led the layer industry to consider routine beak trimming as a necessary intervention.

3 The welfare impact of beak trimming

The avian beak is a specialised sensory organ consisting of bone, dermal and epidermal cellular layers and a smooth, hardened, keratinised layer (the rhamphotheca) (Gentle et al., 1995). The beak is innervated by trigeminal nerve branches (the ophthalmic nerve in the upper beak and the intramandibular nerve in the lower beak; King and McLelland, 1975) and the cellular layers contain mechanoreceptors (Herbst and Grandry corpuscles), blood vessels and free nerve endings including nociceptors (Gentle, 1986b; Kuenzel, 2007). Thus, interference with this complex structure by beak trimming will result in loss of rhampotheca, the bill tip organ (Gentle and Breward, 1986) and possibly bone (depending on the severity of the trim) and will activate mechanoreceptors, nociceptors and other sensory receptors (e.g. thermal receptors) (Breward, 1984). Acute and chronic pain and other welfare impacts such as loss of beak function after beak trimming have been investigated in numerous physiological and behavioural studies. It should be noted that, due to the more recent introduction of IR trimming, the majority of available research is still relevant only to the older HB trimming method. This literature has been previously reviewed (Cunningham, 1992; Hughes and Gentle, 1995; Hester and Shea-Moore, 2003; Jendral and Robinson, 2004; Cheng, 2006) and is briefly summarised below, before a detailed discussion of the available welfare research related to IR trimming.

3.1 HB beak trimming

HB beak trimming involves the placement of the beak on a guide bar or through a hole after which a guillotine-style blade (heated to at least 750°C) simultaneously cuts and cauterises the beak, removing a proportion of both mandibles. The method is usually performed manually, and therefore the skill and consistency of the operator hugely influences the extent and efficacy of the trim. Human error and lack of care and precision can result in overly severe trims, uneven cuts and even facial burns (Jendral and Robinson, 2004).

Based on nerve injury discharge during trimming, Gentle (1991) suggested that initial pain during and immediately following HB beak trimming is short-lived (2–48 s in duration), followed by a pain-free period of up to 26 hours (Gentle et al., 1991). Pain at the time of beak trimming was inferred by Glatz (1987), who found that the heart rates of birds subject to beak trimming took longer to recover to baseline compared to handling alone (though this was not reproduced in later research (Glatz and Lunam, 1994)). There is ample evidence of acute pain in the days following HB trimming in the form of reduction of beak usage for pecking and feeding (Blokhuis et al., 1987; Gentle et al., 1991, 1997; Glatz and Lunam, 1994). In addition, the force generated by pecking was lower in HB trimmed birds (Dennis and Cheng, 2010b) providing evidence of guarding. Reduced use of the injured beak was identified as pain-related behaviour in a study by Glatz et al. (1992), who showed that feed intake was restored to normal levels in birds provided with analgesia. The stressfulness of HB trimming was verified by Davis et al. (2004), who reported increased plasma corticosterone levels within 2 hours of HB trimming at 6 days old. Plasma corticosterone remained elevated for 5 weeks in birds trimmed at 11 weeks of age, suggesting ongoing welfare compromise.

Healing and beak regeneration after HB trimming involves the development of fibroblastic scar tissue covered by a hardened epithelial layer (Gentle, 1986a,b), and there is evidence of absence of sensory receptors in the regenerated beak tip probably associated with reduced sensory feedback (Gentle, 1986b; Dubbeldam et al., 1995; Gentle et al., 1997; Freire et al., 2011). While degeneration of severed nerve fibres may occur initially (Breward and Gentle, 1985), development of neuromas (tangled masses of axons caused by nerve regeneration in multiple directions) has also been associated with HB beak trimming. For example, Breward and Gentle (1985) reported neuroma formation in the beaks of birds 15 days after HB beak trimming at 5 weeks of age. Neuromas are associated with abnormal spontaneous neural activity and possible chronic dysesthesia or pain (Breward and Gentle, 1985; Gentle, 1986a), leading to important welfare concerns if they persist for the bird's lifetime. A few studies have provided data which suggests the presence of chronic pain following HB trimming, primarily in the form of behavioural changes persisting for several weeks after trimming (Gentle, 1986a; Duncan et al., 1989). These changes include reduction in beak usage for non-essential activities such as preening and exploratory pecking, increased inactivity and altered reactions to presentation of warm drinking water (Gentle et al., 1990). Whether these responses are specifically due to the presence of neuromas is not known, but it is notable that these behavioural alterations continued beyond the normal healing period.

The timing of beak trimming appears to be a crucial factor influencing the severity of its welfare consequences. Several studies suggest that neuromas are not present when birds are trimmed at hatch (Dubbledam et al., 1995; Gentle et al., 1997; Cheng, 2006) and are only likely to form when older birds are beak trimmed. This concurs with increased evidence for chronic pain in studies where birds have been beak trimmed as juveniles or adults (e.g. Gentle et al., 1990). Neuroma formation is also more likely when beaks are severely (as opposed to moderately) trimmed (Lunam et al., 1996), defined by Kuenzel (2007) as the removal of 50% or less of the beak. Neuroma formation after moderate trimming at hatch was reported by Lunam et al. (1996), but the persistence of these neural abnormalities throughout adulthood was positively related to the amount of beak tissue removed. Breward and Gentle (1985) suggested that beak-trimmed birds may suffer from a type of 'phantom limb syndrome', where sufferers have a vivid sensory experience of the missing area which may be accompanied by chronic pain. Conversely, Kuenzel (2007) argued that such sensations are unlikely to occur in moderately beak-trimmed birds based on anatomical comparisons of the avian somatosensory cortex with that of primates.

Collectively, the evidence suggests that the likelihood of long-term abnormal sensory and potentially painful consequences resulting from HB trimming relates primarily to age and extent of beak amputation, although acute pain is caused by HB trimming at any age. Wide recognition of the importance of age at trimming forms the basis for applying IR beak trimming exclusively to day old chicks.

3.2 IR beak trimming

IR beak trimming is applied as part of an automated system, which was developed and patented by Nova-Tech Engineering Inc. The procedure is applied while birds are processed at the hatchery, and is usually accompanied by vaccination injections to reduce the handling stress associated with multiple interventions. Chicks are manually placed in individual holders that correctly position the front of the head so that it passes through an IR beam affecting a standardised area of the beak. The head holders are part of a moving carousel so that the process is continuous with birds being loaded on one side and released on the other; the time spent in the carousel is 15–30s (Gentle and McKeegan, 2007). Once the beak is in position, application of a localised, non-contact IR heat source penetrates the hard outer rhamphotheca and damages the underlying dermis and subdermal tissues (see Fig. 1). Resulting necrosis of the affected area causes loss of the beak tip within 2–3

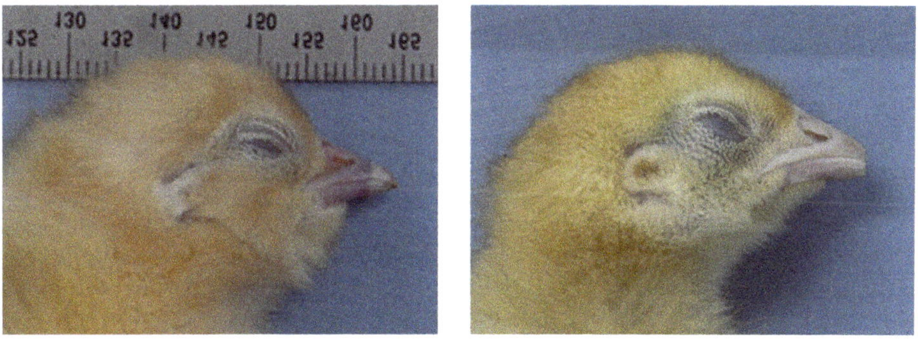

Figure 1 Photographs showing the appearance of the beak following IR beak treatment. The photo on the left shows the beak of a laying hen chick immediately after IR beak trimming at day old – the area affected by the IR beam appears paler than the rest of the beak. The photo on the right shows the appearance of the beak at 4 weeks of age – the affected beak tissue has sloughed off and a blunted beak shape is clearly visible.

weeks following treatment (see Fig. 1). Thus, immediately after IR trimming, the beak is physically intact. The IR beak-trimming process is somewhat flexible, allowing for the use of multiple protocols to accommodate variation in bird strain and age. Head holders of different sizes can be used for different sized birds and the amount of beak tissue exposed to the IR beam can be adjusted via interchangeable guard plates (Dennis and Cheng, 2012). Additionally, the amount of IR energy delivered can be altered via a range of intensity settings (Dennis and Cheng, 2012).

Three clear advantages of IR trimming are immediately apparent, compared to HB beak trimming. Firstly, the beak tip gradually erodes, giving the bird time to adjust to the new beak shape, which may improve pecking accuracy and feeding efficiency. Secondly, the process eliminates the presence of open wounds following trimming, preventing bleeding and reducing the risk of inflammation and infection. Thirdly, the automation and standardisation of the process reduces variability and the risk of human error or inappropriate handling. Despite these benefits, both short- and long-term welfare concerns remain, given the nature of the tissue damage induced.

The acute consequences of IR trimming have been examined in a number of studies, several of which have included direct comparisons with HB trimming. Gentle and McKeegan (2007) examined the behaviour of broiler breeder chicks following IR and HB trimming at day old, compared to control or sham-trimmed birds, including detailed observations in the first hour after treatment. There were no effects on behaviour of beak trimming with either method in the first hour, but information on the initial response to IR trimming is limited to this single study. It is interesting to speculate whether IR trimming may result in a transient pain-free period, characteristic of full thickness burns, as suggested previously for HB trimming (Gentle et al., 1991), which may explain the lack of early behavioural response. Nevertheless, convincing evidence of beak guarding in the days and weeks after IR trimming is provided by a number of studies examining growth and feeding behaviour.

Lower bodyweights in IR trimmed birds have been reported within two days of trimming (Marchant-Forde et al., 2008) and this trend is maintained in the first weeks of life. Gentle and McKeegan (2007) found reduced growth at 14 days and this effect was maintained to 6 weeks (although they noted that absolute differences in bodyweight between trimmed and control birds were small). Disappearance of bodyweight differences has been reported by 3 weeks (Marchant-Forde and Cheng, 2010; Henderson et al., 2009), 4 weeks (Damme and Urselmans, 2013), 5 weeks (Marchant-Forde et al., 2008), 8 weeks (Angevaare et al., 2012), 10 weeks (Marchant-Forde and Cheng, 2010) and 12 weeks after IRBT (Damme and Urselmans, 2013).

Compared to HB trimming, growth rate reduction following IR trimming seems to be in general less pronounced and more short-lived (Gentle and McKeegan, 2007; Henderson et al., 2009). In a study that does not fit this trend, Marchant-Forde et al. (2008) reported lower growth after IR trimming compared with HB trimmed birds until week 4; thereafter IR trimmed birds were similar to HB. However, in later work, Marchant-Forde and Cheng (2010) found that bodyweight was suppressed after IR trimming for only 3 weeks, while birds that underwent HB trimming had slower growth until 9 weeks of age. The difference between these studies probably relates to the fact that the HB trim in the earlier work (Marchant-Forde et al., 2008) was a 'light trim', constituting the removal of a third of beak length, as opposed to a more severe half beak trim in the later experiment (Marchant-Forde and Cheng, 2010). In combination, the results of these two studies suggest that the acute pain associated with IR trimming is intermediate between these HB trimming severities.

Transient negative effects of IR trimming on bodyweight are corroborated by several reports of reduced feeding and pecking at the environment in the days and weeks following trimming. Gentle and McKeegan (2007) observed that sham treated and control birds pecked the environment more than IR trimming birds in weeks 3 and 4, although this effect had disappeared by 6 weeks of age. Marchant-Forde et al. (2008) found that IR-trimmed birds were less active and spent less time eating and drinking than controls, and reduced feed intake until 5 weeks of age was also observed following IRBT in their later study (Marchant-Forde and Cheng, 2010).

These effects on growth and exploratory pecking behaviour are likely to be due to beak guarding, providing circumstantial evidence of acute pain, but there is probably an additional effect on growth caused by reduced feeding efficiency after IR trimming. Angevaare et al. (2012) reported that IR-trimmed birds were less efficient at eating grains, mealworms, grapes, spaghetti and raisins and picked up 63% less food per peck than birds with intact beaks. However, IR-trimmed birds fared better when compared to HB-trimmed birds, spending less time feeding despite being heavier, suggesting that, although feeding efficiency is compromised with both methods of trimming, these effects are less severe following IR trimming (Dennis and Cheng, 2010a). Early body weight reduction (within two weeks of trimming, Gentle and McKeegan, 2007) indicates that welfare impacts occur during the period over which the dead beak tissue is lost, presumably due to acute pain, reduced feeding efficiency and possibly absence of sensory feedback.

In an attempt to more directly identify behavioural evidence of acute pain following beak trimming, Freire et al. (2008) examined self-administration of a non-steroidal anti-inflammatory drug (carprofen, in food) and pecking force in pullets treated with IR or HB trimming at day old. Intact birds and pullets trimmed at 10 weeks were also tested. At 11 weeks of age, pecking force in birds beak trimmed with either method was the same as controls, and there was also no evidence that trimmed birds ate more analgesic feed. Birds trimmed at 10 weeks pecked less forcefully, although birds that ate more analgesic-laced feed had higher maximum force of pecking. The authors noted that the average beak trim of 34% could be considered a light trim, which may not be representative of the effects of more severe trimming, and they concluded that carprofen has no effect on potential neuropathic pain caused by beak trimming. An alternative explanation for the results is that the birds trimmed at day old were not in pain, especially since pecking force was not reduced. In addition, analgesic self-administration methodology can be criticised for rarely determining if the amount of food eaten in a given time period equates to an effective circulating dose. Indeed, this is likely to be difficult to achieve since the required drug concentration may make the food unpalatable. It seems likely that by 11 weeks after trimming, acutely painful effects had subsided, and this is supported by behavioural and bodyweight evidence from other studies of recovery in this timescale.

The apparent variation in the response to IR trimming (e.g. discrepancies in the reported duration of growth suppression) might be explained by the use of different protocols with regard to the intensity of IR energy applied and the amount of beak affected (and therefore subsequently lost). Detailed information on the protocol used is not available for published studies, making comparisons difficult. Only one study has systematically examined the effects of different IR trim settings on welfare and production parameters. Dennis and Cheng (2012) conducted a factorial experiment where chicks were exposed to IR trimming according to two plate sizes (representing a more or less severe trim) and three IR intensities (low, high, moderate), plus a HB control. They reported similar growth curves across treatments, with birds receiving the light IR trim being heaviest, while severely trimmed birds were lightest. Lightly trimmed birds walked and drank more than severely trimmed and HB-trimmed birds initially, but by 10 weeks these differences had gone. The frequency of feeding behaviours did not differ between beak-trimming treatments at any age. They concluded that IR trimming is associated with pain persisting up to 5 weeks of age, but this is reduced in less severely trimmed bird in terms of beak area lost. This appears to be similar to HB trimming where the extent of trimming is related to the severity of welfare consequences.

Longer-term recovery from IR trimming involves beak healing and regeneration, which occurs quickly; Gentle and McKeegan (2007) reported significant beak growth 21 days following IR-trimming. Studies of beak length and condition in rearing and adulthood suggest that IR trimmed birds have less variable beaks in better condition (e.g. absence of abnormalities and scar tissue) than HB-trimmed birds (Glatz, 2009; Dennis and Cheng, 2010a; Carruthers et al., 2012; Damme and Urselmans, 2013). Findings on the effects of IR trimming on beak length are contradictory; Dennis and Cheng (2010a) found that IR-trimmed birds had longer beaks compared to HB, whereas others report shorter beaks compared to HB trimming (Marchant-Forde et al., 2008; Henderson et al., 2009; Carruthers et al., 2012). It has also been noted that uneven beak regrowth (normally the lower beak being longer than the upper) is more common with HB trimming (Marchant-Forde et al., 2008; Glatz, 2009; Dennis et al., 2009), although this has also been reported in IR-trimmed birds (McKeegan and Philbey, 2012). Bill asymmetry has been shown to negatively affect the ability of hens to manipulate and ingest feed pellets (Prescott and Bonser, 2004), so uneven mandible length may have important implications for feeding efficiency.

As with HB trimming, there are concerns about IR trimming inducing neuroma formation which might lead to spontaneous pain or pain during routine beak use. Glatz (2009) reported the presence of persistent traumatic neuromas in birds that had been HB and IR trimmed at day old (Lunam, unpublished). This discovery resulted from detailed histopathological examinations involving seven micrometre sagittal sections of the entire beak and remains the only report of neuroma formation following IR trimming. The presence of neuromas was unexpected, given previous findings in birds HB trimmed at day old, in which neuromas were absent (Gentle et al., 1997). In light of this, it was suggested that either excess heat or tissue removal not typical of routine IR trimming might have been responsible for neuroma formation in those particular experiments, or that the IR method needs further development to reduce incidence of neuromas.

In the only electrophysiological study to examine the welfare consequences of IR trimming, McKeegan and Philbey (2012) compared beak nerve function and anatomy in birds exposed to IR trimming and intact birds at 30 or 50 weeks old. The responses of single sensory nerve fibres were recorded from the intramandibular nerve, which provides sensation to the lower beak. The beaks were also measured and subject to microscopic and radiographic examination. A large number of sensory receptors (mechanoreceptors, thermoreceptors and nociceptors) were detected with receptive field positions across the entire lower beak, regardless of age or treatment. Importantly, there was no evidence of a treatment effect on nociceptive thresholds or spontaneous nociceptor activity at any age, which might have been expected in IR-trimmed birds if neuromas were present. Beak measurements demonstrated that IR trimming affected on average 36% of beak area and resulted in a 44% reduction in beak length by 4 weeks of age. There was no radiographic evidence of adverse or pathological change in relation to IR trimming. Evidence of healing (re-epithelialisation, fibrovascular hyperplasia and bone remodelling) was revealed by microscopic evaluation of the trimmed beaks; by 4 weeks of age, there was limited nerve regeneration in IR-trimmed beaks, including repopulation of mechanoreceptors in some birds. No neuromas or abnormal proliferations of nerve fibres were observed at any age; however, serial sections (as done by Lunam, unpublished) were not carried out. These findings suggest that IR trimming at day old does not have chronic adverse consequences for sensory function or pain. Although the notion that neuromas may be induced by IR trimming is not supported by electrophysiological evidence of neuroma-like activity, further detailed histopathological research is warranted to determine if neuromas are induced following IRBT at day old using standard settings. While age at HB trimming appears to strongly influence the formation of neural abnormalities, this has never been examined in IR trimming, which is always applied in day-old chicks. It is possible that IR trimming gives rise to asymptomatic neuromas, but further electrophysiological research would be needed to investigate this.

The available data on pain, behaviour, beak regeneration and sensory function suggest that the reduced welfare impact of IR trimming (along with its practical advantages) make it the preferred choice when compared to HB trimming. However, IR trimming is almost certainly associated with acute pain, as repeatedly evidenced by reduced beak usage and growth in the first weeks of life. Although there is little evidence that IR trimming has long-term welfare costs, these acute effects are considered to be unacceptable by some. The lack of long-term effects following beak regeneration begs the question – by what mechanism does IR trimming work to reduce injurious pecking? Induction of transient beak guarding and loss of sensation might be crucial, reducing motivation to peck at a sensitive time for learning. Therefore, the acute negative consequences of IR trimming could be unavoidable as an important part of the means by which the process exerts its protective effects, though changes in beak shape are also likely to play a role.

4 Effects of beak trimming on injurious pecking and production

4.1 HB beak trimming

Extensive research has shown that production parameters are generally improved by HB trimming, although reported findings are sometimes inconsistent. One possible reason for this is lack of standardisation between the types and extent of beak trimming performed in different studies. Severe or 'light' trimming involves removing different proportions of the beak, which, along with the application of a range of cauterisation temperatures, may produce diverse outcomes. As outlined above, age at beak trimming also has important effects on subsequent responses in both welfare and production terms. The consequences of HB beak trimming for productivity and its effectiveness for pecking control have been previously reviewed (Hester and Shea-Moore, 2003; Jendral and Robinson, 2004) and will be only briefly outlined here.

Many studies provide evidence of reduced feather pecking, feather pulling and cannibalism following HB trimming (reviewed by Hester and Shea-Moore, 2003), often using plumage condition as a proxy for damaging pecking. Primarily via its effects on damaging pecking, HB trimming therefore usually reduces mortality during the laying phase. In some cases, mortality in untrimmed birds has been dramatically increased due to cannibalism, and figures of 15–20% have been reported in loose housing systems (Blokhuis et al., 2007). The apparent ability of beak trimming to prevent the most destructive pecking outbreaks has been one of the strongest arguments for the practice. Several reports suggest that HB trimming exerts its effects by lowering rates of feather pecking (e.g. Blokhuis et al., 1987) and probably also by reducing the effectiveness of the beak as a tool to peck and pull feathers. Lambton et al. (2010) investigated the development of gentle and severe feather pecking in 111 free range and organic flocks in the United Kingdom, 53.8% of which were beak trimmed within the first week of life (a further 12.6% were retrospectively beak trimmed as a consequence of injurious pecking), mostly by HB, but detailed information was unavailable. Rates of gentle feather pecking were higher in beak-trimmed flocks compared to intact beak or retrospectively beak-trimmed birds. Severe feather pecking rates were highest in intact beak birds, which also had worse plumage scores.

Reports of the effects of HB trimming on egg production are variable. While several studies show that egg production is unaffected by moderate trimming, others report improvements in egg production following HB trimming, with the caveat that it seems to be better for egg production to trim earlier (Hester and Shea-Moore, 2003). This effect is presumably due to early HB trimming being associated with less severe effects on feed intake reduction, which resolve by sexual maturity and peak egg production. Similarly, later HB trimming (after 12 weeks) has been shown to cause smaller eggs and delays in the onset of sexual maturity (Andrade and Carson, 1975).

There is evidence of reduced food wastage in beak-trimmed birds (Craig et al., 1992; Marchant-Forde and Cheng, 2010), which may be related to a reduced rate of ingestion (Persyn et al., 2004). In general, HB trimming has been associated with improved feed efficiency, as indicated by a reduction in feed consumption which does not affect egg size or production rate (Jendral and Robinson, 2004). Superior feather covering due to reduced feather pecking at least partly explains improved feed efficiency; fully feathered birds have improved thermoregulatory capacity and thus require less feed required to maintain body temperature. Indeed, feed requirements increase significantly with feather loss (Leeson and Morrison, 1978). Tauson and Svenson (1980) showed that hens with poor feather cover had 46% higher energy maintenance needs, resulting in 27% larger feed requirements.

In a recent study, Vezzoli et al. (2015) compared the responses of intact or HB-trimmed birds to experimental infestation with lice and mites. Time spent preening before, 6 and 9 weeks after infestation was recorded and mite/lice numbers were scored weekly. HB-trimmed birds spent more time preening than before infestation with lice, but intact birds did not, and HB-trimmed birds had 17 times more lice and higher mite loads than intact birds. This strongly suggests that an intact beak is important for dealing with ectoparasite infestations. Unusually for such a recent study, this research examined HB trimming only (despite this method becoming obsolete), so it is not known if these results also apply to IR beak trimming.

4.2 IR beak trimming

There are limited data concerning the production implications of IR trimming, compounded by the fact that several studies compare HB and IR trimming without an intact control. For such a widespread practice with a clear goal, a surprisingly small number of studies have investigated whether IR trimming has protective effects against damaging pecking. Dennis et al. (2009) compared production parameters in hens exposed to IR trimming at hatch or HB trimming at 7–10 days of age, and measured feather scores and focal observations of pecking behaviour at 30 weeks. The feather scores of HB-trimmed birds were worse than IR-trimmed birds (particularly on the head and back), but there was no comparison to intact birds. They also reported that IR birds showed reduced 'aggressiveness' under high light intensity compared to HB; however, this was based on a simplistic categorisation of pecking behaviour (with no separation between gentle and severe pecks, for example) and a small behavioural sample. In later work, Dennis and Cheng (2012) concluded that a range of IR trimming protocols were as effective as HB trimming in reducing plumage damage, but again a control group was absent. Damme and Urselmans (2013) examined feather condition at 72 weeks in white and brown laying strains, treated with HB, IR trimming or no trimming. In the white strain, 54.8% of the IR-trimmed birds had intact feather cover compared to 35.0% after HB trimming and only 15.7% of the untreated controls. In the brown strain, HB trimming was more effective than IR trimming to prevent feather damage, with the untreated control having the most plumage damage. These results suggest that both methods of trimming reduced feather pecking, but that the effectiveness of different methods of beak trimming may be influenced by genetic factors.

A recent study examined plumage damage in intact versus IR-trimmed flocks kept in aviary systems on commercial farms (Sepeur et al., 2015). At the end of rearing, plumage damage scores were higher in untrimmed birds and this effect was further amplified in the laying phase, although a difference between intact and trimmed flocks was not exhibited on all farms (again highlighting the unpredictable nature of feather pecking). Hartcher et al. (2015) reported improved plumage condition in hens at 43 weeks following IR trimming. Compared to intact birds, beak-trimmed birds also performed less ground pecking, less severe feather pecking and more gentle feather pecking, but this study was confounded by the fact that the birds were given a second trim by HB at 11 weeks. Lambton et al. (2013) reported increased gentle feather pecking in IR-trimmed birds compared to intact controls, and higher rates of severe feather pecking in organic flocks (which presumably were not beak trimmed). Comparing rates of gentle and severe feather pecking in 12 flocks that were IR beak-trimmed and 22 that were not, Gilani et al. (2013) reported that non-trimmed flocks had a 14.56 times higher chance of showing severe feather pecking during rearing than beak-trimmed ones. However, during lay, the only effect of IR trimming was that intact birds gentle feather pecked significantly less than beak-trimmed ones.

Findings concerning the effects of IR trimming on egg production are contradictory. No differences between IR and HB trimming at 7–10 days of age on bodyweight, egg production or HL ratio at 30 weeks were reported by Dennis et al. (2009), but there was no intact control. Glatz (2009) also reported no effect on egg production throughout the laying phase between birds HB trimmed at 10 days and IR trimmed at day old, but did note that the body weight of IR-trimmed birds was higher while egg weights were lower. Damme and Urselmans (2013) described increased egg production in IR-trimmed birds compared to both HB and intact controls with a reduced cumulative mortality of up to 50% associated with IR trimming. By contrast, Angevaare et al. (2012) found that IR-trimmed birds had lower egg productivity indicator scores at 17 weeks, and to a lesser extent at 21 weeks, compared to intact birds. The positive effects of HB trimming on feed efficiency are mirrored in a couple of IR trimming studies. Damme and Urselmans (2013) showed that IR-trimmed hens required up to 48 g less feed per kg egg mass, reflecting a better feed conversion ratio than both HB trimmed and intact controls. Reduced feed waste following IR trimming was also reported by Marchant-Forde and Cheng (2010) and Dennis and Cheng (2012).

It is clear that significant production advantages arise from beak trimming by both HB and IR methods, and in some cases IR trimming has been shown to outperform traditional HB trimming, adding to its appeal. While a growing number of studies support the notion that IR trimming is effective as a preventative measure against injurious pecking, data on this are still sparse and direct evidence of its effects on pecking behaviour in adulthood in particular is lacking. It is therefore not clear to what extent long-term behavioural changes are induced by IR trimming, and whether its protective effects are primarily related to changes in beak length and shape. Whatever the mechanism triggered by IR trimming, the direct welfare benefits to the birds arise only from the reduced risk of feather pecking injury, not other production gains, and this must be weighed against the acute welfare costs that it imposes. Obviously, damaging pecking can have very severe welfare consequences and prevention of the most severe outbreaks is a priority, although high levels of feather pecking can also occur in beak-trimmed flocks (e.g. Green et al., 2000). Recent work following intact-beak flocks in free-range housing in the United Kingdom concluded that, when feather pecking does occur, the consequences in intact beak flocks are rapid and far more serious than in beak-trimmed flocks in terms of tissue damage, subsequent infection and bird welfare (Defra Project final report, 2015). In addition, reduced mortality, positive effects on egg production and improved feed efficiency are strong economic drivers that also support the use of beak trimming in the current era of sustainable intensification and heightened concern over food security.

5 Alternative strategies to reduce the need for beak trimming

Several risk factors have been identified in both the rearing and laying phases, which increase the likelihood of injurious pecking, providing a starting point for the development of ameliorative management strategies. These include manipulation of diet, lighting, litter quality, use of the range and provision of environmental enrichment (reviewed in Nicol et al., 2013; Rodenburg et al., 2013). Given the apparently complex causation of damaging pecking, it has been suggested that multifactorial approaches to addressing the issue are most likely to be effective (Rodenburg et al., 2013).

A traditional response to outbreaks of severe feather pecking and cannibalism has been to reduce light intensity or alter light wavelength, with the aim of reducing overall activity levels as well as the visibility of conspecifics (and in the case of red light, the conspicuousness of blood and tissue damage). While low light intensities are associated with reduced injurious pecking (Tauson, 2005), this method risks negative effects on eye development and function (Prescott et al., 2003), and is thus not a straightforward solution. Encouraging work on the provision of dark brooders has shown reduced feather pecking in both small-scale (Jensen et al., 2006) and commercial-scale experiments (Gilani et al., 2012), and is thought to exert its effects by separating active and inactive chicks and enhancing behavioural synchronisation.

While dietary factors such as protein level and type, energy content and minerals have been shown to affect feather pecking, especially when deficiencies are present (Rodenburg et al., 2013), a completely preventative diet is not available. Dietary dilution with high fibre levels leads to increased feeding time and has been shown to reduce damaging pecking (Van Krimpen et al., 2009). However, implementation of dietary dilution to reduce the risk of severe feather damage may have negative production implications. Feed form is also important; mashed feed (which takes longer to eat) has been associated with reduced plumage damage in commercial flocks (e.g. Lambton et al., 2010). Given the important role of foraging behaviour in the development of injurious pecking, provision of friable substrate is a key factor; provision of litter (and scattered food) during rearing can have long-term protective effects (Nicol et al., 2013).

Environmental enrichment in the form of 'pecking distractions' such as straw bales or nets and pecking objects have not been shown to be particularly effective in controlling feather pecking. Sepeur et al. (2015) reported that intact beak commercial flocks had significantly poorer plumage condition than IR beak-trimmed flocks, despite the provision of sand baths, blocks for pecking, bales of straw and plastic objects. Hartcher et al. (2015) also found no beneficial effect of enrichment (string pecking devices and scattering of whole oats in the litter) on pecking damage. Provision of hay bales was found to reduce gentle feather pecking, but not severe pecking in small group housed hens (Daigle et al., 2014).

Bringing a range of approaches together based on a literature review, Lambton et al. (2013) examined the effects of bespoke management regimes in free-range flocks in the United Kingdom. Forty possible strategies were identified and tailored to each farm such that on average, 5.5 strategies to reduce pecking were implemented in each flock. Lower rates of feather pecking (gentle and severe) were related to the number of strategies employed, and in particular, reduced severe feather pecking was linked to the provision of aerated breeze blocks and shelters to encourage range use. However, most of the flocks were beak trimmed and so it is not known if these control methods would be similarly effective with intact beak hens. It is worth noting that many of the practices implemented with the aim of prevention and control of damaging pecking may also have other welfare benefits.

Recently, a major UK research project investigated the feasibility of keeping intact beak flocks of laying hens commercially and the effectiveness of a range of management strategies previously shown to reduce injurious pecking. The study followed 20 flocks (19 free range and one barn) ranging in size from 1200 to 16000 hens to the end of lay. No pecking problems were observed during rearing, but the behaviour of the flocks during lay was highly variable with regard to damaging pecking and mortality. Although 18 of the 20 flocks had acceptable levels of mortality at 40 weeks, two of the flocks had severe outbreaks of injurious pecking, leading to high mortality and significant welfare compromise. It was noted that these were both large (16000 bird) flocks, but a direct link between flock size and injurious pecking was not apparent. Improvement in end of lay plumage scores and mortality was related to whether the previous flock had been beak trimmed or not, suggesting that experience with intact beak flock and increased uptake of management strategies may have positive effects in a cumulative way (DEFRA, 2015). This study shows that even the application of multiple management strategies unfortunately cannot guarantee the prevention of severe pecking damage in intact beak flocks.

There has been much interest in the prospect of genetic manipulations that may produce birds that exhibit reduced rates of injurious pecking. The genetic basis for differing propensities towards injurious pecking has been known for some time, in particular between brown and white laying strains (e.g. Damme and Urselmans, 2013). Within strains, individual selection for damaging pecking behaviour appears to be feasible with heritability between 0.05 and 0.65 (Jendral and Robinson, 2004). Despite success in experimental contexts (e.g. Kjaer et al., 2001), progress in this area from commercial breeding companies remains limited, as their programmes have focused on individual selection for egg laying performance and bird survival. More recently, there has been considerable effort devoted to selection for traits related to social behaviour and damaging pecking. It is now recognised that using survival as an isolated trait for selection may actually make matters worse by leading to the birds that peck being selected for breeding. Since each bird's survival is greatly affected by the injurious pecking behaviour of its group mates, information on group performance must be incorporated (Muir, 1996; Ellen et al., 2007). Such an approach resulted in a 10% decrease in mortality in the first generation (Rodenburg et al., 2010) and holds considerable promise for genetic improvement of laying hens (Nicol et al., 2013). Genomic selection technology is currently being used in selection programmes aimed at bird survival and may accelerate genetic advances in this area. Despite this potential for progress, the strong influence of multiple environmental factors on the development of feather pecking suggests that a 'genetic fix' will never completely eliminate these tendencies in laying hens.

6 Conclusions and future directions for research

Injurious pecking in laying hens remains a complex and intractable problem. The unpredictability of pecking outbreaks and their extremely severe implications for welfare make a reliable prevention strategy imperative. Currently, although much progress has been made in the identification and practical implementation of a range of management strategies, it seems that alone or in combination, these are not sufficient to consistently prevent severe pecking outbreaks. Beak trimming, although not absolutely effective, is currently the most reliable way of preventing severe feather pecking and cannibalism. Despite this, consumer and lobby group pressure to eliminate beak trimming continues, and many governments view continuing to allow beak trimming as an interim measure in a long-term plan to ban the practice. The UK's Beak Trimming Action Group takes this view, stating 'deferring of any ban on beak trimming of laying hens should only be an interim step' and 'We believe there is considerable scope for further improvement in the control of injurious pecking through greater efforts by producers to implement further mitigating management interventions' (DEFRA, 2015). Further research should continue to investigate the commercial viability of managing intact beak flocks in a sustained effort to move away from beak trimming.

It is undeniable that IR beak trimming is a traumatic process for chicks, which is associated with acute pain. Its long-term welfare consequences appear to be minimal, but there are gaps in our knowledge. Much of the available research has examined IR trimming in comparison with HB trimming, and more comparative work with intact beak flocks would inform us on the absolute and not the relative, welfare costs that it imposes. An obvious area for future research is collection of more data on the long-term behavioural consequences of IR beak trimming, to elucidate its underlying mechanisms and potentially provide reassurance that chronic welfare effects are minimal or absent. More research on the likelihood of neuroma formation (and whether these are symptomatic and persistent) following IR trimming at day old is also needed. Although beak trimming confers production and therefore economic advantages unrelated to feather pecking, these are not sufficient to justify this undesirable practice. However, while it remains the case that we cannot reliably control injurious pecking in flocks with intact beaks under commercial conditions, current evidence suggests that the welfare costs of IR beak trimming are probably justified.

7 Where to look for further information

FeatherWel Project. http://www.featherwel.org/injuriouspecking.

DEFRA (2005), A guide to the practical management of feather pecking and cannibalism in free-range laying hens. https://www.gov.uk/government/uploads/system/uploads/attachment_data/file/69374/pb10596-feather-pecking-050309.pdf.

Nicol, C. J., Bestman, M., Gilani, A.-M., De Haas, E. N., De Jong, I.C., Lambton, S., Wagenaar, J. P., Weeks, C. A. and Rodenburg, T. B. (2013), The prevention and control of feather pecking: application to commercial systems. *World's Poultry Science Journal* 69: 775–87.

Rodenburg, T. B., Van Krimpen, M. M., De Jong, I. C., De Hass, E. N., Kops, M. S., Riedstra, B. J., Nordquist, R. E., Wagenaar, J. P., Bestman, M. and Nicol, C. J. (2013), The prevention and control of feather pecking in laying hens: identifying the underlying principles. *World's Poultry Science Journal* 69: 361–74.

8 References

Andrade, A. N. and Carson, J. R. (1975), The effect of age and methods of debeaking on future performance of White Leghorn pullets. *Poultry Science* 54: 666–74.

Angevaare, M. J., Prins, S., van der Staay, F. J. and Nordquist, R. E. (2012), The effect of maternal care and infrared beak trimming on development, performance and behaviour of Silver Nick hens. *Applied Animal Behaviour Science* 140(1–2): 70–84.

Bennewitz, J., Bogelein, S., Stratz, P., Rodehutscord, M., Piepho, K. P, Kjaer, J. B. and Bessei, W. (2014), Genetic parameters for feather pecking and aggressive behaviour in a large F2-cross of laying hens using generalized linear mixed models. *Poultry Science* 93(4): 810–17.

Blokhuis, H. J., van der Haar, J. W. and Koole, P. G. (1987), Effects of beak trimming and floor type on feed consumption and body weight of pullets during rearing. *Poultry Science* 66: 623–5

Blokhuis, H. J. (1989), The effect of a sudden change in floor type on pecking behaviour in chicks. *Applied Animal Behaviour Science* 22: 65–73.

Blokhuis, H. J., Fiks-van Niekerk, T., Bessei, W., Elson, A., Guemene, D., Kjaer, J. B., Levrino, G. A. M., Nicol, C. J., Oester, H., Tauson, R., Weeks, C. A. and De Weerd, H. A. V. (2007), The LayWel project: welfare implications of changes in production systems for laying hens. *World's Poultry Science Journal* 63: 101–14.

Breward, J. (1984), Cutaneous nociceptors in the chicken beak. *Journal of Physiology*, 346, 56.

Breward, J. and Gentle, M. J. (1985), Neuroma formation and abnormal afferent nerve discharges after partial beak amputation (beak trimming) in poultry. *Experentia* 41: 1132–4.

Carruthers, C., Gabrush, T., Schwean-Lardner, K., Knezacek, T. D., Classen, H. L. and Bennett, C. (2012), On-farm survey of beak characteristics in White Leghorns as a result of hot blade trimming or infrared beak treatment. *Journal of Applied Poultry Research* 21(3): 645–50.

Cheng, H. (2006), Morphopathological changes and pain in beak trimmed laying hens. *World's Poultry Science Journal* 62: 41–52

Craig, J. V., Craig, J. A. and Milliken, G. A. (1992), Beak trimming effects on beak length and feed usage for growth and egg production. *Poultry Science* 71: 1830–41.

Cunningham, D. L. (1992), Beak trimming effects on performance, behavior and welfare of chickens: a review. *Journal of Applied Poultry Research* 1(1): 129–34.

Daigle, C. L., Rodenburg, T. B., Bolhuis, J. E., Swanson, J. C. and Seigford, J. M. (2014), Use of dynamic and rewarding environmental enrichment to alleviate feather pecking in non-cage laying hens. *Applied Animal Behaviour Science* 161: 75–85.

Damme, K. and Urselmans, S. (2013), Infrared beak treatment – a temporary solution? *Lohman Information* 48(2): 36–44.

Davis, G. S., Anderson, K. E. and Jones, D. R. (2004), The effects of different beak trimming techniques on plasma corticosterone and performance criterial in Single Comb White Leghorn hens. *Poultry Science* 83(10): 1624–8.

DEFRA (2015), The Beak Trimming Action Group's Review. https://www.gov.uk/government/publications/beak-trimming-action-group-review.

Dennis, R. L., Fahey, A. G. and Cheng, H. W. (2009), Infrared beak treatment method compared with conventional hot blade trimming in laying hens. *Poultry Science* 88(1): 38–48.

Dennis, R. L. and Cheng, H. W. (2012), Effects of different infrared beak treatment protocols on chicken welfare and physiology. *Poultry Science* 91(7): 1499–505.

Dennis, R. L. and Cheng, H. W. (2010a), A comparison of infrared and hot blade beak trimming in laying hens. *International Journal of Poultry Science* 9(8): 716–19.

Dennis, R. L. and Cheng, H. W. (2010b), Effects of beak trimming on pecking force. *International Journal of Poultry Science* 9(9): 863–6.

Dubbeldam, J. L., De Bakker, M. A. G. and Bout, R. G. (1995), The composition of trigeminal nerve branches in normal adult chickens and after debeaking at different ages. *Journal of Anatomy* 187: 619–27

Duncan, I. J. H., Slee, G. S., Seawright, E. and Breward, J. (1989), Behavioural consequences of partial beak amputation (beak trimming) in poultry. *British Poultry Science* 30: 479–88

Ellen, E. D., Muir, W. M. and Bijma, P. (2007), Genetic improvement of traits affected by interactions among individuals: sib selection schemes. *Genetics* 176: 489–99.

Freire, R., Eastwood, M. A. and Joyce, M. (2011), Minor bear trimming in chickens leads to loss of mechanoreception and magnetoreception. *Journal of Animal Science* 89(4): 1201–6.

Freire, R., Glatz, P. C. and Hinch, G. (2008), Self-administration of an analgesic does not alleviate pain in beak trimmed chickens. *Asian-Australasian Journal of Animal Sciences* 21(3): 443–8.

Gentle, M. J. (1991), The acute effects of amputation on peripheral trigeminal afferents in *Gallus gallus* var *domesticus*. *Pain* 46: 97–103.

Gentle, M. J. (1986a), Neuroma formation following partial beak amputation (beak trimming) in the chicken. *Research in Veterinary Science* 41(3): 383–5.

Gentle, M. J. (1986b), Beak trimming in poultry. *World's Poultry Science Journal* 42: 268.

Gentle, M. J and Breward, J. (1986), The bill tip organ of the chicken (*Gallus gallus var. domesticus*). *Journal of Anatomy* 145: 79–85.

Gentle, M. J., Hughes, B. O., Fox, A. and Waddington, D. (1997), Behavioural and anatomical consequences of two beak trimming methods in 1- and 10-day old chicks. *British Poultry Science* 38: 453–63

Gentle, M. J., Hunter, L. N. and Waddington, D. (1991), The onset of pain related behaviours following partial beak amputation in the chicken. *Neuroscience Letters* 128: 113–16.

Gentle, M. J. and McKeegan, D. E. F. (2007), Evaluation of infrared beak trimming in broiler breeder chicks. *Veterinary Record* 160(5): 145–8.

Gentle, M. J., Thorp, B. H. and Hughes, B. O. (1995), Anatomical consequences of partial beak amputation (beak trimming) in turkeys. *Research in Veterinary Science* 58: 158.

Gentle, M. J., Waddington, D., Hunter, L. N. and Jones, R. B. (1990), Behavioural evidence for persistent pain following partial beak amputation in chickens. *Applied Animal Behaviour Science* 27: 179–57.

Gilani, A.-M., Nicol, C. J. and Knowles, T. G. (2012), The effect of dark brooders on feather pecking on commercial farms. *Applied Animal Behaviour Science* 142: 42–50.

Gilani, A.-M., Knowles, T. G. and Nicol, C. J. (2013), The effect of rearing environment on feather pecking in young and adult laying hens. *Applied Animal Behaviour Science* 148: 54–63.

Glatz, P. C. (2009), Effect of hot blade and infrared beak trimming on beak condition, production and mortality of laying hens. *Proceedings of the 8th European Symposium on Poultry Welfare*, Cervia, Italy, 18–22 May. World's Poultry Science Association.

Glatz, P. C. (1987), Effects of beak trimming and restraint on heart rate, food intake, body weight and egg production in hens. *British Poultry Science* 28: 601–11.

Glatz, P. C. and Lunam, C. A. (1994), Production and heart rate responses of chickens beak trimmed at hatch or at 10 or 42 days of age. *Australian Journal of Experimental Agriculture* 34: 443–7

Glatz, P. C., Murphy, L. B. and Preston, A. P. (1992), Analgesic therapy of beak trimmed chickens. *Australian Veterinary Journal* 69: 18.

Grams, V., Bogelein, S., Grashorn, M. A., Bessei, W. and Bennewitz, J. (2015), Quantitative genetic analysis of traits related to fear and feather pecking in laying hens. *Behavior Genetics* 45(2): 228–35.

Green, L. E., Lewis, K. Kimpton, A. and Nicol, C. J. (2000), Cross-sectional study of the prevalence of feather pecking in laying hens in alternative housing systems and its associations with management and disease. *Veterinary Record* 147: 233–8.

Hartcher, K. M., Tran, K. T. N., Wilkinson, S. J., Hemsworth, P. H., Thomson, P. C. and Cronin, G. M. (2015), The effects of environmental enrichment and beak trimming during the rearing period on subsequent feather damage due to feather pecking in laying hens. *Poultry Science* 94(5): 852–9.

Henderson, S. N., Barton, J. T., Wolfenden, A. D., Higgins, S. E., Higgins, J. P., Kruenzel, W. J., Lester, C. A., Tellez, G. and Hargis, B. M. (2009), Comparison of beak-trimming methods on early broiler breeder performance. *Poultry Science* 88: 57–60.

Hester, P. Y. and Shea-Moore, M. (2003), Beak trimming egg-laying strains of chickens. *World's Poultry Science Journal* 59: 458–74.

Huber-Eicher, B. and Wechsler, B. (1997), Feather pecking in domestic chicks, its relation to dustbathing and foraging. *Animal Behaviour* 54: 757–68.

Hughes, A. L. and Buitenhuis, A. J. (2010), Reduced variance of gene expression at numerous loci in population of chickens selected for high feather pecking. *Poultry Science* 89: 1858–69.

Hughes, B.O. (1973), The effect of implanted gonadal hormones on feather pecking and cannibalism on pullets. *British Poultry Science* 14: 341–8.

Hughes, B. O. and Gentle, M. J. (1995), Beak trimming of poultry – its implications for welfare. *World's Poultry Science Journal* 51(1): 51–61.

Jendral, M. J. and Robinson, F. E. (2004), Beak trimming in chickens: historical, economical, physiological and welfare implications, and alternatives for preventing feather pecking and cannibalistic activity. *Avian and Poultry Biology Reviews* 15(1): 9–23.

Jensen, A. B., Palme, R. and Forkman, B. (2006), Effect of brooders on feather pecking and cannibalism in domestic fowl (*Gallus gallus domesticus*). *Applied Animal Behaviour Science* 60: 25–41.

Johnsen, P. F., Vestergaard, K. S. and Norgaard Neilsen, G. (1998), Influence of early rearing conditions on the development of feather pecking and cannibalism in domestic fowl. *Applied Animal Behaviour Science* 60: 25–41.

Keeling, L. (1994), Feather pecking – who in the group does it, how often and under what circumstances? In *Proceedings of the 9th European Poultry Conference*, 288–99.

King, A. S. and McLelland, J. (1975), Nervous system. In Outlines of Avian Anatomy, Cassel and Collier Macmillan Publishers Ltd., London, 106.

Kjaer, J. B. and Vestergaard, K. S. (1999), Development of feather pecking in relation to light intensity. *Applied Animal Behaviour Science* 62: 243–53.

Kjaer, J. B., Sorensen, P. and Su, G. (2001), Divergent selection on feather pecking behaviour in laying hens (*Gallus gallus domesticus*). *Applied Animal Behaviour Science* 71: 229.

Kuenzel, W. J. (2007), Neurobiological basis of sensory perception: welfare implications of beak trimming. *Poultry Science* 86: 1273–82.

Lambton, S. L., Knowles, T. G., Yorke, C. and Nicol, C. J. (2010), The risk factors affecting the development of gentle and severe feather pecking in loose housed laying hens. *Applied Animal Behaviour Science* 123: 32–42.

Lambton, S., Nicol, C. J., Friel, M., Main, D. C. J., McKinstry, J. L., Sherwin, C. M., Walton, J. and Weeks, C. A. (2013), A bespoke management package can reduce the levels of injurious pecking in loose house laying hen flocks. *Veterinary Record* 172: 423–30.

Leeson, S. and Morrison, W. D. (1978), Effect of feather cover on feed efficiency in laying birds. *Poultry Science* 57: 1094.

Lunam, C. A., Glatz, P. C. and Hsu, J. H. (1996), The absence of neuromas in beaks of adult hens after conservative trimming at hatch. *Australian Veterinary Journal* 74: 46–9.

Marchant-Forde, R. M. and Cheng, H. W. (2010), Different effects of infrared and one-half hot blade beak trimming on beak topography and growth. *Poultry Science* 89(12): 2559–64.

Marchant-Forde, R. M., Fahey, A. G. and Cheng, H. W. (2008), Comparative effects of infrared and one-third hot blade trimming on beak topography, behaviour and growth. *Poultry Science* 87(8): 1474–83.

McKeegan, D. E. F. and Philbey, A. W. (2012), Chronic neurophysiological and anatomical changes associated with infra-red beak treatment and their implications for laying hen welfare. *Animal Welfare* 21(2): 207–17.

McKeegan, D. E. F. and Savory, C. J. (1999), Feather eating in layer pullets and its possible role in the aetiology of feather pecking damage. *Applied Animal Behaviour Science* 65: 73–85.

Muir, W. M. (1996), Group selection for adaptation to multiple hen cages: selection program and direct responses. *Poultry Science* 75: 447.

Newberry, R. C. (2004), Cannibalism. In G. C. Perry (ed.), *Welfare of the Laying Hen*, 239–58. Wallingford, United Kingdom, CABI Publishing.

Newberry, R. C., Keeling, L. J., Estevez, I. and Bilcik, B. (2007), Behaviour when young as a predictor of severe feather pecking in adult laying hens: The redirected foraging hypothesis revisited. *Applied Animal Behaviour Science* 107: 262–74.

Nicol, C. J., Bestman, M., Gilani, A.-M., De Haas, E. N., De Jong, I. C., Lambton, S., Wagenaar, J. P., Weeks, C. A. and Rodenburg, T. B. (2013), The prevention and control of feather pecking: application to commercial systems. *World's Poultry Science Journal* 69: 775–87.

Nicol, C. J., Lindberg, A. C., Phillips, A. J., Pope, S. J., Wilkins, L. J. and Green, L. E. (2001), Influence of prior exposure to wood shavings on feather pecking, dustbathing and foraging in adult laying hens. *Applied Animal Behaviour Science* 73: 141–55.

Persyn, K. E., Xin, H., Nettleton, D., Ikeguchi, A. and Gates, R. S. (2004), Feeding behaviors of laying hens with or without beak trimming. *Transactions of the ASAE* 47(2): 591–6.

Prescott, N. B and Bonser, R. H. C. (2004), Beak trimming reduces feeding efficiency of hens. *Journal of Applied Poultry Research* 13(3): 468–71.

Prescott, N. B., Wathes, C. M. and Jarvis, J. R. (2003), Light, vision and the welfare of poultry. *Animal Welfare* 12: 269–88.

Rodenburg, T. B. and Koene, P. (2003), Comparison of individual and social feather pecking tests in two lines of laying hens at ten different ages. *Applied Animal Behaviour Science* 81: 133–48.

Rodenburg, T. B., Bijma, P., Ellen, E. D., Bergsma, R., De Vries, S., Bolhuis, J. E., Kemp, B. and Ven Arendonk, J. A. M. (2010), Breeding amiable animals? Improving farm animal welfare by including social effects in breeding programmes. *Animal Welfare* 19: S77–82.

Rodenburg, T. B., Buitenhuis, A. J., Ask, B., Uitdehaag, K. A., Koene, P., van der Poel, J. J., van Arendonk, J. A. M. and Bovenhuis, H. (2004), Genetic and phenotypic correlations between feather pecking and open-field response in laying hens at two different ages. *Behaviour Genetics* 34: 407–15.

Rodenburg, T. B., Van Krimpen, M. M., De Jong, I. C., De Hass, E. N., Kops, M. S., Riedstra, B. J., Nordquist, R. E., Wagenaar, J. P., Bestman, M. and Nicol, C. J. (2013), The prevention and control of feather pecking in laying hens: identifying the underlying principles. *World's Poultry Science Journal* 69: 361–74.

Savory, C. J. (1995), Feather pecking and cannibalism. *World's Poultry Science Journal* 51: 215–19.

Savory, C. J. and Griffiths, J. D. (1997), Individual variation in rates of giving and receiving feather pecks in bantams, and some behavioural correlates. In *5th European Symposium on Poultry Welfare*, Wageningen, The Netherlands.

Sepeur, S., Spindler, B., Schulze-Bisping, M., Habig, C., Andersson, R., Beyerbach, M. and Kemper, N. (2015), Comparison of plumage condition of laying hens with intact and trimmed beaks kept on commercial farms. *European Poultry Science* 79. doi: 10.1399/eps.2015.116.
Simonsen, H. B., Vestergaard, K. S. and Willeburg, P. (1980), Effect of floor type and density on the integument of egg layers. *Poultry Science* 59: 2202–6.
Tauson, R. (2005), Management and housing systems for layers – effects on welfare and production. *World's Poultry Science Journal* 61: 477–90.
Tauson, R. and Svenson, S. A. (1980), Influence of plumage condition on the hen's feed requirement. *Swedish Journal of Agricultural Research* 10: 35.
Van Krimpen, M. M., Kwakkel, R. P, Reuvekamp, B. F. J., van der Peet-Schwering, C. M. C., Den Hartog, L. A. and Verstegen, M. W. A. (2005), Impact of feeding management on feather pecking in laying hens. *World's Poultry Science Journal* 61: 663–85.
Van Krimpen, M. M., Kwakkel, R. P., Van der Peet-Schwering, C. M. C., Den Hartog, L. A. and Verstegen, M. W. A. (2009), Effects of nutrient dilution and nonstarch polysaccharide concentration in rearing and laying hen diets on eating behaviour and feather damage of rearing and laying hens. *Poultry Science* 88: 759–73.
Vezzoli, G., Mullens, B. A. and Mench, J. A. (2015), Relationships between beak condition, behaviour and ectoparasite infestation levels in laying hens. *Poultry Science* 94(9): 1997–2007.

Chapter 7

Maintaining the health of laying hens: a practical approach

Richard M. Fulton, Michigan State University, USA

1 Introduction

No treatise on maintaining the health of any poultry, including egg-laying hens, can be written without discussing biosecurity, which is the protection of the birds from agents that may cause disease. Biosecurity also prevents the spread of disease-causing agents such as bacteria, viruses and parasites from one group of birds to another. Biosecurity is a way of life that must be adopted and practised by everyone who comes in contact with poultry. Although there are many complex strategies to maintain biosecurity, these strategies can be classified as

- separation
- people
- traffic control
- equipment and vehicles
- air, feed and water

http://dx.doi.org/10.19103/AS.2016.0012.27

By providing proper biosecurity, the most common mechanisms of disease spread can be prevented and controlled. This chapter reviews the range of biosecurity measures required to maintain the health of laying hens. It also reviews disease identification and detection, including the identification of specific diseases and vaccination regimes for prevention or treatment.

2 Biosecurity measures: separation

Separation means keeping the birds protected and separate from each other and sorted by age. This is especially important for chickens that are yet to attain the egg-producing age (pullets) and for egg-laying hens. Separation, as for other elements of biosecurity, is best applied prior to building bird housing facilities. Young chickens are typically more susceptible to disease than old chickens. When constructing a farm that will have both pullets and laying hens, the pullet facilities should be built upwind of the layer farm. Since some agents can be borne on the wind, a minimum distance of 0.5 miles (0.8 km) is recommended. The more the distance between the facilities, the better.

Separation is also maintained by providing a break or buffer zone in facilities where workers must change their footwear, outer clothing and head covering (hair nets or hats) before entering the compartment where the birds are housed. Ideally, it is a seperate building where workers arrive to work on the farm. The workers enter the dirty side (the section just inside the exterior opening), remove their clothing and shoes, take a shower, put on the provided clothing, and enter farm or bird room if shower facility is attached to bird-raising facility. On the clean side (the section adjacent to the entrance of the bird room), workers will don clothing, head coverings and footwear provided by the farm before entering the farm. A separate facility at the edge of the farm is best.

At a minimum, this can be accomplished with an anteroom constructed just inside of an exterior door but before the area where the birds are kept. An anteroom placed prior to the bird room could be divided into a 'dirty' and 'clean' side. A real or imaginary break between these two sides may be delineated by a painted red line or by benches that stretch across the entire anteroom. People enter the anteroom; remove covering garments (hats, coats, jackets, etc.); step or swivel across the red line or bench; and put on building-specific coveralls, boots and head coverings on the 'clean' side of the anteroom. It is best if workers wash their hands before entering the bird room. Ideally, coveralls, hats and footwear should be designated for use in only one facility. Colour coding of coveralls, footwear and head cover attracts the immediate attention of observers, who can report this infraction to their superiors if items of the wrong colour are observed in areas other than those assigned.

Separation of birds of different ages is commonly referred to in the poultry industry as practising 'all in, all out' poultry management. For broiler chickens, one building or all buildings on a farm are stocked with chickens of the same age. When those chickens reach the market age, the entire building or the entire farm has to be processed before adding extra birds. 'All in, all out' management prevents disease from spreading between flocks, since the birds are not maintained in the building or on the farm continuously, and the buildings can be cleaned and disinfected prior to stocking with new chickens. Providing 'down time' when there is no poultry on a farm will help reduce the load of some disease-causing agents or eliminate them altogether. Unfortunately, because of scalability, land availability and costs of construction, many commercial egg layer farms house chickens of multiple ages. Birds of each individual age are typically housed in a single unit, but one often finds a conveyor that passes through multiple houses to transport the eggs to the central egg-processing building. In addition, the bird houses on some layer farms are constructed in close proximity to each other. Often, the air exhaust vents of one building are located directly across from the air intake vents of another building.

Separation from pets and pests including mice, rats, raccoons, woodchucks, possums and wild birds is also important to prevent the spread of disease. Cats and rats are known to harbour in their mouth *Pasteurella multocida*, the bacteria causing fowl cholera. *Mycoplasma gallisepticum* has been found in wild birds and is thought to have caused problems in turkeys (Ferguson, 2003). Bird facilities and their surroundings must, therefore, be constructed in such a way that prevents pests from being attracted to the buildings or from entering them. In addition, wild-bird screening must be in place on all openings of the buildings. The common rule of thumb for pest exclusion is that if a pest can get its head through an opening, it can get its entire body through the opening. At a minimum, a 10-foot (3 metre) zone surrounding the bird facility should be kept free of grass and weeds. This provides a buffer zone to prevent mice from entering the facility. Tall weeds and/or grass close to the buildings act as a hiding place for mice, allowing them to evade aerial predators. It also provides food source, including insects and weed seeds, for mice. Piles of unused equipment or trash should be kept at least 100 feet (30 metre) away from the buildings, since rats like to live in piles of rubbish and will travel 100 feet for food and water. Rodent bait stations should be placed every 25 feet (7.6 metres) along the exterior foundations as well as immediately beside any exterior openings on the interior of buildings. The type of rodent bait that is used should be changed every 3 months. Since most rodents have continuously growing incisor teeth, they must gnaw on things. A chew bait, therefore, is often a better choice than a seed bait or paste.

3 Other biosecurity measures

People must always be mindful of biosecurity and put it into practice. At first, biosecurity may not be natural behaviour, but with time, practice and a well-designed facility, it can become second nature. Only those people who have a definite need to visit the bird facilities should be allowed to do so. A visitor logbook should be maintained, where the names of the persons, the company they represent, the purpose for the visit, the areas that are visited, and the date and time they enter and leave the facility should all be recorded. A person should not be allowed into a poultry facility unless a period of at least 48 hours has elapsed since they last visited other poultry facilities. Outside contractors should be encouraged to come to the facility on Mondays since this will allow at least 48 hours between poultry visits. Ideally, visitors coming from other countries should allow at least seven days between contact with poultry and visiting poultry facilities.

People are known to spread disease from one group of poultry to another. They can carry disease agents on their clothing, shoes, hands and headwear. Most, if not all, poultry companies require their workers to sign a form stating that they will not own any poultry, visit anybody who owns poultry, visit cock fights, visit pet stores or poultry exhibits and hunt waterfowl or upland game birds. If hunting is allowed, a period of 48–72 hours is required before they can return to work. They will also be asked to declare whether or not they live with anybody who works at another poultry farm. Employees can be spot-checked for compliance, although this is difficult to enforce. Most, if not all, egg companies outline, in writing, the biosecurity practices used to protect their birds. All new workers are trained on biosecurity practices at the farm. Some companies require that their workers retrain in biosecurity once a year.

Traffic control describes how visitors and workers progress through the farm or perform their jobs. They should go from young birds to old birds and from healthy birds to sick birds. Following this routine will help prevent the spread of disease. Farm workers should not be shared with multiple farms. If workers must be shared, it is best if they remove their headwear, outer clothing and shoes and then take shower and don clean clothing and disinfected footwear before going to another farm. Some companies that lack appropriate facilities make arrangements with local motels for their workers to carry out this activity. When workers have to be with sick flocks, they are not allowed to go to other flocks that same day.

Traffic control also includes the control of vehicles and equipment. All unessential vehicles should be excluded from a farm. If possible, vehicles that visit multiple farms, like feed trucks or rendering trucks, should be excluded from the farm. If it is not possible to totally exclude them, at least their distance from the bird buildings should be maximized. An ideal way for excluding feed trucks from a farm is to construct a road at the edge of the property. A fence should be placed along the road to separate the road from the farm. Feed tanks should be located next to the fence on the side of the building. Feed trucks and their drivers need not have to enter the farm.

If separation is not possible, vehicles that are allowed to enter the farm should have their tyres and undercarriages washed and sprayed with a disinfectant. Doors of the poultry facility should always be locked. Feed truck and other vehicle drivers should never enter the buildings. A container, such as a rural delivery mailbox, should be affixed to or be kept near feed tanks for the drivers to deposit feed delivery tickets. Vehicles should be parked at least 30 feet (10 metres) away from the buildings. It is best that vehicles visiting farms be parked at the farm gate and not enter the farm. It is best that no equipment be shared between multiple poultry facilities. If a necessity arises that equipment must be shared between facilities, it should be cleaned to remove all organic material (feed, manure, feathers, litter, etc.) and disinfected just before transporting it to a particular farm and cleaned and disinfected when it arrives at the next farm. Stations where rendering trucks pick up farm waste, including routine mortalities, should be constructed at the edge of the farm property so as to prevent the trucks from entering the farm. Care should be taken to ensure that there is no cross traffic contamination of vehicles or people from the farm while carrying materials to the rendering station.

Control of air, feed and water should also be addressed to prevent the spread of disease. Air is best controlled by placement of facilities (see separation discussion above). Research facilities should use filtered air at intakes, and the air leaving bird facilities should be filtered prior to exiting. Following the Spring 2015 outbreak of Eurasian H5N2 avian influenza in the Unites States, air filtration has become a priority in commercial facilities. Intake air may be drawn in from the roof peak and air exhausted from the ends or sides, and production facilities must consider this system rather than the current system in which air enters intakes located near the ground on the sides or ends of buildings. Feed should be kept in closed containers (and one has to make sure feed bin lids are always closed) at all times to prevent wild bird and other pests from intruding. Feed spills should be cleaned up immediately to prevent wild birds or rodents from getting attracted to the facilities. Feed mills and other feed manufacturing facilities should be designed and utilized in such a way that cross-contamination of finished feed with raw feed ingredients is prevented.

Sanitation of water has already caused a dramatic improvement in disease prevention in meat-type poultry and is also gaining acceptance in egg production. The use of halogens such as chlorine or iodine and/or hydrogen peroxide has already become common practice in commercial poultry. These compounds, when added to drinking water, will control the spread of bacteria and viruses.

4 Disease identification and detection

Identifying disease in poultry may sometimes prove difficult. Often, sick birds tend to hide their illness from flock mates and thus from the casual observer. When other birds in the flock observe a sick flock mate, they pester them for long periods which may even lead to the death of the sick birds. The most common presenting history provided on diagnostic cases submitted to the author by either caged pet bird owners or small flock owners is, 'Bird was fine yesterday. Found dead this morning.' When these birds are examined, they are often found to be very thin, with marked atrophy of breast muscle and lack of body fat. This change occurs over several weeks or even several months. When sick, no matter the cause, birds will be reluctant to move, stand with their head tucked close to their body, appear sleepy and often have their feathers fluffed up. They may have a hunched or puffed up appearance. In some circumstances, there may be respiratory signs such as a snick (like a sneeze), rattling when they breathe or cough when a respiratory disease is present. There may be a paralysed leg, a dropped wing or a twisted neck if the bird has a disease of the nervous system. Watery diarrhoea may be present with either enteric disease or renal disease. Often there are no other signs. Many a time, sick birds pass a large amount of foul-smelling faeces and die soon afterwards. Owners of small flocks observe this phenomenon in multiple birds that die weeks apart and they often think that this happens because of a specific disease, illness or syndrome. That activity, however, is considered normal bird behaviour associated with illness and death. When an individual bird dies, it is often just normal attrition. But when multiple birds in a flock die the same day and there are signs of illness, it becomes important to investigate what is going on in the flock.

Flock disease may be detected by having multiple birds within a flock with the above presentations. More commonly, disease in a flock is best detected by monitoring daily mortality, feed and/or water intake, and egg production in egg-laying chickens. With large numbers of birds, it is likely that some will die every day due to natural causes. This is especially true of egg-laying chickens, since many factors associated with egg production can go wrong and cause hens to die. In the poultry industry, this phenomenon is referred to as daily mortality and it is closely monitored. Some diseases may cause increase in the daily mortality rate. Some diseases may cause the daily mortality to rise rapidly, while some others cause this mortality to rise slowly. The daily mortality rate may continue to rise until steps are taken to stop it, or, in some cases, the rate may peak and then decline until it reaches the normal level. The normal mortality rate for egg-laying chickens varies with breed and management conditions, but it is typically 3% from hatching until 60 weeks of age or is 0.05% per week (Hy-Line, 2015). Paying attention to more subtle changes in feed or water consumption and/or egg production may also provide early detection of disease.

Feed and water intake varies with the strain of the chicken. Most if not all layer breeder companies provide information about these statistics for their customers in their published production guides. Owners can use these guides to determine whether or not their birds reach standards recommended for a particular strain. These production guides are updated frequently by breeder companies, and copies can be requested from them or can even be downloaded from their websites. For example, Hy-Line International makes the production guide for their most popular leghorn strain, Hy-Line W-36, available on the Internet (Hy-Line, 2015b). For example, 18- to 90-week-old Hy-Line W-36 chickens during egg production consume approximately 22 pounds (9.8 kg) of feed per day for 100 hens. Water consumption increases with age and also depends on the environmental temperature. When it is hot, birds, which are warm blooded like humans, decrease their feed consumption and increase their water intake. As a general rule, from 18 weeks of age onwards, water consumption is at the rate of 2.5–5 gallons (10–20 litres) per day for 100 hens. During periods of moderate temperatures, water intake generally parallels feed consumption. Birds that are sick may decrease or even stop eating. Their water intake will decrease to parallel their lack of feed intake. Birds will continue to drink water up to and until their time of death, but at a much-reduced rate. Chickens that lay brown eggs typically eat more feed than those that lay white eggs. For instance, Hy-Line International's Hy-Line Brown chicken (Hy-Line, 2015a) from 18 to 90 weeks of age consumes approximately 24.2 pounds (10.9 kg) of feed per day per 100 hens. Water consumption for the same strain of bird is approximately 4.0–6.1 gallons (15–23 litres) per day per 100 hens.

Unexplained drops in egg production may also provide an early indication of a disease problem. The pattern of the drop in egg production, and whether or not egg production returns to normality, will also vary with each disease. Although feed consumption and egg production can be an early indicator of flock disease, it may also reflect issues that are directly related to management or feed errors rather than to the disease itself.

5 Vaccination of poultry

Vaccination of poultry usually requires mass application, since the number of birds within a unit of housing is typically in the thousands, and even in the hundreds of thousands. Vaccination is usually applied through drinking water or by spray. Application by injection or eye drop is seldom used because of labour expenses. Injection and eye drop methods are used in some vaccinations when they are required for successful vaccination.

Most egg-laying chickens receive vaccines in the hatchery through either subcutaneous injection – as is the case with Marek's disease vaccine, ILT- or infectious-bursal-disease-vectored Marek's disease vaccine – or spray vaccination – as is the case with infectious bronchitis (IB), Newcastle disease, salmonellosis and coccidiosis vaccines. Once the chicks are placed in pullet-rearing facilities, vaccines are given via drinking water or by spray for those that have not been vaccinated previously, or booster vaccine is given to those that have been vaccinated before to boost the efficacy of the vaccination. Booster vaccines are given every two to three weeks for a series of two or three vaccinations. Pox combined with AE vaccines are typically injected after six weeks of age. Prior to being relocated to the laying facility, the pullets may be given an injection of a killed vaccine containing IB, Newcastle disease and *Salmonella enteritidis*. Instead of administering a killed vaccine, many companies will use a live-boost-in-lay vaccine, which contains modified live viruses of IB and Newcastle disease.

The reader may notice that some agents that are included in the vaccination protocol are not discussed previously. *Salmonella enteritidis* is given in an attempt to prevent eggs from being contaminated with that bacteria, since it is considered to be a food safety issue. *Salmonella enteritidis* is not a disease of egg-laying chickens and hence not discussed herein.

6 Common diseases of egg-laying chickens: respiratory diseases

This discussion will provide a synopsis of diseases that are common to egg-laying chickens, starting with respiratory diseases. If the reader requires a full discussion of the topic, he or she should read discussions about the disease found in what is referred to as the 'bible of poultry diseases', *Diseases of Poultry* (Swayne, 2013).

Respiratory diseases are the most common diseases in egg-laying chickens. One of the most common viral respiratory diseases of egg-laying chickens is infectious bronchitis (IB). Infectious bronchitis is caused by a coronavirus. Being an RNA virus, IB virus has a tendency to mutate and there are approximately ten different serotypes. There is little if any cross protection among the serotypes, and thus, no single vaccine will protect against all other serotypes. IB infects only chickens, and it can occur at any age. If very young female chickens get infected, some segments of their reproductive tract will not develop properly, thus affecting their egg-producing capacity. IB is typically a respiratory disease which causes birds to gasp, cough and have tracheal rales. Increased mortality rates may be seen with concomitant or secondary infections. In chickens actively producing eggs, there can be a drop in production up to 70%. Usually, it is a 5–10% production drop, with a return to normal production levels at 2–3 weeks postinfection. Changes in egg quality are often reported, with pallor of egg shells in brown eggs, wrinkled egg shells and runny thick albumen being the most common. In commercial settings, it is often discovered when customers complain about the runny egg whites. Liquefaction of the albumen can occur normally if eggs are stored for long periods of time. Dissection of birds that die from IB may reveal a slight reddening of 1/3 of the trachea closest to the thoracic inlet.

Vaccination for IB is performed with many modified live vaccines given during the pullet stage. Some companies will vaccinate in the hatchery at one day of age. Others will delay first vaccination until two or three weeks of age. Booster vaccines are given every two or three weeks until a total of three or four vaccinations are received. To provide protection against IB viruses that commonly circulate on multiple age egg layer farms, some companies use a killed IB vaccine around 16 weeks of age, while others provide protection in egg-producing hens via a live-boost-in-lay programme. Live-boost-in-lay is where a modified live IB vaccine is given every four to six weeks while birds are in egg production. Some strains of IB cause kidney problems, with birds dying from kidney stones (urolithiasis) and visceral gout. These strains may cause a gradual rise in daily mortality in affected flocks, but are not common. It is the author's experience that many egg quality and gout problems have been blamed on IB, but when further investigated, IB was not proven to be the cause. Many egg-quality and gout problems can be traced to nutrition issues.

The second most common respiratory disease of egg-laying chickens is Newcastle disease. It is caused by a paramyxovirus. Like IB, it is an RNA virus. The Newcastle disease virus however changes very little when compared to IB virus. Newcastle disease can occur in other avian species, with turkeys being the most commonly infected commercial poultry species next to chicken. Newcastle disease virus (NDV) typically spreads from house to house in large, multi-age egg-laying facilities, yet it seldom if ever causes clinical disease. NDVs are classified on the basis of their ability to cause disease and mortality. The mildest strains are called lentogenic, the moderate strains are called mesogenic and the highly virulent strains are called velogenic. In the United States, lentogenic and a few mesogenic strains of NDV are most common. Occasionally, viscerotrophic velogenic Newcastle disease virus (vvNDV) occurs within the United States, and the disease it causes is referred to as Exotic Newcastle disease, since it is not commonly found within that country and therefore is considered to be an exotic disease. However, vvNDV is found in many countries throughout the world. NDV often causes disease of the central nervous system as well as respiratory disease, referred to as pneumoencephalitis.

Clinical signs in non-vaccinated chickens are similar to those of IB, but with facial oedema. With mesogenic strains, there may be a mixture of respiratory signs and nervous system signs such as a twisted neck (torticollis), head tremors and neck paralysis. vvNDV infections typically cause death within 48 hours postinfection. Gross dissection of birds with mesogenic strains may reveal reddening of the entire trachea and red, wet and heavy lungs. Gross dissection of birds that die from vvNDV usually reveals haemorrhages in the oesophagus and trachea as well as the gastrointestinal system. Gross dissection is not able to differentiate vvNDV from highly pathogenic avian influenza. Avian influenza rarely occurs in commercial egg-laying chickens and the reader is referred to *Diseases of Poultry* for a detailed discussion (Swayne, 2013).

Egg production in commercial egg-laying chickens is typically not affected by Newcastle disease, since egg-laying chickens are well vaccinated against this disease. There is little to no mortality with Newcastle disease. However, vvNDV is a catastrophic illness and may cause an extremely high and rapid death loss in an infected flock, sometimes up to 100% mortality within three or four days. Vaccination against Newcastle disease is usually the same as for IB, since these two viruses are often combined in the same vaccine vial. Humans may get conjunctivitis when exposed to either the wild type or vaccine NDV. The disease typically disappears after three days.

Infectious laryngotracheitis (ILT) is a viral respiratory disease of chickens caused by a herpes virus. ILT occurs sporadically in egg-laying chickens and often occurs in outbreak scenarios with large death loss. This disease is often transmitted by people and equipment. It may also occur when vaccinated birds are comingled with non-vaccinated birds. Since ILT is a herpes virus infection, when a bird is infected with herpes virus it is infected for life; birds that have been vaccinated can be asymptomatic carriers. Under periods of stress, the herpes virus evades the immune system, is shed and causes clinical disease in non-vaccinated chickens.

Clinical signs of ILT are those of difficulty breathing (dyspnea), often with 'pump handle breathing', unusual vocalization and with shaking of the head which spreads blood on and near the affected bird's environment and flock mates. 'Pump handle breathing' is when an infected bird raises its head to inhale and drops its head to exhale. This type of breathing is due to partial obstruction of the trachea by an inflammatory exudate. Daily mortality rises rapidly, and it may double or triple with each passing day as the disease spreads to the other birds within a facility. It typically starts as a death loss in a particular area of the building (often where the virus was introduced by people and/or equipment) and spreads from that area, affecting other areas of the facility. Gross dissection of birds that die from ILT may reveal haemorrhage and/or yellow material (fibrin) in the trachea. ILT can be differentiated from NDV and highly pathogenic avian influenza, since ILT produces haemorrhage and/or fibrin in the trachea alone and not in other tissues.

Vaccination for ILT in some individual states of the USA is controlled by the State Veterinarian's office. The use of modified live ILT vaccine should only occur in the face of an outbreak. ILT spreads slowly enough in a flock that is going through an outbreak of ILT such that vaccination of that flock can stop the spread of the disease and the associated death loss. Chicken embryo origin (CEO) ILT vaccine is the best one to use in an outbreak scenario. It is typically applied by spraying all of the birds on the farm. It is important when vaccinating large poultry facilities against ILT in an outbreak scenario to not allow a time when vaccination is halted for a day or two, since the author is aware of an occasion where ILT spread to adjacent nonvaccinated buildings when a vaccine crew quit for the weekend during a response to an outbreak. A plan should be developed to eliminate the use of modified live ILT vaccine containing the whole virus on a farm after the outbreak is under control. Switching to a tissue culture origin (TCO) modified live ILT vaccine could be the next reasonable step and this should be used on replacement pullets being added to the layer flock. Once all the layer flocks on the farm are vaccinated with the TCO vaccine, a genetically modified vaccine can be used to exclude the whole virus from the farm. As of this writing, there are a pox-virus vectored ILT and a turkey herpes-virus vectored ILT vaccine which only contain the protective part of the ILT virus. The use of the vectored vaccines allows protection against ILT without the danger of vaccinated birds shedding vaccine virus and causing additional outbreaks. Turkey herpes-virus vectored ILT vaccine is given at the hatchery. This vaccine also provides protection against Marek's disease. Pox-virus vectored vaccine is given at six to eight weeks of age and provides protection against pox as well as ILT.

Mycoplasma gallisepticum is a bacterium that typically causes a mild respiratory disease when non-vaccinated birds or naïve birds are exposed to this agent. *Mycoplasma gallisepticum* can spread through the egg to the chick, also known as vertical transmission, as well as horizontally to flock mates. Most chicks and pullets are free of *M. gallisepticum* since infection with this agent is frequently screened for in breeder flocks and breeder flocks that become infected are typically destroyed. Some layer facilities are chronically infected with *M. gallisepticum* since they are never without poultry flocks. Introduction may be by people, wild birds or infected poultry.

Clinical signs are those of mild respiratory signs such as a snick or tracheal rales. Birds typically do not die from this infection, but may, if also, have gotten infected with *Escherichia coli*. Birds that die or are sacrificed during the acute infection may have small amounts of fibrin within their air sacs. Once the acute infection is completed, birds that die from other causes will have no gross lesions associated with *M. gallisepticum* infection. In commercial egg-laying operations, the farmer can expect the *M. gallisepticum*-infected birds to lay 12 less eggs per year than those chickens that are not infected with *M. gallisepticum*. For commercial operations, it is evident that there is an issue. *Mycoplasma* infections in small (backyard) flocks are common. In these flocks, the infection is silent and the owners never know their flocks are infected. In commercial operations that become infected with *M. gallisepticum*, there can be a drop in the egg production of up to 20%. Egg production will gradually rise over the next two to three weeks, but it will never return to normal levels.

Treatment with antibiotics approved for use in the United States, such as tylosin and chlortetracycline, can ease some of the clinical signs and help egg production but they do not provide a sterilizing treatment (i.e. ways for getting rid of the mycoplasma). Currently, there is no egg withdrawal mandated for chlortetracycline or tylosin use in laying hens. Vaccination is practised in flocks that are infected or in pullets that will be taken to an infected facility. Vaccination typically consists of at least two doses of vaccines with a modified live vaccine. Vaccination for *M. gallisepticum* does not eliminate the egg production loss, but will ameliorate it.

Colibacillosis caused by the bacterium *Escherichia coli* produces many syndromes in egg-type chickens. There can be umbilical infections commonly known as mushy chick disease or more formally as omphalitis. It can produce an oviductal infection known as salpingitis, and, as it relates to respiratory disease, it can produce laryngeal plugs and airsacculitis/peritonitis. The respiratory form of airsacculitis/peritonitis is regional in distribution, with some farms and companies more severely affected than others.

Clinical signs are absent and the condition is typically found as a gradually increasing daily mortality rate, with a cumulative mortality rate higher than target rates. Laryngeal plugs are often found in birds dying in a specific location, often adjacent to intake or exhaust fans. Birds die soon after a worker blows the fan blades and fan shroud to remove the accumulated dust. Dissection of those birds reveals that the opening to the trachea (larynx) is partially to completed occluded by a plug of easily crumbled, yellow material (fibrin). Dissection of birds that die from airsacculitis/peritonitis reveals the presence of fibrin on the capsular surface of the liver, sometimes with fibrin on the surface of the heart or within the air sacs.

Treatment with chlortetracycline can be effective if it is given early in the infection, typically within the first 7–10 days of a rising mortality. Providing treatment afterwards typically fails to cause any effect. Many of the isolates from such cases are resistant to most if not all of the approved antibiotics. A modified live *E. coli* vaccine is available. The first vaccine should be applied at hatch, and a booster vaccine at 12–14 weeks of age. It is thought by some that *Mycoplasma synoviae* may contribute to the development of *E. coli* peritonitis. Although *Mycoplasma synoviae* is not believed to cause respiratory disease, there have been reports of strain variations. Many of the egg-laying flocks in the United States are infected with *Mycoplasma synoviae,* with little or no known effects.

Infectious coryza is a respiratory disease caused by the bacterium *Avibacterium* (previously *Haemophilus*) *paragallinarum*. Most strains of *Av. paragallinarum* require the addition of V-factor for culture, but there are also some V-factor-independent strains in Mexico and South Africa. There are three serovars of *Av. paragallinarum* referred to as A, B and C, with little or no cross protection between the serovars. Thus, if vaccination is to be performed, the vaccine being used should contain the offending serovar. In the United States, infectious coryza is most prevalent in California egg-laying flocks and is seen in the southern part of the country. Outbreaks of infectious coryza can cause a mortality rate as high as 34% (Swayne, 2013). Sources of infection are chronically ill and asymptomatic carriers, people and equipment.

Clinical signs consist of swollen faces, conjunctivitis, nasal discharge, coughing, sneezing, rattling, while breathing and a characteristic foul odour within the flock have been reported. Often, there has been a drop in feed consumption and a correlated drop in egg production. Gross dissection reveals swollen faces with oedema and inflammation of the tissue under the skin of the face, a yellow exudate within sinuses and possibly fibrin within air sacs.

Treatment is suggested with the use of an approved antibiotic such as erythromycin or chlortetracycline in the drinking water until it can be added to the feed. Relapse may occur after antibiotic treatment is discontinued. Vaccination with the offending serovar containing killed vaccine (bacterin) is recommended with at least two doses of vaccines given three to four weeks apart, with the last dose of vaccine being given four weeks prior to the birds being placed in a contaminated facility. Cleaning, disinfection and complete depopulation of a facility is required to eliminate the infection.

Although not truly a respiratory disease, fowl pox virus occasionally causes problems in the trachea of egg-laying chickens. In some cases, tracheal pox infection is complicated by ILT infection as well, and it can be difficult to separate the mortality caused by tracheal pox from that caused by ILT. Pox viral infection in the trachea is called diphtheritic or wet pox since it affects mucous membranes (wet skin). Another more common form of pox viral infection in chickens is called cutaneous or dry pox since it affects normal (dry) skin of the face and comb. In dry pox, egg production may be decreased as much as 10% and feed consumption may be decreased. There is little or no increased mortality with dry pox. In wet pox, mortality rate will be increased.

Clinical signs in dry pox are those of the visible lesions of raised areas with scabs and crusts. Clinical signs in wet pox are those of gasping for air and sometimes of 'pump handle' breathing as described for ILT. Gross dissection will reveal papules, pustules and scabs on the skin in the dry form. No internal lesions are associated with dry pox. Polypoid masses are formed on the tracheal epithelium when the respiratory tract is affected. Pox virus is found within the scabs produced by the infection, and when shed, the virus can persist for months, or even years, within the environment. Infection typically develops when the virus enters a break in the skin by biting insects, equipment or by contaminated needles. The tracheal infection may develop without a break in the tracheal epithelium.

Since pox is a viral disease, there is no treatment for pox infection. In an outbreak, however, the spread of the virus is slow and vaccination during an outbreak may prevent the disease from spreading. Outbreaks in commercial flocks have been linked to vaccine application failures (personal experience). Fowl pox vaccine is typically used and should be given after six weeks of age. Vaccine is typically given by piercing the skin between the shoulder and carpus, known as the wing web, with needles dipped in the vaccine. Variant forms of fowl pox virus containing retrovirus gene segments have, in some cases, been responsible for losses in commercial laying hens that were previously vaccinated with fowl pox vaccine alone (Garcia, 2003). So far, the addition of pigeon pox vaccine with fowl pox vaccine in vaccine programmes has been able to prevent infections by variant strains.

7 Common diseases of egg-laying chickens: nervous system diseases

There are numerous nervous system diseases that affect commercial poultry. The ones that most commonly affect egg-laying chickens are Marek's and Newcastle diseases. The others are sporadic to rare in occurrence. Newcastle disease was described under respiratory diseases, since respiratory system is the most affected part in egg-laying chickens. Most strains of NDV in the United States do not produce nervous system disease. Marek's disease is, however, common in laying hens.

Marek's disease is probably the most common viral disease of chickens. If someone raises chickens long enough (no matter the scale of production), they will see Marek's disease and/or coccidiosis (an intestinal disease). Marek's disease is caused by a herpes virus that is different from ILT virus. The mature virus develops in the feather follicles and is shed in the dander produced by the infected chicken. Chicks are exposed to wild type Marek's disease virus within the first week of life from breathing the dander or dust-contaminated air found in poultry houses.

The incidence of Marek's disease has decreased quite dramatically due to the discovery and development of vaccines. In the 1950s and 1960s, egg-laying chicken farmers could expect to lose 30–60% of their flock due to Marek's disease. But the discovery of a vaccine virus dramatically decreased the mortality due to Marek's disease. Over time, common field isolates of Marek's disease virus were able to break through vaccination and developed a greater ability to cause disease. Additional vaccine discoveries ameliorated the field isolates affects, but there is always a risk that a new, more pathogenic field isolate may develop. To further decrease the incidence of Marek's disease in today's egg-laying chickens, some egg-type chicken breeders are now selecting breeding stock to build up resistance for Marek's disease. As such, the most common cause of Marek's disease in egg-type chickens today is vaccination errors.

Marek's disease most commonly occurs in one of two forms, either paralysis or lymphomas. The classical presentation for Marek's disease is a 12- to 14-week-old pullet that is paralysed with one leg forward and one leg backward. Paralysis can happen at any age, but it can most commonly occur before sexual maturity. In some forms of Marek's disease paralysis, the bird will get up and walk in about three days after the onset of paralysis. In the most common form, the bird will never be able to walk again and will die from disease complications. Lymphomas in chickens are caused by either Marek's disease virus or avian leukosis virus, a retrovirus. Being a retrovirus, avian leukosis virus inserts itself into the host DNA, allowing itself to be passed to the chick from the hen through the egg. Egg-laying breeders typically test their breeding stock for avian leukosis virus and eliminate those found to be positive. Thus, the incidence of avian leukosis is extremely low or completely absent in today's commercial egg-laying chickens. Marek's disease lymphomas can occur at any age. Since it is a viral disease, there is no treatment for Marek's disease, with only prevention available through vaccination. Vaccines for Marek's disease are typically given at hatch or *in ovo* in egg-laying chickens. There are many different types of Marek's disease vaccines. The choice of which vaccine to use is dependent on the pathogenicity of the wild type virus.

8 Common diseases of egg-laying chickens: intestinal system diseases

Coccidiosis is the most common intestinal disease of egg-laying chickens, as well as meat-type chickens and turkeys. Coccidiosis in chickens is caused by a single-cell protozoan parasite of the genus *Eimeria*. Coccidia are species specific in that they do not cross species barriers; chicken coccidia do not infect turkeys, ducks or gamebirds. There are many different species of coccidia that infect chickens. Specific species can be determined by observing the location within the intestinal tract of the lesions they produce and by determining the size of the oocyst (eggs) produced by that species. Determining the specific species is not necessary in all cases. It is important to determine the species of coccidia causing the infection to either ensure that the coccidial vaccine contains the species that is causing the problem or determine if the coccidiosis is not responding to treatment. Some species are more resistant than others to commonly used anti-coccidial drugs. Coccidiosis typically occurs in young chickens that have not developed immunity to the parasite or in older chickens when they have not been exposed previously to coccidia. Coccidiosis is more common in poultry species that are raised on litter or have access to manure. The parasite enters chickens when they peck at litter or manure that contains the coccidial oocysts (eggs). The parasite once inside of the intestines leaves the oocyst and enters the cells that line the intestines (enterocytes). That stage of coccidia multiplies to hundreds or even thousands of a different stage. Those stages then burst out of the cell and invade hundreds or even thousands of cells. The cycle continues again. After undergoing asexual reproduction and once these stages are freed from their host cell by bursting, those stages produce sexual stages which form the oocysts which then leave the body and mature on the litter or passed manure. The cycle begins all over again.

Clinical signs in naive chickens are those of reluctance to move, a puffed-up appearance, the head held close to the body and a sleepy appearance. There may be a bloody diarrhoea. Mortality rate can be high, reaching 30% in some cases. Gross dissection will reveal a variety of changes to the intestines. Some species of coccidia cause white stripes on the mucosal surface (inside) of the intestine. Some species cause ballooning of the intestine with or without blood and some cause yellow material (fibrin and death) of the lining of the intestine. Some species of coccidia cause more damage and therefore produce a higher mortality rate. Death loss may also be increased if there are complications with coccidiosis such as superinfection with *Clostridium perfringens*. This condition is commonly referred to as necrotic enteritis. Diagnosis is made by scraping the intestine, putting the material gained from scraping on a microscope slide and examining the material with a microscope for the presence of oocysts. The combination of lesions and the presence of oocysts provides the diagnosis of coccidiosis. A word of caution is necessary. Some species of coccidia produce their lesions in the middle of the small intestine and form oocysts only in the caeca. Therefore, it is important to always look at cecal material for oocysts. The discovery of oocysts in manure of adult chickens or chickens without intestinal lesions of coccidiosis does not warrant a diagnosis of coccidiosis since it only indicates that some of those parasites are present in the bird and that this presence is not causing any problems. The term for this condition is coccidiasis.

Prevention of coccidiosis is always better than treatment. The largest group of drugs used in poultry production consists of coccidiostats. The most commonly used coccidiostats includes polyether ionophores, carbanilides, febrifugine, quinolones, pyridones and vitamin analogs. However, most are not used in the egg type of chickens. The most commonly used coccidiostats in egg-laying chicken are ionophores and amprolium, a thiamine analog. In caged egg-laying chickens, coccidiostats are used whenever there is clinical coccidiosis or during times when exposure of naive birds are suspected. Those times are when pullets are first moved from the pullet facility to the laying facility, especially if the houses are constructed such that newly placed pullets have access to the manure belt from the tier of cages above. Coccidiosis appears to be in some housing facilities on a recurring basis. Cage facilities without manure belts will have some areas of the house where coccidiosis always breaks, whereas in manure belt houses, it can occur anywhere in the house, but most commonly in the first cages that have access to the manure belt on the return side after the manure has been dumped or scraped from the belt. As noted above, litter houses will have a higher incidence of coccidiosis than cage houses. Cage-free organic egg-laying chickens are not allowed to use medications including coccidiostats and therefore must rely on vaccination to prevent coccidiosis. Cage-free conventional egg-laying chickens will use coccidiostats more routinely than caged laying hens.

Vaccination against coccidiosis is becoming more common in egg-laying chickens. It is typically applied as a spray in the hatchery. The effectiveness of coccidiosis vaccination in chickens kept in cages has been variable (personal experience). Coccidiosis vaccination is a controlled infection where a multitude of coccidial species oocysts are given in reduced numbers. The chickens get a mild, sub-clinical infection and pass the vaccine oocysts in their manure. According to vaccine manufacturers, the vaccine must complete three cycles of infection, excretion, re-infection in order to be effective. This is difficult to accomplish in caged chickens, since the manure and oocysts fall through the cage floor.

Focal duodenal necrosis (FDN) is a recent phenomenon in egg-laying chickens. Clinical signs are those of failing to reach target case weights, by up to 2 pounds (0.9 kg) per case. Egg production may be decreased by up to 10%. (Hy-Line, 2013) FDN is thought to be caused by the bacterium *Clostridium colinum*. The condition can only be discovered by dissecting freshly euthanized laying hens. Gross dissection reveals grey to black areas on the serosal (outside) surface of the duodenum. The mucosa (inside) of the duodenum may show blackened areas to areas with fibrin adhered to the surface. Treatment with bacitracin, tylosin, tetracycline or chlortetracycline is recommended until target case weights have been reached; however, the condition may recur after discontinuing treatment.

Intestinal parasitism in caged laying hens is extremely rare. The most common intestinal parasite found in laying hens is tapeworm. There are eight species of tapeworms that occur in chickens in the United States. (Swayne, 2013). There are no approved treatments for tapeworms in egg-laying chickens. Prevention should be focused on control of flies and beetles. Cage-free and those egg-laying chickens having access to their own manure and litter have a much higher incidence of intestinal parasitism. The most common nematodes are *Ascaridia galli*, known simply as an ascarid, and *Heterakis gallinarum*, known simply as the caecal worm. There are no approved treatments for use in laying hens.

9 Common diseases of egg-laying chickens: skeletal system, urogenital system and unclassified diseases

9.1 Diseases of the skeletal system

Osteomalacia is the major skeletal system disease in egg-laying chickens and is related to the large demand of calcium metabolism related to producing an egg. The hen needs approximately 2.4 grams of calcium supplied to the oviduct each day to produce the egg shell. (NRC, 1994) The average white leghorn produces about 240 eggs per year. Calcium that is not made available in the diet is extracted from the bird's bones. This condition was first described as 'caged layer fatigue', where egg-laying chickens that were kept in cages had easily broken bones. They could not stand but could eat and drink. The condition is more like osteomalacia, which means 'softening of the bones', than like osteoporosis, which means 'loss of bone'. The origin of this disease is mostly nutritional in nature and the reader is referred to that section. This condition may occur in spite of there being an appropriate amount of calcium, phosphorus and vitamin D in the diet and can occur when there is an intestinal upset, which prevents the absorption of calcium from the diet. Osteomalacia is best prevented by routinely monitoring the flock for skeletal integrity. Routine monitoring of skeletal integrity without the use of sophisticated and expensive equipment unfortunately requires the humane euthanasia of a sample of the flock and gross dissection. Treatment by adding additional calcium to the ration in the form of dicalcium phosphate or lime stone while providing the water-soluble vitamin D3 in the drinking water may provide some effect. Adding to the diet a more active form of vitamin D, such as vitamin D3, than what is routinely added is sometimes effective in combatting this condition.

9.2 Urogenital system diseases

Gout is the most common result of kidney dysfunction in poultry. Birds excrete nitrogen primarily as uric acid. The low solubility of uric acid can cause problems of precipitation if it reaches high enough levels in the blood. (Scanes, 2015) These levels can be reached by conditions such as water deprivation (Julian, 1982, personal experience) and by kidney damage when layer feed containing 4.0% calcium is accidentally fed to pullets that normally receive feed containing 1.0% (personal experience). Clinical signs in cases of gout are failure to meet egg production targets and a steadily increasing daily mortality rate or a rapid rise in daily mortality, depending on the age of the bird and severity of the error. Water deprivation or calcium feed errors are worse in young chickens. Some egg producers have tried using urine acidifier such as ammonium sulphate, ammonium chloride or potassium chloride with mixed results. As expected, using these compounds will cause the manure to be wetter than normal.

In addition to gout, a condition known as urolithiasis occurs in egg-laying chickens. Urolithiasis is the formation of stone-like material within the ureters of the kidney. The causes are the same as for gout, but can also include some strains of IB (see respiratory diseases above).

9.3 Conditions not classified

Fatty liver syndrome is usually an individual bird problem within a flock but may cause an increase in mortality if enough chickens within the flock are affected. This condition is related to the birds consuming too many calories. The condition typically occurs in obese-laying hens. Lipids accumulate within the liver in response to the hormones of egg production. Affected livers may have a 70% lipid content. (Swayne, 2013) Affected birds die from haemorrhage when the liver ruptures. Gross dissection reveals a healthy bird with a pale comb and abundant subcutaneous and coelomic (abdominal) fat stores. The liver is often enlarged. There may be large amounts of blood within the coelom with or without clotting. Often, a blood clot is attached to the liver. The liver is easily punctured with a finger. The fingers will feel greasy upon handling the affected liver. If a piece of affected liver is placed in water, it will float rather than sink.

Vent prolapse and cannibalism is another condition that occurs in leghorn chickens to a much greater degree than other poultry species. When the egg is laid, the oviduct is protruded from the vent. Being red, it is a natural attractant to cage or flock mates. Other chickens peck at the oviduct and/or cloaca which can cause bleeding, another attractant. Further pecking can lead to ulceration of the oviduct and/or cloaca with prolapse of intestine through the resulting ulcer. Once protruded, other chickens will peck at the intestine, and even eat it. Consequently, the bird dies from evisceration. Leghorns are more cannibalistic than other commercial poultry excluding gamebirds. Dead birds may be found that have prolapsed vents alone. Vent prolapse is due to egg production and with a higher frequency in obese birds. The contribution of low blood levels of calcium (hypocalcaemia) to this condition is not known, but it should be considered.

Northern fowl mite, *Ornithonyssus sylviarum*, is a mite that can be a problem in egg-laying chicken flocks. It spends its entire life cycle on the chicken and can be found around the bird's vent. Their numbers can be in the tens of thousands and appear as black to brown staining of the feathers around the vent. Closer inspection of the feather staining will reveal small (0.5 mm) red to brown to black specks that move. These mites are common on wild birds and it is thought to have been introduced into a flock from wild birds. Treatment is difficult because of the complexity of modern egg-laying facilities. Prevention by excluding wild birds from poultry operations is important.

A mite commonly referred to as the chicken, red or roost mite, *Dermanyssus gallinae*, is a mite known to suck blood. This mite is typically found on chickens only at night when it gets a blood meal. The mite spends most of its time in the cracks and crevices of poultry housing. These mites are not commonly found in caged egg-laying chickens in the northern regions of the United States. This trend may change as the egg production in that country shifts to cage-free systems. Severe infection in young birds can lead to anaemia and death, and in adult birds it can cause a drop in egg production.

Egg production decline due to avian encephalomyelitis (AE) virus infection can occur in flocks that are not vaccinated. AE is caused by a Tremorvirus. AE rarely occurs in commercial egg-laying chickens due to the use of vaccination to prevent drop in production. AE is more of a problem in chicks originating from infected breeder flocks.

10 Where to look for further information

Many books and pamphlets are written about the subject. Two of the most helpful books are given below:

- American Association of Avian Pathologists (AAAP). A Practical Guide for Managing Risk in Poultry Production, 2011.
- United States Department of Agriculture. Biosecurity Guide for Poultry and Bird Owners, April 2014.

The 'bible of poultry diseases' is Swayne, D. E., J. R. Glisson, L. R. McDougald, L. K. Nolan, D. L. Suarez and V. Nair. *Diseases of Poultry*, 13th Edition, John Wiley & Sons, Inc. Ames, Iowa, 2013.

11 References

American Association of Avian Pathologists (AAAP). *Biosecurity in the Poultry Industry*, 1995.

American Association of Avian Pathologists (AAAP). *A Practical Guide for Managing Risk in Poultry Production*, 2011.

Ferguson, N. M., D. Hermes, V. A. Leiting and S. H. Kleven. Characterization of a naturally occurring infection of *Mycoplasma gallisepticum* house finch-like strain in turkey breeders. *Avian Dis.* 47: 523–30, 2003.

Garcia, M., N. Narang, W. M. Reed and A. M. Fadly. Molecular characterization of reticuloendotheliosis virus insertions in the genome of field and vaccine strains of fowl poxvirus. *Avian Dis.* 47: 343–54, 2003.
Technical Update: An Overview of focal duodenal necrosis (FDN). Hy-Line International, 2013. http://www.hyline.com/UserDocs/Pages/TB_FDN_ENG.pdf (accessed 8 January 2016).
A Hy-Line Brown Commercial Layers 2015. http://www.hyline.com/UserDocs/Pages/BRN_COM_ENG.pdf (accessed 22 December 2015).
B Hy-Line W-36 Commercial layers management guide 2015. http://www.hyline.com/UserDocs/Pages/36_COM_ENG.pdf (accessed 22 December 2015).
Julian, R. Water deprivation as a cause of renal disease in chickens. *Avian Path.* 11: 615–17, 1982.
National Research Council. Signs of nutritional deficiencies in chickens and turkeys: Mineral deficiencies in Nutrient requirements of Poultry, 9th Edition. National Academy Press, Washington, DC., p. 56, 1994.
Sturkie's Avian Physiology, C. G. Scanes, editor, Academic Press, London, England, 2015.
Swayne, D. E., J. R. Glisson, L. R. McDougald, L. K. Nolan, D. L. Suarez and V. Nair. Diseases of Poultry, 13th Edition, John Wiley & Sons, Inc. Ames, Iowa, 2013.
United States Department of Agriculture. Biosecurity Guide for Poultry and Bird Owners, April 2014.

Chapter 8

Managing laying hen flocks with intact beaks

Thea van Niekerk, Wageningen Livestock Research, The Netherlands

1 Introduction

Worldwide, the majority of laying hens are beak trimmed to prevent injurious pecking. Although this procedure has been refined over the past few years from hot blade trimming (in Europe mostly at 6 weeks of age) to infrared treatment in the hatchery (which is less painful and therefore is referred to as 'treatment' instead of 'trimming'), it is still a mutilation causing pain and discomfort to the birds (Cheng 2006, Marchant-Forde et al. 2008). Therefore, in various countries a ban on beak treatments has come into force (e.g. Sweden) or is being discussed (e.g. the United Kingdom, Germany and the Netherlands). In addition, the European legislation prohibits beak treatment of organic flocks. When it is no longer allowed to treat the beaks, farmers are facing the challenge of preventing injurious pecking behaviour in their flocks. This requires specific management skills.

Injurious pecking comprises any pecking that leads to feather or skin damage. It leads to bald flocks, which have a lower feed efficiency, and it may lead to wounds and mortality. Birds with trimmed beaks also perform injurious pecking behaviour, but the blunt beaks do not cause as much damage as intact beaks do. This makes the necessity to prevent this behaviour less urgent in flocks with trimmed beaks.

http://dx.doi.org/10.19103/AS.2016.0012.32

2 Types of feather pecking

There are various types of injurious pecking behaviour: (1) aggressive pecking, forceful pecks directed at the head or neck of the recipient to establish and maintain a dominance hierarchy, but usually not resulting in much feather damage; (2) gentle feather pecking, usually directed at the tips of the feathers without causing any damage; (3) severe feather pecking, consisting of forceful pecks and pulling out of feathers that are frequently eaten and resulting in feather loss on the back, vent and tail area; (4) tissue pecking in denuded areas, a form of cannibalistic pecking where pecking continues on the skin, leading to wounds; (5) vent pecking, possibly leading to the pulling out of the inner organs (Rodenburg et al. 2004, 2013, Savory 1995). Feather pecking is the pecking at feathers of conspecifics, as described under type 2 and 3 of injurious pecking behaviour. Sometimes a feather pecking bird may also peck at its own feathers. As gentle feather pecking does not harm the feathers, it is usually not regarded as problematic behaviour. However, gentle feather pecking can develop into severe feather pecking when the severity of the pecks increases and, therefore, is often regarded as a first sign in the onset of severe feather pecking (Rodenburg et al. 2004). Not all scientists support this hypothesis, as a relationship between gentle and severe feather pecking has not always been observed (Newberry et al. 2007). Aggressive pecking is part of the agonistic behavioural repertoire of chickens and most likely has a different motivational basis than feather pecking (Savory 1995). Tissue pecking and vent pecking are also distinct types of pecking behaviour, although they may be triggered by damage caused by severe feather pecking (Rodenburg et al. 2004).

Gentle feather pecking can already be seen in day-old chicks. Severe feather pecking can be seen in the rearing phase, but is mostly not very frequent. The first signs of severe feather pecking usually appear early in the laying period (Rodenburg et al. 2004, 2013). Rodenburg et al. (2004) found 15% of the hens performing severe feather pecking at 30 weeks of age. According to farmers' observations in the United Kingdom, feather pecking started on average at 23 weeks of age, with a range of 14–40 weeks of age (Lambton et al. 2015).

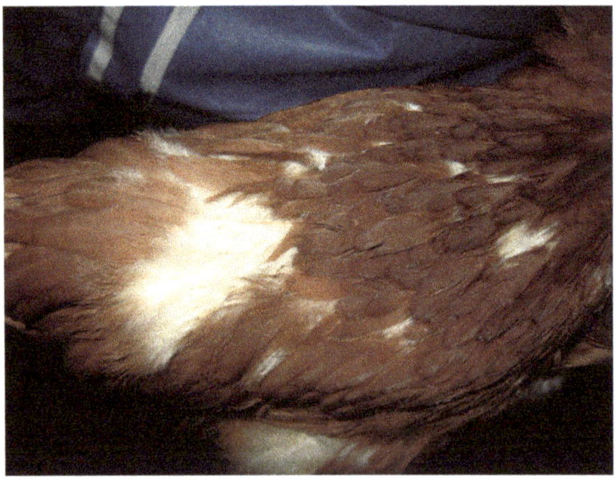

Figure 1 Signs of feather pecking at the base of the tail.

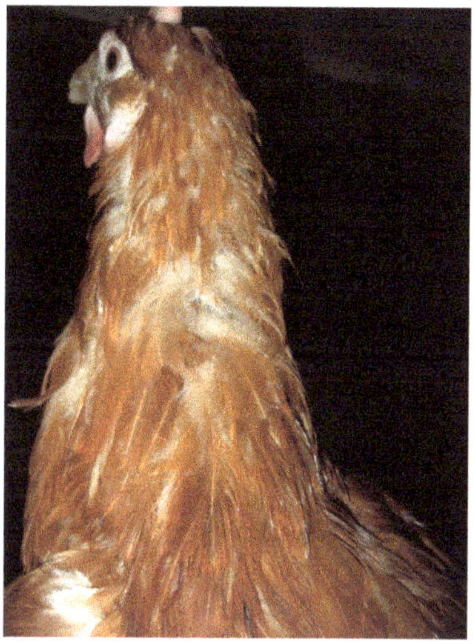

Figure 2 Rough feather cover due to moulting.

Not all feather damage is caused by feather pecking. Feather damage may also be caused by abrasion at feed troughs (front of the neck), wire side partitions (wings) or artificial grass mats in nest boxes (breast/belly) (Heerkens et al. 2015, Rodenburg et al. 2004, Savory 1995). Figure 1 presents clear signs of feather pecking, which often (but not exclusively) starts at the base of the tail. Sometimes feather pecking damage is hard to distinguish from rough feather cover due to moulting (Fig. 2).

3 The origin of feather pecking behaviour

In (semi-)wild conditions, chickens spend about 60% of their time foraging. Pecking behaviour is an element of foraging and part of the natural behavioural repertoire of chickens. Blokhuis (1986) showed that the motivation for non-aggressive pecking at conspecifics varies along with ground-pecking motivation, supporting the view of a common regulating mechanism. On the basis of this, Blokhuis developed the theory that feather pecking is redirected ground-pecking.

Vestergaard et al. (1993) indicated a relationship between feather pecking and absence of suitable dustbathing substrate. There is also evidence that early feather pecking could be interpreted as social exploration (McAdie and Keeling 2002, Rodenburg et al. 2004). Feather pecking has a hereditary component with a reported heritability ranging from 0.04 to 0.50 (Bessei 1996, Kjaer and Hocking 2004, Rodenburg et al. 2004). Selection for higher levels of feather pecking may also result in more fearfulness, higher stress levels and lower immune response (Bolhuis et al. 2009, Buitenhuis et al. 2004, De Haas et al. 2013). De Haas et al. (2014b) found a correlation for stress and feather damage between parents and offspring. They measured feather damage and stress in parent flocks and found that flocks with high levels of maternal feather damage and stress produced offspring with higher stress levels and more feather damage. This was particularly the case for white genotypes; for brown genotypes, the relations were not significant. This indicates that feather pecking may be a problem that needs to be addressed throughout the whole production chain.

4 Prevention of feather pecking: rearing conditions

Apart from the genetic predisposition and maternal influences, many factors have been identified as affecting the onset and severity of feather pecking. They can be led back to the theory of redirected ground pecking (Blokhuis 1986, Vestergaard et al. 1993) or to the findings that stress and fear may trigger feather pecking. All factors add up until a threshold is surpassed; at that point, feather pecking will start. Good management can reduce the amount each factor adds to reaching the threshold at which feather pecking starts. This is most successful in preventing the onset of feather pecking but, once it has started, it also helps to prevent escalation. Lambton et al. (2013) indeed showed that a management package does help to reduce levels of feather pecking, although, once started, feather pecking is very hard to stop (Bestman et al. 2009, De Haas et al. 2014a). This section will concentrate on how rearing conditions can be managed to minimise feather pecking. Section 5 then covers the prevention of feather pecking during the laying period.

Bestman et al. (2009) showed that the rearing period has a major influence on the incidence of feather pecking later in life. If (organic) birds did not show feather pecking in the rearing period, there was a 71% chance they would not start feather pecking in the laying period. Flocks that did perform feather pecking in the rearing period also did this in the laying period in 90% of cases. These findings were confirmed by other authors (De Jong et al. 2013, Gilani et al. 2013, Janczak and Riber 2015). Several authors have reported risk factors for feather pecking in the rearing period. One of the major factors is the presence or absence of litter. Although Nicol et al. (2001) indicated that birds could compensate for an earlier absence of litter, most authors find that early presence of litter influences the incidence of feather pecking later in life and disruptions in litter availability during rearing may lead to feather pecking in both the rearing and laying periods (Bestman et al. 2009, De Haas and Rodenburg 2014, Gunnarsson et al. 1999, Huber-Eicher and Sebö 2001, Vestergaard et al. 1997). Higher stocking densities during rearing are also associated with a higher risk for feather pecking. Bestman et al. (2009) found more feather pecking at a density of 34 birds/m^2 compared to 21 birds/m^2 (both during the first 4 weeks of rearing). Huber-Eicher and Audige (1999) found more feather pecking in flocks reared at densities of more than 10 birds/m^2. Stocking density, as such, only influences the space per bird. However, apart from the space availability, it is also important what the birds can do with it and thus what the design of the space is.

Various studies reported positive effects of a more complex environment on bird welfare (Colson et al. 2008, Janczak and Riber 2015). More complex refers to the presence of perches, elevated platforms with feed and water supplied on them, foraging material and other enrichments. Other enrichments may be provided as objects stimulating pecking behaviour, like pecking blocks and string devices. These reduce feather pecking, especially when introduced at a young age, preferably as early as day 1 (De Haas and Rodenburg 2014, McAdie et al. 2005). A more complex environment keeps birds busy and thus prevents them from developing feather pecking. This not only applies to the rearing period, but may also have a long-lasting effect throughout the laying period. For instance, perches in the rearing period reduce the incidence of feather pecking in the laying period (Huber-Eicher and Audige 1999) and may also reduce vent pecking (Donaldson and O'Connell 2012, Gunnarsson et al. 1999). Special measures in the rearing period may, therefore, be one of the best ways to prevent feather pecking throughout the whole life of the laying hen. One of these special measures is the so-called dark brooders. These are heaters with a dark enclosure underneath, placed on the litter in the rearing house, mimicking the brooding hen. Especially in the first week of rearing, when the chicks are held under 24-h light, the brooders are used by the chicks to hide and seek warmth. Gilani et al. (2012) found direct effects on the level of feather pecking during rearing, but also found less feather damage in the laying period in flocks that had been reared with dark brooders. Only a few trials on commercial farms have been carried out with dark brooders, but they seem to have a long-term preventive effect on feather pecking (Gilani et al. 2012, Jensen et al. 2006).

In terms of increasing environmental complexity, free range is probably the most enriched environment. Although providing additional ranging area is very uncommon in rearing birds, free range is available in organic rearing. Owing to climatic reasons, this only comprises the second half of the rearing, when the pullets are able to maintain their body temperature and have a good feather cover to withstand outside temperatures. Only a few studies have been conducted regarding the use of outdoor range for pullets (Janczak and Riber 2015) and all find similar results. Early use of, or access to, a free-range area reduced fearfulness, even if the duration of the outdoor access was only one week. In addition, birds that had access to free range during rearing tended to use the outdoor area in the laying period better (Krause et al. 2006). Other management measures that reduce stress and fear in both the rearing and laying phases, for example, a radio and frequent contact with humans, may also reduce the incidence of feather pecking (De Haas and Rodenburg 2014, Jones 2004, Lambton et al. 2013).

An often underestimated effect on the onset of feather pecking is the climate in the poultry house. Bad climatic conditions (high CO_2, ammonia) can result in stress and feather pecking (Drake et al. 2010). In a large survey on commercial farms, Drake et al. (2010) found an average of 20.9 ppm ammonia in rearing houses between 15 and 17 weeks of age, with the lowest being 0 and the highest being 100 ppm. Each 15 ppm higher ammonia level raised the chance of feather pecking by 10%. Average levels of CO_2 were 586, with the lowest being 43 and the highest being 2000 ppm. Each 200 ppm more CO_2 represented an almost 15% higher risk for feather pecking. Maintaining a good climate may be easier in multi-tier systems with manure belts compared to floor systems with a manure pit (Dekker et al. 2011).

Most authors report more feather pecking in flocks kept under higher light intensities (Drake et al. 2010, Janczak and Riber 2015, Kjaer and Vestergaard 1999), but some find the opposite (Bestman et al. 2009, Gunnarsson and Hermansson 2011). However, light sources in the different studies vary greatly, making the spectrum of the light completely different. Studies reporting daylight often do not mention if this is filtered by a glass window, taking out the UV component. Most artificial light sources hardly contain the ultraviolet spectrum, which can be seen by the birds and which may play an important role in their perception of the environment (De Jong et al. 2013, Gunnarsson et al. 2008). UV does not play a role in reproduction, but it does have an influence on the synthesis of vitamin D3 and thus on skeletal and skin health (Lewis et al. 2000). Moreover, it seems to play a role in the recognition of conspecifics and in vision of the environment. The exact mechanism is not known, but UV light seems to have an influence on feather pecking, where there is less feather pecking in the presence of UV (Lewis and Gous 2009). In recent years, LED has become increasingly popular. As this type of light does not seem to have a limitation in spectral range, it can be adapted to the needs of the birds. Although a lot is known about the vision of poultry (Lewis and Morris 2006), the optimal spectrum and best illumination in relation to the onset of feather pecking is not yet completely understood. In general, it is agreed that low-frequency light sources should be avoided, as poultry can see the flickering and may respond in a more stressful way (Lisney et al. 2012). Not only does this apply to fluorescent lights, but also LED may have a flickering frequency due to the dimming electronics. Dimming is often realised by rapid on–off switching of the LED. In some LEDs with low frequency of on–off switching, at a certain stage of dimming, this may be perceived as flickering (Niu and Takagaki 2016).

Feed has an important influence on the prevalence of feather pecking in both the rearing and laying periods. It is well known that deficiencies in minerals, proteins or amino acid levels (methionine, arginine) will lead to feather pecking (Krimpen et al. 2007). Structure also influences the prevalence of feather pecking. Small particle size will keep the birds occupied for a longer time and thus prevent the development of feather pecking behaviour. Not only research findings but also analysis of results on commercial farms confirmed this finding (Lambton et al. 2010). Low-energy diets require birds to eat more and thus may have the same result. Diets with high-fibre content will provide a longer satiation to the birds, may improve gut health and postpone the onset of feather pecking (Krimpen et al. 2005, Lambton et al. 2015, Van Krimpen et al. 2009). With regard to feed composition, Lambton et al. (2010) indicated that changes in formulation may lead to feather pecking. This seems to be in conflict with phase feeding strategies, where different formulations are used in different periods of the life of the laying hen in order to meet the demands of the birds as best as possible. Further research is needed to fine-tune these two seemingly conflicting interests. If the changes in formulation are limited, the risk for feather pecking may be lower than the benefit of an optimal feed formulation.

Finally, any change in husbandry may induce stress and lead to feather pecking (De Jong et al. 2013, Drake et al. 2010, Janczak and Riber 2015, Van de Weerd and Elson 2006). Changes are most pronounced in the transition from rearing to the laying house. This potentially stressful event can easily lead to feather pecking. The more similar the rearing and laying environments, the lower will be the level of stress. Similarities may concern the housing system (layout, feeder system, drinkers and light) and management (e.g. lighting conditions and schedule, feeding times and enrichment) (Janczak and Riber 2015). It is therefore advisable for laying hen keepers to have close contact with the rearer to prepare the pullets in the best way for their stay in the laying house. This also enables the laying hen keeper to start the laying period with a management package that is as much similar to the rearing management as possible.

5 Prevention of feather pecking: laying period

The prevention of feather pecking in the laying period does not differ to a large extent from the rearing period, as similar management and housing factors are involved and should be addressed. Some additional elements related to egg laying are involved, putting other demands on the birds. For some flocks, the free range may be an additional factor. Finally, the laying period comprises a longer period and problems at the start may have long-lasting effects. Hence it is important to monitor performance and behaviour of the birds thoroughly from the beginning. Although the propensity to develop feather pecking behaviour may originate in the rearing period, feather damage generally is not a problem before 20 weeks of age, and the severity increases with age (De Haas and Rodenburg 2014, Huber-Eicher and Sebö 2001, Lambton et al. 2010).

Although the possibilities to provide enrichments are limited in furnished cages, feather quality in general is better compared to non-cage systems (Nicol et al. 2013). This could be related to group size, as smaller groups are less prone to initiate feather pecking (Rodenburg and Koene 2007). Both group size and stocking density have effects on feather pecking, but often experimental setups fail to distinguish between the two. Higher stocking densities and larger groups seem to result in higher levels of fear and stress, and this may trigger feather pecking (De Haas et al. 2013, Nicol et al. 2013, Rodenburg et al. 2013, Zimmerman et al. 2006).

In addition, the type of housing system has an influence on feather damage. Birds in single-tier non-cage systems tend to have poorer feather cover than birds in aviaries (Sherwin et al. 2010). This may be because multi-level systems allow birds to escape from conflicts more easily. This is in line with the finding that elevated perches result in better feather cover if positioned well (Forkman 2005, Sandilands et al. 2009). Well positioned not only means easily accessible to prevent keel bone damage but also refers to the height of the perch. Perches on a height of 45 cm may lead to vent pecking, whereas perches at 70 cm height will not have this risk (Bilcik and Keeling 2000). The type of multi-tier system also has an influence on the feather cover. Flocks in aviaries with wire floors tend to have better feather cover compared to flocks in aviaries with plastic floors (Heerkens et al. 2015). This might be related to the presence of red mites, as in the same study infestations with red mites were higher in systems with plastic slats. In large quantities, red mites not only lead to anaemia and higher mortality in affected flocks, but also lead to higher levels of stress (Kowalski et al. 2006, Mul et al. 2009). Both health and stress issues increase the risk for feather pecking.

Enrichment in the hen house is important to prevent the development of feather pecking. Several researchers report the positive effects of enrichment materials, such as roughage, hay bales, string devices or pecking stones (Daigle et al. 2014, De Haas and Rodenburg 2014, Holcman et al. 2008, Lambton et al. 2013, McAdie et al. 2005). Dixon et al. (2010) reported the lowest level of feather pecking when foraging materials were provided and the highest level when no additional material was present. The presence of dustbaths and novel objects gave intermediate results. In general, enrichment material is thought to be more effective if it contains edible particles, as it keeps hens interested in the material (Lambton et al. 2010).

Free range is an important enrichment factor as well, with a known positive effect on feather pecking. Several studies have shown that a higher usage of the free range was associated with less feather damage due to feather pecking (Bestman and Wagenaar 2003, Green et al. 2000, Lambton et al. 2010). Hens may be encouraged to use the free range by providing suitable shelter, which can be artificial (roofs and nettings) or natural (trees and bushes). The size of the range is less important, but the degree of shade and shelter (i.e. quality of cover) does have a relation with the use of the range and thus the preventive effect on feather pecking (Bright et al. 2016).

Feed has been identified as a major factor in the prevention and control of feather pecking. Wild jungle fowl allocate 60% of their daytime to foraging (Dawkins 1989).

Laying hens still have this urge for foraging. Feed that is high in insoluble fibres, such as non-starch polysaccharides (NSPs), increases feeding time, and this leads to a later onset of feather pecking (Krimpen et al. 2007). Van Krimpen et al. (2009) increased the NSP concentration of the rearing diet from 124 to 184 g/kg, and in the laying diets from 72 to 115 g/kg, and found a significant better feather cover. Lambton et al. (2010) indicate that fibre content should be above 80 g/kg to reduce feather pecking. Fibres also have a positive effect on gut motility and health, which may reduce the incidence of feather pecking (Rodenburg et al. 2013). Finally, small particle sizes like mash feeding instead of pellets can prolong the feeding time and lead to less damage due to feather pecking (Lambton et al. 2010). Lambton et al. (2010) stated that, in addition to particle size and fibre content, protein content and the prevention of deficiencies are also important in preventing feather pecking (increased feather pecking was associated with the following: crude protein <125 g/kg, lysine <8.2 g/kg, methionine + cysteine <5.1 g/kg). Although it is often mentioned that feather pecking has increased after animal origin protein was removed from the feed (as a measure to reduce the risk of BSE-like health problems), Krimpen et al. (2011) could not confirm this effect. The ban on processed animal proteins in poultry feed made it necessary to use other protein sources. This resulted in a major change in feed formulation, which makes it hard to find a precise cause for the possibly higher levels of feather pecking. Lambton et al. (2010) found differences in feather pecking between flocks fed from different feed companies, indicating that feed formulation may have a major influence. This was also confirmed by Green et al. (2000) in a large survey of flocks in alternative housing systems. These authors found that three or more diet changes in the laying phase were associated with a higher risk for feather pecking.

Not only changes in feed but also changes in management may cause stress to the birds (Hemsworth 2009). Stress is known to have a relationship with feather pecking in that more stressed and fearful birds tend to perform more feather pecking (Rodenburg et al. 2010), which itself may also result in more fearful birds (Vestergaard et al. 1993). The level of stress and fearfulness varies between genetic lines (Bolhuis et al. 2009, Kjaer and Hocking 2004). Rodenburg et al. (2013) indicated that genetic lines that differ in the tendency to show feather pecking often also differ in fearfulness. The authors found White Leghorns to be more fearful than brown egg layers; this was confirmed by De Haas et al. (2013). On commercial farms, white hens indeed tend to be more fearful than brown hens. However, it also tends to be easier to keep white layers with intact beaks in non-cage systems compared to brown hens. This is not in compliance with the findings that higher fearfulness more easily leads to feather pecking. As white and brown hens have genetically different founder lines, there are probably more factors that may play a role and interact with fearfulness and feather pecking. For instance, white hens tend to be more susceptible for influences of the parent stock: stressful events in the parent stock causes the offspring to be more stressful and more likely to start feather pecking. Brown egg layers are more sensitive to the presence of good foraging material (De Haas et al. 2014b). Reduction of stress in laying hens can also be done through management measures such as knocking on the door before entering the house, walking calmly through the flock and turning on a radio (De Haas et al. 2014a, Hemsworth 2004). Flocks can be trained to react less to changes in management. If daily routines are varying over time, birds get used to different situations, which then do not cause stress anymore. In this perspective, one can think of wearing overalls, of different colours or changing the inspection route through the hen house frequently (not always from left to right, but also from right to left and sometimes starting in the middle of the house). In flocks that are extremely fearful from the start this may be a problem, but in other flocks this is a good measure to prevent stress in unforeseen cases (e.g. reparations in the house, vet visits).

Fearfulness and stress can be influenced by light intensity. Hens tend to react more calmly to novel objects in brighter light (Van Niekerk, non-published data), whereas dim lights may impair vision, making the birds less able to monitor their environment (Prescott and Wathes 2002). Negotiating distances may be more difficult in dimmed lighting, making it more difficult for birds to move through aviary systems (Moinard et al. 2004). In addition, hens are stimulated to forage under brighter light intensities (Prescott and Wathes 2002). A survey of 46 commercial farms in the Netherlands revealed that feather cover was better in flocks that were exposed to more daylight (with UV, not filtered by glass) (Van Niekerk, non-published data). This seems to be in conflict with the advice to dim the lights to prevent or reduce feather pecking. However, results on commercial farms are not always better in houses with dim lights. The quality of the light, and more specifically the spectrum, may be a key factor in this. There is still a lot unknown regarding the optimum light (spectrum and light intensity) for laying hens.

As mentioned earlier, health status does have an influence on feather pecking. Fibre content in the feed may have an influence on gut health. Consequently, a malfunction of the digestive and metabolic processing of the nutrients may result in deficiencies leading to feather pecking (Krimpen et al. 2005). Infectious diseases or worm infestations may cause a similar effect, although this is not always apparent or easy to prove (Gauly et al. 2007, Green et al. 2000). Hoglund and Jansson (2011) indicated that parasites reproduce more easily in hosts with a suboptimal condition, which could mean that birds that are under stress of feather pecking may be more susceptible to parasite infections. Meyer et al. (2013) found a correlation between feather eating and intestinal microbes, suggesting a relationship between feather eating and intestinal health. On the other hand, feather pecking may also be a result of suboptimal health. Buitenhuis and Kjaer (2008) found a relationship between selection against feather pecking and immune response, indicating that feather pecking may start in birds with a lower immune response, for example, as a result of a disease problem. Keeping the flock healthy and stress-free, therefore, may be one of the best measures to prevent feather pecking.

One of the measures to maintain a healthy poultry flock is good ventilation. This not only regulates temperature but also reduces ammonia and dust levels, and thus reduces the presence of endotoxins, airborne bacteria, fungi and other particles that may challenge bird health (Matkovic et al. 2009, 2010). Maintaining a good climate will also improve litter quality which, in turn, will encourage birds to be engaged with the litter instead of pecking at the feathers of conspecifics.

Lambton et al. (2013) investigated the effect of management on the prevalence of feather pecking and found that a bespoke management package indeed was protective against injurious pecking. This means that, even if feather pecking started in a flock, management can control it and thus is worth the effort.

6 Summary

Beak trimming of laying hens, as a measure to prevent injurious pecking behaviour, is a painful procedure that in several countries is or will be banned. Therefore, other strategies need to be applied to successfully keep laying hens with intact beaks. This phenomenon already starts in the rearing period with the provision of suitable litter material, providing perches and feed in small particles. If no feather pecking occurs in the rearing period, there is an increased chance it will not develop in the laying period either. The transition from rear to lay is important as it implies a stressful period for the birds, which can give rise to feather pecking. The more similar the environment, equipment, feed and management is between rearing and lay, the lower the risk for the development of feather pecking. During the laying period, suitable litter, additional foraging material, feed with high-fibre content and small particle size are important factors in preventing the start of feather pecking. Maintaining good health and preventing or counteracting stressors will further reduce the risk, as does a free-range area with enough cover. If feather pecking does occur, a package of management factors, similar to the preventive measures, can be used to counteract the problem. This may even reduce the pecking behaviour.

7 Future trends in research

Drake et al. (2010) reported less feather pecking if birds were reared in the same house that was used for laying. Further research is needed to confirm this and to investigate how to realise and optimise this.

Feed formulation is one of the key factors in preventing the onset of feather pecking. Research has been conducted to find optimum formulations with regard to fibre, protein and energy. Feed is often formulated in various phases, so it meets the specific demands of the laying hen in the different phases of the rearing and laying period. However, changes in feed formulation are indicated as risk factor for feather pecking as well. Therefore, research should focus on phase feeding formulations that meet the requirements of the birds, but are subtle and thus not triggering feather pecking.

Light is a factor that has not yet been thoroughly examined. Although a lot is known about the vision of poultry, little is known about the ideal spectrum with regard to production and the prevention of feather pecking. Further research is needed so that manufacturers of LED light sources can build the optimal light source for laying hens. Also the role of natural daylight should be studied in more detail to find the specific aspects that play a role in either triggering or preventing feather pecking.

8 Where to look for further information

Several excellent review papers have been published on the causation and prevention of feather pecking (De Jong et al. 2013, Drake et al. 2010, Lambton et al. 2013, Nicol et al. 2013, Rodenburg et al. 2004, Van de Weerd and Elson 2006). A lot of more farmer-oriented information about feather pecking can be found on the project websites of FeatherWel (www.featherwel.org) and Hennovation (www.henhub.eu).

9 References

Bessei, W. (1996), Genetics of feather pecking, in Sörensen, P. (ed.), *Proceedings 2nd European Poultry Breeders Roundtable*, 6–8 September 1995, Danish Institute for Animal Sciende, Research Centre Foulum: Report 73, pp. 9–21.

Bestman, M., Koene, P. and Wagenaar, J. P. (2009), Influence of farm factors on the occurrence of feather pecking in organic reared hens and their predictability for feather pecking in the laying period. *Applied Animal Behaviour Science*, *121*(2), pp. 120–5.

Bestman, M. W. P. and Wagenaar, J. P. (2003), Farm level factors associated with feather pecking in organic laying hens. *Livestock Production Science, 80*(1–2), pp. 133–40.

Bilcik, B. and Keeling, L. J. (2000), Relationship between feather pecking and ground pecking in laying hens and the effect of group size. *Applied Animal Behaviour Science, 68*(1), pp. 55–66.

Blokhuis, H. J. (1986), Feather-pecking in poultry: Its relation with ground-pecking. *Applied Animal Behaviour Science, 16*(1), pp. 63–7.

Bolhuis, J. E., Ellen, E. D., Van Reenen, C. G., De Groot, J., Ten Napel, J., Koopmanschap, R. E., Reilingh, G. D. V., Uitdehaag, K. A., Kemp, B. and Rodenburg, T. B. (2009), Effects of genetic group selection against mortality on behavior and peripheral serotonin in domestic laying hens with trimmed and intact beaks. *Physiology & Behavior, 97*(3–4), pp. 470–5.

Bright, A., Gill, R. and Willings, T. H. (2016), Tree cover and injurious feather-pecking in commercial flocks of free-range laying hens: a follow up. *Animal Welfare, 25*(1), pp. 1–5.

Buitenhuis, A. J. and Kjaer, J. B. (2008), Long term selection for reduced or increased pecking behaviour in laying hens. *Worlds Poultry Science Journal, 64*(4), pp. 477–87.

Buitenhuis, A. J., Rodenburg, T. B., Wissink, P. H., Visscher, J., Koene, P., Bovenhuis, H., Ducro, B. J. and Poel, J. J. V. D. (2004), Genetic and phenotypic correlations between feather pecking behavior, stress response, immune reponse, and egg quality traits in laying hens. *Poultry Science, 83*(7), pp. 1077–82.

Cheng, H. (2006), Morphopathological changes and pain in beak trimmed laying hens. *Worlds Poultry Science Journal, 62*(1), pp. 41–52.

Colson, S., Arnould, C. and Michel, V. (2008), Influence of rearing conditions of pullets on space use and performance of hens placed in aviaries at the beginning of the laying period. *Applied Animal Behaviour Science, 111*(3–4), pp. 286–300.

Daigle, C. L., Rodenburg, T. B., Bolhuis, J. E., Swanson, J. C. and Siegford, J. M. (2014), Use of dynamic and rewarding environmental enrichment to alleviate feather pecking in non-cage laying hens. *Applied Animal Behaviour Science, 161*(0), pp. 75–85.

Dawkins, M. S. (1989), Time budgets in Red Junglefowl as a baseline for the assessment of welfare in domestic fowl. *Applied Animal Behaviour Science, 24*(1), pp. 77–80.

De Haas, E. N., Bolhuis, J. E., de Jong, I. C., Kemp, B., Janczak, A. M. and Rodenburg, T. B. (2014a), Predicting feather damage in laying hens during the laying period. Is it the past or is it the present? *Applied Animal Behaviour Science, 160*(0), pp. 75–85.

De Haas, E. N., Bolhuis, J. E., Kemp, B., Groothuis, T. G. G. and Rodenburg, T. B. (2014b), Parents and early life environment affect behavioral development of laying hen chickens. *Plos ONE, 9*(3), p. e90577.

De Haas, E. N., Kemp, B., Bolhuis, J. E., Groothuis, T. and Rodenburg, T. B. (2013), Fear, stress, and feather pecking in commercial white and brown laying hen parent-stock flocks and their relationships with production parameters. *Poultry Science, 92*(9), pp. 2259–69.

De Haas, E. N. and Rodenburg, T. B. (2014), Feather pecking in laying hens during rearing and laying in non-cage systems, management practices and farmers opinions. In *Proceedings of the 48th Congress of the International Society for Applied Ethology (ISAE)*, Wageningen Academic Publishers, Wageningen, The Netherlands, pp. 159.

De Jong, I. C., Rodenburg, T. B. and Niekerk, T. G. C. M. V. (2013), Management approaches to reduce feather pecking in laying hens. *CAB Reviews: Perspectives in Agriculture, Veterinary Science, Nutrition and Natural Resources, 8*(036), pp. 1–8.

Dekker, S. E. M., Aarnink, A. J. A., Boer, I. J. M. d. and Groot Koerkamp, P. W. G. (2011), Emissions of ammonia, nitrous oxide, and methane from aviaries with organic laying hen husbandry. *Biosystems Engeneering, 110*(2), pp. 123–33.

Dixon, L. M., Duncan, I. J. H. and Mason, G. J. (2010), The effects of four types of enrichment on feather-pecking behaviour in laying hens housed in barren environments. *Animal Welfare, 19*(4), pp. 429–35.

Donaldson, C. J. and O'Connell, N. E. (2012), The influence of access to aerial perches on fearfulness, social behaviour and production parameters in free-range laying hens. *Applied Animal Behaviour Science, 142*(1–2), pp. 51–60.

Drake, K. A., Donnelly, C. A. and Dawkins, M. S. (2010), Influence of rearing and lay risk factors on propensity for feather damage in laying hens. *British Poultry Science, 51*(6), pp. 725–33.
Forkman, B. (2005), An overview of advantages, disadvantages of construction of perches for laying hens. *Dansk Veterinaertidsskrift, 88*(2), pp. 20–3.
Gauly, M., Duss, C. and Erhardt, G. (2007), Influence of Ascaridia galli infections and anthelmintic treatments on the behaviour and social ranks of laying hens (Gallus gallus domesticus). *Veterinary Parasitology, 146*(3–4), pp. 271–80.
Gilani, A.-M., Knowles, T. G. and Nicol, C. J. (2012), The effect of dark brooders on feather pecking on commercial farms. *Applied Animal Behaviour Science, 142*, pp. 42–50.
Gilani, A.-M., Knowles, T. G. and Nicol, C. J. (2013), The effect of rearing environment on feather pecking in young and adult laying hens. *Applied Animal Behaviour Science, 148*(1–2), pp. 54–63.
Green, L. E., Lewis, K., Kimpton, A. and Nicol, C. J. (2000), Cross-sectional study of the prevalence of feather pecking in laying hens in alternative systems and its associations with management and disease. *Veterinary Record, 147*(9), pp. 233–8.
Gunnarsson, S., Heikkila, M., Hultgren, J. and Valros, A. (2008), A note on light preference in layer pullets reared in incandescent or natural light. *Applied Animal Behaviour Science, 112*(3–4), pp. 395–9.
Gunnarsson, S. and Hermansson, A. (2011), *Lighting System For Laying Hens – Pre-Testing of New Technique in Sweden,* Tribun EU, Brno.
Gunnarsson, S., Keeling, L. J. and Svedberg, J. (1999), Effect of rearing factors on the prevalence of floor eggs, cloacal cannibalism and feather pecking in commercial flocks of loose housed laying hens. *British Poultry Science, 40*(1), pp. 12–18.
Heerkens, J. L. T., Delezie, E., Kempen, I., Zoons, J., Ampe, B., Rodenburg, T. B. and Tuyttens, F. A. M. (2015), Specific characteristics of the aviary housing system affect plumage condition, mortality and production in laying hens. *Poultry Science, 94*(9), pp. 2008–17.
Hemsworth, P. H. (2004), Human animal interactions, in Perry, G. (ed.), *Welfare of the Laying Hen*, Cabi Publishing, Cambridge, pp. 329–43.
Hemsworth, P. H. (2009), *Impact of Human-Animal Interactions on the Health, Productivity and Welfare of Farm Animals, Sustainable Animal Production,* Wageningen Academic Publishers, Wageningen.
Hoglund, J. and Jansson, D. S. (2011), Infection dynamics of Ascaridia galli in non-caged laying hens. *Veterinary Parasitology, 180*(3–4), pp. 267–73.
Holcman, A., Gorjanc, G. and Stuhec, I. (2008), Porous concrete block as an environmental enrichment device increases activity of laying hens in cages. *Poultry Science, 87*(9), pp. 1714–19.
Huber-Eicher, B. and Audige, L. (1999), Analysis of risk factors for the occurrence of feather pecking in laying hen growers. *British Poultry Science, 40*(5), pp. 599–604.
Huber-Eicher, B. and Sebö, F. (2001), The prevalence of feather pecking and development in commercial flocks of laying hens. *Applied Animal Behaviour Science, 74*(3), pp. 223–31.
Janczak, A. M. and Riber, A. B. (2015), Review of rearing-related factors affecting the welfare of laying hens. *Poultry Science, 94*(7), 1454–69.
Jensen, A. B., Palme, R. and Forkman, B. (2006), Effect of brooders on feather pecking and cannibalism in domestic fowl (Gallus gallus domesticus). *Applied Animal Behaviour Science, 99*(3–4), pp. 287–300.
Jones, R. B. (2004), *Environmental Enrichment: The Need For Practical Strategies to Improve Poultry Welfare, Welfare of the Laying Hen,* Cabi Publishing, Cambridge.
Kjaer, J. B. and Hocking, P. M. (2004), *The Genetics of Feather Pecking and Cannibalism, Welfare of the Laying Hen,* Cabi Publishing, Cambridge.
Kjaer, J. B. and Vestergaard, K. S. (1999), Development of feather pecking in relation to light intensity. *Applied Animal Behaviour Science, 62*(2–3), pp. 243–54.
Kowalski, A., Sokol, R. and Jedlinska-Krakowska, M. (2006), Influence of red mite Dermanyssus gallinae invasions on corticosterone and haematological levels and immunological indices in egg-laying hens. *Medycyna Weterynaryjna, 62*(10), pp. 1188–90.

Krause, E. T., Naguib, M., Trillmich, F. and Schrader, L. (2006), The effects of short term enrichment on learning in chickens from a laying strain (Gallus gallus domesticus). *Applied Animal Behaviour Science, 101*(3–4), pp. 318–27.

Krimpen, M. M. v., Kwakkel, R. P., Andre, G., Peet-Schwering, C. M. C. v. d., Hartog, L. A. d. and Verstegen, M. W. A. (2007), *Impact of Nutritional Factors on Feather Pecking Behaviour of Laying Hens in Non-Cage Housing Systems,* World's Poultry Science Association (WPSA), Beekbergen.

Krimpen, M. M. van, Kwakkel, R. P., Reuvekamp, B. F. J., Peet-Schwering, C. M. C. van der, Hartog, L. A. der and Verstegen, M. W. A. (2005), Reduction of feather pecking behaviour in laying hens by feeding management – a review. *Animal Science Papers and Reports, 23*(Suppl. 1), pp. 161–74.

Krimpen, M. M. van, Veldkamp, T., Binnendijk, G. and Veer, R. der (2011), Effect of four processed animal proteins in the diet on behavior in laying hens. *Applied Animal Behaviour Science, 132*(3/4), pp. 138–45.

Lambton, S. L., Knowles, T. G., Yorke, C. and Nicol, C. J. (2010), The risk factors affecting the development of gentle and severe feather pecking in loose housed laying hens. *Applied Animal Behaviour Science, 123*(1–2), pp. 32–42.

Lambton, S. L., Knowles, T. G., Yorke, C. and Nicol, C. J. (2015), The risk factors affecting the development of vent pecking and cannibalism in free-range and organic laying hens, *Animal Welfare, 24*, pp. 101–11.

Lambton, S. L., Nicol, C. J., Friel, M., Main, D. C. J., McKinstry, J. L., Sherwin, C. M., Walton, J. and Weeks, C. A. (2013), A bespoke management package can reduce levels of injurious pecking in loose-housed laying hen flocks. *Veterinary Record, 172*(16), p. 423.

Lewis, P. and Morris, T. (2006), *Poultry Lighting: The Theory And Practice,* Northcot, Andover.

Lewis, P. D. and Gous, R. M. (2009), Responses of poultry to ultraviolet radiation. *Worlds Poultry Science Journal, 65*(3), pp. 499–510.

Lewis, P. D., Perry, G. C. and Morris, T. R. (2000), Ultraviolet radiation and laying pullets. *British Poultry Science, 41*(2), pp. 131–5.

Lisney, T. J., Ekesten, B., Tauson, R., Håstad, O. and Ödeen, A. (2012), Using electroretinograms to assess flicker fusion frequency in domestic hens Gallus gallus domesticus. *Vision Research, 62*(0), pp. 125–33.

Marchant-Forde, R. M., Fahey, A. G. and Cheng, H. W. (2008), Comparative effects of infrared and one-third hot-blade trimming on beak topography, behavior, and growth. *Poultry Science, 87*(8), pp. 1474–83.

Matkovic, K., Vucemilo, M., Vinkovic, B. and Benic, M. (2009), *Air Quality and Welfare of Cage Housed Laying Hens,* Tribun EU, Brno.

Matkovic, K., Vucemilo, M., Vinkovic, B., Benic, M. and Frizon, E. (2010), *Dust and endotoxins in animal housing,* Korunić d.o.o, Zagreb.

McAdie, T. M. and Keeling, L. J. (2002), The social transmission of feather pecking in laying hens: effects of environment and age. *Applied Animal Behaviour Science, 75*(2), pp. 147–59.

McAdie, T. M., Keeling, L. J., Blokhuis, H. J. and Jones, R. B. (2005), Reduction in feather pecking and improvement of feather condition with the presentation of a string device to chickens. *Applied Animal Behaviour Science, 93*(1–2), pp. 67–80.

Meyer, B., Zentek, J. and Harlander-Matauschek, A. (2013), Differences in intestinal microbial metabolites in laying hens with high and low levels of repetitive feather-pecking behavior. *Physiology & Behavior,* 110–11, pp. 96–101.

Moinard, C., Statham, P., Haskell, M. J., McCorquodale, C., Jones, R. B. and Green, P. R. (2004), Accuracy of laying hens in jumping upwards and downwards between perches in different light environments. *Applied Animal Behaviour Science, 85*(1–2), pp. 77–92.

Mul, M., Van Niekerk, T., Chirico, J., Maurer, V., Kilpinen, O., Sparagano, O., Thind, B., Zoons, J., Moore, D., Bell, B., Gjevre, A.-G. and Chauve, C. (2009), Control methods for Dermanyssus gallinae in systems for laying hens: results of an international seminar. *World's Poultry Science Journal, 65*(04), pp. 589–600.

Newberry, R. C., Keeling, L. J., Estevez, I. and Bilcik, B. (2007), Behaviour when young as a predictor of severe feather pecking in adult laying hens: The redirected foraging hypothesis revisited. *Applied Animal Behaviour Science, 107*(3–4), pp. 262–74.

Nicol, C. J., Bestman, M., Gilani, A.-M., De Haas, E. N., De Jong, I. C., Lambton, S., Wagenaar, J. P., Weeks, C. A. and Rodenburg, T. B. (2013), The prevention and control of feather pecking: application to commercial systems. *World's Poultry Science Journal, 69*(04), pp. 775–88.

Nicol, C. J., Lindberg, A. C., Phillips, A. J., Pope, S. J., Wilkins, L. J. and Green, L. E. (2001), Influence of prior exposure to wood shavings on feather pecking, dustbathing and foraging in adult laying hens. *Applied Animal Behaviour Science, 73*(2), pp. 141–55.

Kozai, T., G. Niu and Takagaki, M., Eds (2016). Light. In *Plant Factory: An Indoor Vertical Farming System for Efficient Quality Food Production*, Academic Press, San Diego, pp. 115–28.

Prescott, N. B. and Wathes, C. M. (2002), Preference and motivation of laying hens to eat under different illuminances and the effect of illuminance on eating behaviour. *British Poultry Science, 43*(2), pp. 190–5.

Rodenburg, T. B., Haas, E. N. der, Nielsen, B. L. and Buitenhuis, A. J. (2010), Fearfulness and feather damage in laying hens divergently selected for high and low feather pecking. *Applied Animal Behaviour Science, 128*(1/4), pp. 91–6.

Rodenburg, T. B., Hierden, Y. M. v., Buitenhuis, A. J., Riedstra, B., Koene, P., Korte, S. M., Poel, J. J. van der, Groothuis, T. G. G. and Blokhuis, H. J. (2004), Feather pecking in laying hens: new insights and directions for research? *Applied Animal Behaviour Science, 86*(3/4), pp. 291–8.

Rodenburg, T. B. and Koene, P. (2007), The impact of group size on damaging behaviours, aggression, fear and stress in farm animals (Special issue: Too many, too few: the effects of group size and density in captive animals). *Applied Animal Behaviour Science, 103*(3/4), pp. 205–14.

Rodenburg, T. B., Van Krimpen, M. M., De Jong, I. C., De Haas, E. N., Kops, M. S., Riedstra, B. J., Nordquist, R. E., Wagenaar, J. P., Bestman, M. and Nicol, C. J. (2013) The prevention and control of feather pecking in laying hens: identifying the underlying principles. *World's Poultry Science Journal, 69*(02), pp. 361–74.

Sandilands, V., Moinard, C. and Sparks, N. H. C. (2009), Providing laying hens with perches: fulfilling behavioural needs but causing injury? *British Poultry Science, 50*(4), pp. 395–406.

Savory, C. J. (1995), Feather pecking and Cannibalism. *Worlds Poultry Science Journal, 51*(2), pp. 215–19.

Sherwin, C. M., Richards, G. J. and Nicol, C. J. (2010), Comparison of the welfare of layer hens in 4 housing systems in the UK. *British Poultry Science, 51*(4), pp. 488–99.

Van de Weerd, H. A. and Elson, A. (2006), Rearing factors that influence the propensity for injurious feather pecking in laying hens. *World's Poultry Science Journal, 62*(04), pp. 654–64.

Van Krimpen, M. M., Kwakkel, R. P., Van der Peet-Schwering, C. M. C., Den Hartog, L. A. and Verstegen, M. W. A. (2009), Effects of nutrient dilution and nonstarch polysaccharide concentration in rearing and laying diets on eating behavior and feather damage of rearing and laying hens. *Poultry Science, 88*(4), pp. 759–73.

Vestergaard, K. S., Kruijt, J. P. and Hogan, J. A. (1993), Feather pecking and chronic fear in groups of red junglefowl: their relations to dustbathing, rearing environment and social status. *Animal Behaviour, 45*(6), pp. 1127–40.

Vestergaard, K. S., Skadhauge, E. and Lawson, L. G. (1997), The Stress of Not Being Able to Perform Dustbathing in Laying Hens. *Physiology & Behavior, 62*(2), pp. 413–19.

Zimmerman, P. H., Lindberg, A. C., Pope, S. J., Glen, E., Bolhuis, J. E. and Nicol, C. J. (2006), The effect of stocking density, flock size and modified management on laying hen behaviour and welfare in a non-cage system. *Applied Animal Behaviour Science, 101*(1–2), pp. 111–24.

Part 2

Sustainability

Chapter 9

Waste management in egg production

Ruihong Zhang, University of California at Davis, USA; and Hamed. M. El- Mashad, University of California at Davis, USA and Mansoura University, Egypt

1 Introduction

Waste management is an essential component of an egg production system. It affects the environmental impact of production, food safety, worker safety, and animal health and safety. Manure, animal deaths, and egg washing and processing wastewater are the major types of waste generated in an egg production facility. Manure mostly contains nutrients fed to the hens and is an excellent source of energy and nutrients if managed properly. It is also the main source of various environmental pollutants and disease organisms. Common air pollutants are ammonia, greenhouse gases (GHG), volatile organic compounds and particulate matter. Most of the nitrogen in poultry manure is present in urea, uric acid and undigested protein. Uric acid represents 65–75% of nitrogen in manure (Pan et al., 2009; Roberts et al., 2007). Ammonia is generated when these compounds are decomposed via enzymatic and/or microbial reactions during storage and application as a fertilizer. More than 60% of ammonia in manure can be released under farm conditions (Mahimairaja et al., 1994), causing environmental problems and loss of fertilizer value.

Ammonia release depends on dietary composition, manure properties, housing type and management, bird stocking density, manure handling schemes, production stage and activities of birds, and environmental conditions (Zhang et al., 2009). Liang et al. (2006) reported that ammonia emissions from commercial laying hen houses in Iowa and Pennsylvania were 0.9 ± 0.09 g NH_3/hen-d for high-rise houses, 0.054 ± 0.005 g NH_3/hen-d for manure belt houses with daily manure removal and 0.094 ± 0.019 NH_3/hen-d for manure

http://dx.doi.org/10.19103/AS.2016.0012.30

belt houses with manure removal every three or four days. The average emission rate of ammonia from cage-free layer houses was 0.29 g/hen/d (Lin et al., 2017). Shepherd et al. (2015) measured ammonia emission in conventional cage, enriched colony and aviary houses to be 0.16, 0.30 and 0.29 g/hen/d, respectively. Manure storage accounted for 60–70% of the farm-level emissions. Various methods are available for reducing ammonia losses, including applying chemicals to reduce manure pH or inhibit the enzymatic hydrolysis, or drying quickly to reduce the moisture content and, therefore, reduce the ammonia volatilization potential.

Bird mortality could cause negative environmental impact and disease transmission if not properly managed. The extent of mortality depends on housing type and management, bird stocking density, and hygiene and health of hens. In addition, the wastewater generated from egg washing could also be a source of environmental pollution. Proper management of manure, hen mortality and egg washing wastewater is very important for achieving sustainability in egg production. There are several methods and technologies that can be applied for the treatment of waste streams on egg production farms. Anaerobic digestion and composting are common biological treatment methods used for treating manure and mortalities. Incineration, gasification and pyrolysis are thermochemical treatment methods. Rendering plants use caustic cooking to treat mortalities. Egg washing wastewater can be treated using aerobic and anaerobic biochemical processes and other physical and chemical processes. In this chapter, the characteristics of manure, poultry mortality, and egg washing wastewater and various methods and technologies available for treating and managing these waste streams are presented.

2 Characteristics of waste and wastewater

2.1 Laying hen manure

The quantity and chemical and physical characteristics of laying hen manure generated from an egg production farm depend on hen strain and age, diet, housing, and manure collection and storage methods. Typically, daily manure production ranges from 64 to 94 g per hen (Rahman et al., 2012). The effect of the laying hen strain on manure characteristics and amount are attributed to the differences among different strains in feed consumption, body weight, duration of egg production and egg size. Moreover, laying hen genetics could affect manure characteristics due to the different nutrient requirements. Table 1 shows the quantity and characteristics of laying hen manure for different strains and ages (Rahman et al., 2012).

The housing type can significantly affect the characteristics of manure. In manure belt houses, manure is allowed to accumulate for two to three days inside the houses before it is transferred to storage or land application. Lin et al. (2015) measured the nutrient distribution of feed intake among eggs, manure and losses in houses in three types of houses (conventional cage, enriched colony and aviary). Mature laying hens aged 31 weeks to 78 weeks were housed during the study period. The distribution of the feed total solid (TS), nitrogen (N) and carbon (C) among the eggs, manure and losses is shown in Fig. 1. As can be seen, most TS in the feed is used for egg production, while the majority of N intake is excreted in manure. The losses of TS and C are mainly in the form of carbon dioxide while N losses are in the form of ammonia. A small proportion (<1%) of feed intake is used for chicken growth. All three houses used belts to collect and remove manure except that the aviary had manure collection on wood shavings litter as well as belts.

Table 1 Quantity and characteristics of laying hen manure for different strains and ages

Colour	Variety	Daily manure production (g/d as-is)		Moisture (%, w.b.)		pH		Daily nitrogen production (g/d as-is)	
		27–28 wks	35–36 wks	27–28 wks	35–36 wks	27–28 wks	35–36 wks	27–28 wks	35–36 wks
	ISA	95.8	86.6	78.1	77.6	7.81	7.5	0.87	0.79
	Lohmann	92.4	77.4	77.4	75.9	7.88	7.4	0.91	0.75
	Bovans	90.8	82.2	77.5	75.6	7.88	7.6	0.92	0.82
Brown	Hy-Line	95.3	87.2	78.0	78.9	7.80	7.4	0.88	0.73
	Hy-Line W-98	84.3	65.1	76.6	72.2	7.67	7.5	0.85	0.80
	Lohmann LSL Lite	76.1	63.7	72.4	71.6	7.72	7.4	0.91	0.78
	Hy-Line W-36	62.8	50.7	70.9	67.2	7.53	7.3	0.93	0.67
White	Bovans	76.0	75.3	74.1	73.0	7.78	7.4	0.89	0.78

Source: Rahman et al., 2012.

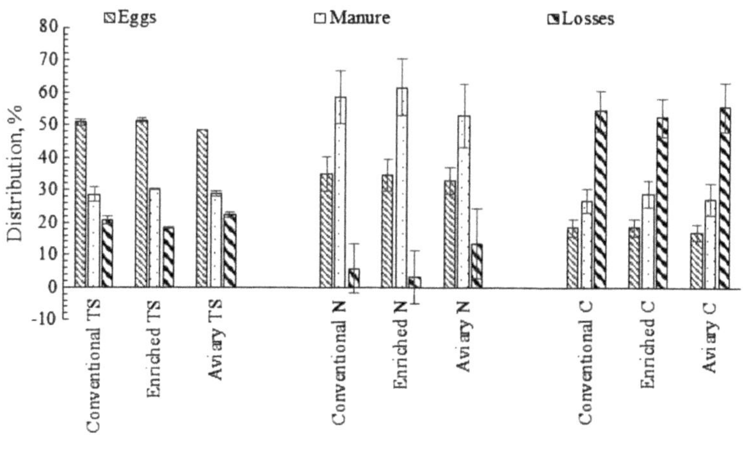

Figure 1 Distribution of the intake TS, N and C in eggs, manure and losses.

Manure belts are shown in Fig. 2. The manure on the belts was cleaned out every three or four days and moved to manure storage (Fig. 3). The litter in the aviary house was cleaned out at the end of the production cycle. The characteristics of manure in the three studied houses are shown in Table 2.

Manure under bird cages in high-rise laying hen houses is scraped off from dropping boards onto the first floor that is usually used as a manure storage (Fig. 4). Manure was sampled in two high-rise houses in California (Cortus et al., 2010). The buildings were tunnel-ventilated with single-speed fans. The manure was collected in two, triangular-shaped piles each of 115 m length (Fig. 4). Depending on storage time, pH, TS and

Figure 2 Manure belts on poultry farms.

Figure 3 Manure shed in manure belt housing.

Table 2 Manure characteristics (%, w.b.) with different manure ages on belt from different houses

Manure type and age on belt	House	TS (%)	N (%)	C (%)	S (%)	P (%)	K (%)
Manure, 3d	Conventional cage	45.02	2.51	16.57	0.37	0.64	1.04
Manure, 3d	Enriched colony	51.54	2.97	19.03	0.40	0.69	1.10
Manure, 3d	Aviary	47.08	2.60	17.52	0.36	0.65	0.97
Manure, 4d	Conventional cage	47.86	2.66	17.43	0.38	0.67	1.09
Manure, 4d	Enriched colony	57.25	3.14	21.15	0.45	0.77	1.28
Manure, 4d	Aviary	49.61	2.63	18.19	0.38	0.71	1.15
Litter	Aviary	82.47	3.25	30.48	0.66	0.97	1.81

Source: Lin et al., 2015.

Figure 4 Manure storage in high-rise housing.

ammonia of the manure were measured to be 7.18–8.56, 35.6–88.9% and 0.15–0.52% (w.b.). Manure was stored for six to eight months. Loadout manure was also analysed. Samples were taken from random locations in the manure piles during manure loadout. The TS and N contents of manure were 2.45–2.79% and 1.78–3.48% (w.b.).

Manure storage is necessary in many cases (Ogejo and Collins, 2009): (1) for growers who do not have enough land to use all the available manure on their farm, (2) when cleaning of poultry houses does not coincide with available open cropland and (3) when field conditions do not permit operation of manure application equipment. The manure storage time depends on the laying hen egg production housing system and local manure land application practices and regulations. The manure storage time, configuration of hen manure stack, housing type and moisture content are among the factors affecting the characteristics of manure during storage. During storage, manure can be chemically treated or treated physically by reducing the surface area to volume to reduce the emissions of pollutant gases (Roberts et al., 2007).

2.2 Laying hen mortality

Animal mortality needs to be managed safely, practically and economically. Mortality can occur due to disease or non-disease reasons. The non-disease reasons could be fires, collapse of farm structures due to tornados and floods, and poisons or poor housing management such as ventilation failure (Glanville and Harmon, 2007). Other reasons include outbreaks of *Pasteurella*, predatory attacks and piling. Van Horne (1996) reported the mean yearly mortality rate as 6.7% and 9.2%, respectively, for aviaries and battery cages. In a healthy, well-managed conventional cage housing, mortality is less than 0.1% per week (Lay et al., 2011). Lin et al. (2015) used farm inventories to calculate the mortality percentages and measured their composition in three houses located in Iowa. The average mortality rates were 4.3%, 4.8% and 11.6% on a yearly basis for conventional cage, an enriched colony and an aviary, respectively. The occupied area was 516, 752 and 1166 cm^2/hen, respectively. Mortality increased with the cage sizes. For the three studied houses, the TS, N, C, S, P and K contents of chicken mortalities sampled were

34.72–37.02%, 3.02–3.22%, 17.47–18.71%, 0.23–0.25%, 0.61–0.69% and 0.21–0.24%, respectively. Hegelund et al. (2006) determined mortality rate in seven commercial Danish farms with a total of 18 organic egg-producing flocks. The flocks are loosely housed with daily access to an outdoor run with a minimum area per hen of 4 m^2. Mortality rate ranged from 9 to 62% with an average of 22%. The disease-caused mortality should be disposed of according to local regulations. For example, Iowa regulations require that the dead animals from non-transmissible or non-infectious disease can be either burned or sent to a licensed rendering facility within 24 hours (Glanville and Harmon, 2007).

2.3 Egg washing wastewater

In many countries, eggs are normally washed before packaging. Kiepper and Ritz (2012) reported that water consumption in commercial shell egg processing facilities varied from 5.7 to 11.4 L per a packaged case of 360 table eggs. Egg washing wastewater is treated on site in more than 80% of the commercial table egg facilities in Georgia. The characteristics of egg washing wastewater vary substantially at different facilities, depending on several factors such as amount of water used for egg washing, frequency of replacement of water in the washing tank, egg breakage in washer, cleanliness of eggs and the application of oil on egg shells before washing (Northcutt et al., 2005). The characteristics of egg washing wastewater generated from several plants in California and Georgia are shown in Table 3. When treating egg washing wastewater, besides the need for reducing the organic matter, hormones should be removed from egg wash wastewater. Egg wash wastewater contained 17 β-estradiol equivalents (E_2 Eqs) ranging from 9 to 18 ng/L (Shappell, 2013). The E_2 Eqs is calculated based on multiplying the measured concentration of each hormone present in the wastewater by its corresponding relative potency factor relative to 17 β-estradiol (Furuichi et al., 2004).

Table 3 Characteristics of egg washing wastewater. All values are in mg/L except pH

	Measured mean range	
Parameter	Hills and Hauser (1979)	Northcutt et al. (2005)
Chemical oxygen demand	5,050–6,560	1,764–7,288
Biological oxygen demand	2,610–3,120	n.d.*
Total solids	7,630–10,550	n.d.
Total volatile solids	3,030–3,440	n.d.
Suspended solids	610–680	306–796
Kjeldahl nitrogen	220–300	81–302
Phosphorus	90–230	n.d.
Chlorine	n.d.	0.9–4.5
Iron	n.d.	0.3–1.5
pH	10.11–10.97	10.0–11.40

*Not determined

3 Methods for treating waste and wastewater: anaerobic digestion

Anaerobic digestion is a biochemical conversion of organic wastes into biogas in the absence of oxygen. The produced biogas can be used, after upgrading, as a transportation fuel, for heating purposes on farms or co-generation of electricity and heat (combined heat and power, CHP). The costs of the anaerobic digestion system depend on the types and characteristics of feedstock, climate conditions, digestion system configuration, end-use of biogas and digester effluent management. Anaerobic digestion involves complex enzymatic and microbial reactions (Zhang and El-Mashad, 2007). Temperature is a very important parameter that affects the performance and stability of the anaerobic digestion process. Anaerobic digestion can occur at a temperature as low as 4°C, but the degradation rates of organic material increase with increased temperature. It can be applied under one of three different temperature regimes: psychrophilic (<25°C), mesophilic (25–45°C) and thermophilic (>45°C). There are several anaerobic digestion systems that are suitable for solid and liquid substrates. Detailed descriptions of these systems have been presented by many researchers (e.g. Zhang and El-Mashad, 2007). Anaerobic digesters are commonly designed to operate in either mesophilic or thermophilic temperature range. This is because digesters operated at higher temperatures can handle higher organic loading rates and require shorter solids retention time and are therefore more space-efficient. Thermophilic conditions have many advantages over other temperature conditions such as (1) faster degradation rates as a result of the higher growth rates (but low growth yield) of thermophilic microorganisms, (2) increased destruction of pathogens and (3) improved solid–liquid separation using mechanical dewatering technologies (Hobson et al., 1981). The residence time under thermophilic conditions may approach one-third of those applied under mesophilic conditions (Ahring, 1995). On the other hand, thermophilic digestion requires more energy for keeping the digester at high temperatures. Recently, in collaboration with our industry partner, CleanWorld, we developed a three-stage thermophilic anaerobic digestion system on the University of California, Davis campus.

Figure 5 UC Davis high rate anaerobic digestion system.

The anaerobic digestion system treats 20 000 tonnes of mixed wastes per year, including food waste and animal manure. The system is a three-stage anaerobic digestion system. The biogas produced is used in CHP generation. The electricity generated (approximately 16000 kWh/day) is provided to the university.

For a successful anaerobic digestion process, there should be a balance between the carbon and nitrogen content of the feedstock. Moreover, anaerobic digesters should contain micro and macro nutrients necessary for the growth of microorganisms. Most organic matter contains sufficient micro and macro nutrients. Carbon-to-nitrogen ratio (C/N) is an important parameter affecting the performance of anaerobic digestion. The desirable C/N ratio is in the range of 15–30, depending on the biodegradability of feedstock. The C/N ratio of laying hen manure and mortality were measured to be 6.6 and 5.8, respectively (Lin et al., 2015). Because of low C/N ratio, the organic loading rate of anaerobic digesters is limited by the ammonia concentration in the digesters. In order to limit the ammonia concentration below 4000 mg/L, manure and mortality need to be diluted with water or mixed with other types of waste that have higher C/N ratio. High ammonia concentration can cause inhibition or even toxicity to the microorganisms in anaerobic digesters. Process inhibition is usually indicated by decreasing methane production rate and increasing VFA concentrations, while toxicity causes a total cessation of the methanogenic activity (e.g. Kroeker, 1979). The negative effects of ammonia on the performance of anaerobic digestion depend on ammonia concentration, substrate biodegradability, retention time, pH, temperature and, most importantly, the extent of adaptation of the micro-flora to the applied ammonia concentration (El-Mashad, 2003). Koster (1986) noted that the adaptation of methanogenic population to high ammonia concentrations could take up to six months during the digestion of synthetic wastewater at 30°C using granular sludge as inoculum. Different measures have been applied to reduce the negative effects of high ammonia concentrations on the degradation of poultry manure: digestion at mesophilic temperatures, dilution of manure before digestion, immobilization of microorganisms with activated carbon zeolite, clay, addition of adsorbents and ion exchangers such as natural zeolite, and codigestion of chicken manure and other substrates (Chen et al., 2008).

Anaerobic digestion has been applied for treating poultry manure, mortality and egg washing wastewater. Safley et al. (1987) reported the digestion of manure from a farm with 70000 caged layers. The anaerobic digester was operated at 35°C and 22 days hydraulic retention time. The average biogas yield was 0.39 m^3/kg volatile solids (VS) and methane content was 58.3%. Chen and Huang (2006) treated poultry mortality in a temperature-phased anaerobic digestion system that consisted of one upflow anaerobic sludge blanket (UASB) operated at 55°C and three leach-bed reactors operated at ambient temperatures. Methanogenesis was inhibited due to high concentrations of ammonia. When the temperature was lowered to 35°C, the stability of the system improved. In a laboratory-scale anaerobic digester treating egg washing wastewater at 35°C and hydraulic retention times of 12 and 18 days, Hauser and Hills (1981) determined a reduction in VS and chemical oxygen demand (COD) of about 45% after reaching steady-state conditions. Gwyther et al. (2011) reviewed several methods for the disposal of animal mortality and concluded that anaerobic digestion and composting are preferable on-farm disposal methods due to the environmental, practical, economical and biosecurity benefits. A large anaerobic digestion facility has been in operation since 2008 on an egg production farm located near Beijing, China, which has 3 million hens. The facility was originally developed and owned by Beijing DQY Agricultural Technology Company. Currently, it is operated by Helee Bio-Energy Company. The facility uses two-stage anaerobic digesters operated at around 40°C.

Figure 6 Beijing DQY Biogas Energy Facility located in Yanqing, China.

On an annual basis, approximately 70 000 tonnes of manure and 120 000 tonnes of wastewater (egg processing wastewater and sewage from offices and worker dorms) are processed. The biogas production is approximately 7.3 million m^3. The biogas was initially used to generate 1400 MWh electricity for the farm. Recently, a biogas upgrading system was added to produce biomethane that is then taken to villages nearby for use as cooking and heating fuel. The anaerobic digesters, biomethane upgrading system and fuelling station are shown in Fig. 6. The digestate produced in the digesters is applied to the nearby farms for food production.

4 Methods for treating waste and wastewater: aerobic treatments

4.1 Composting of laying hen manure and carcasses

Aerobic biodegradation is a biochemical process that occurs in the presence of oxygen (i.e. air) to convert organic matter into carbon dioxide, water, new microbial cells and energy. Aerobic treatment has been applied for reducing contamination and for stabilization of wastewaters and composting of solid organic materials. Composting has been applied as an effective method for solid waste management while producing a nutrient-rich product that is used as a soil amendment and organic-based fertilizer. Composting improves the characteristics of waste, reduces the weight and eliminates pathogenic microorganisms. In addition to extract production from laying hen manure, aerobically treated chicken manure could be used as feed supplements. Martin and Loehr (1983) reported that essential amino acid content of laying hen manure increased by 22% after aerobic treatment for three days retention time due to the degradation of protein in solids.

Compost can have a positive impact on the physical, biological and chemical properties of soils. It could stimulate the growth of soil microorganisms by providing the required nutrients for their growth. The preferred conditions for composting are 25–30 for C/N, 50–60% for moisture, 5–15% for oxygen, 6.5–8.0 for pH and 54–60°C for temperature (Rynk et al., 1992). During composting, the low C/N ratio of hen manure causes high nitrogen loss, which reduces the fertilizer value of the produced compost. Mixing manure with carbon-rich materials (such as crop residues and paper waste), elemental sulphur and zeolite could reduce nitrogen losses during composting of poultry waste (Bolan et al., 2010). The litter produced from aviary houses, where bedding materials such as wood shavings are used (Lin et al., 2015), is a good material for composting. At oxygen concentration below 15%, anaerobic conditions could be developed causing methane production and odours. The high temperature achieved during composting could destroy the pathogens and therefore reduce the risk of water and crop contamination. Composting could effectively reduce the concentrations of the majority of chemicals and pesticides in manure (Kawata et al., 2006). Chemicals such as rabon, zoalene, nicarbazin and nitrofurazone are used in layer diets to control insects and other organisms in litter (Bolan et al., 2010). Larvicides such as cyromazine and pyridine are used for treating manure. Composting was found to be effective for the degradation of some antibiotics in animal manure such as chlortetracycline, monensin and tylosin but not effective for others such as sulfamethazine (Dolliver et al., 2008). The degradation of antibiotics was attributed to biotic factors such as microbial activity and abiotic factors such as chemical transformations under high temperatures. Composting systems need to be properly designed and managed in order to obtain compost with consistent characteristics, and to reduce the emissions of various gases (including GHG and VOC) and ammonia and nutrients leachate and run off.

Composting can be carried out in windrows with natural aeration or enclosed bins and vessels with forced aeration. Windrows that are turned periodically with machines are the most common systems used for manure composting. These systems can be labour-intensive and produce air emissions. There is a space requirement for operating and storing pile-turning equipment. Open windrow composting requires less capital investment than forced-aeration composting. In the forced-aeration composting, air is supplied through perforated bases, on which composted materials are placed, or perforated tubes that are inserted in the compost pile. The forced air systems need smaller areas than windrows. From a process control standpoint, the in-vessel systems are preferred because aeration and temperature can be well controlled. Moreover, applications of the in-vessel systems could reduce risk of leaching and run-off nutrients and other pollutions that occur in windrow composting systems. To reduce the chance of rain percolation into compost piles, and to reduce emissions of GHG, ammonia and other gases from windrows, impermeable covers should be used. To reduce the nutrients leachate and run off, impermeable liners should be used underneath the composting materials.

Composting has been applied as a cost-effective treatment for animal mortality and manure. Composting at thermophilic temperatures (50–70°C) could help in reducing the pathogen content of chicken mortality so that biosafety could be attained on farms. To achieve the biosafety on farms, a certain temperature should be maintained for a certain period: 53°C for 5 days, 55°C for 2.6 days or 70°C for 30 min. At temperatures above 70°C, the activity of microorganisms declines (Haug, 1993). Sivakumar et al. (2008) studied the co-composting of dead poultry with layer manure and farmyard manure. The latter is usually composed of manure, bedding and field residues. Composting experiments were carried out in a small-scale composter during hot and cold months. The peak temperatures were reached by the second week of composting and started declining. Higher temperatures (above 60°C) were achieved during hot months and maintained for 3–4 weeks. However, the temperature did not exceed 55°C during cold months. Results also showed that noxious odour or fly menace were not detected under any of the experimental conditions. Higher degradation rates of organic matter were achieved for the mixtures containing the farmyard manure due to the comparatively higher moisture retention in the mixtures containing farmyard manure. The weight reduction was measured to range from 31.8 to 58.7% for the mixtures containing farmyard manure and from 19.3 to 48.6% for the mixtures containing layer manure. Composted manure has shown a higher value than raw manure. It can be dried and made into pelletized fertilizer products as shown in Fig. 7. The drying and pelletizing equipment used to make such product are shown in Fig. 8 and 9.

Figure 7 Pelletized fertilizer made from composted manure.

Figure 8 Equipment used for drying composted manure.

Figure 9 Equipment used for pelletizing composted manure to produce fertilizer product.

The emissions of GHG, ammonia, hydrogen sulphide and other gases from composting are affected by the composition of composted materials, aeration method and oxygen supply. Oliveira et al. (2012) measured the emissions of ammonia (NH_3) and nitrous oxide (N_2O) from composting of a mixture of layer manure, sawdust, waste eggs and dead birds. The nitrogen losses were 22.8% and 25.8% of the original N content of the mixture, with and without inoculation, respectively. Biological inoculum (Humidibiol) was used in the experiments. No details were provided on the characteristics of Humidibiol. The composting experiments lasted for 28 days: 7 days in batch reactors wherein thermophilic condition was achieved, and 21 days in weekly turned piles wherein mesophilic conditions could be attained. The nitrogen losses could decrease the value of the compost as fertilizer compared with non-composted manure. However, the produced compost contains more stable organic matter and can improve the soil properties (Tiquia and Tam, 2002). An adequate supply of O_2 to the composting process is important for reducing GHG emissions. Park et al. (2011) reported that swine manure composting with forced aeration had lower GHG emission as compared with the composting with passive aeration. Chowdhury et al. (2014) reported that addition of biochar and barley straw during the composting of layer manure could reduce the emissions of GHG and ammonia but not N_2O. The low emission of ammonia could be attributed to the high C/N ratio and the adsorption of ammonium to biochar. However, there were no data presented on the economics of applying biochar. Patni et al. (1992) studied the effect of adding peat and chopped straw on the composting of layer manure in a static pile passive aeration composting system. Initial moisture content of two mixtures of laying manure and peat were 73% and 80% in the peat piles and the mixture with straw was 76%. Composting time increased with the increase in the initial moisture content in the peat piles. The acidic peat was effective in reducing the loss of ammonia emissions from the composting pile. Formation of struvite by adding magnesium and phosphate during the composting of broiler litter could reduce ammonia emissions by 40% to 80% (Zhang and Lau, 2007).

4.2 Aerobic stabilization of egg washing wastewater

Egg wash wastewater is treated in municipal wastewater treatment plants when available. The most common method for disposal of this wastewater is irrigation directly after discharge or after lagoon storage. Storing egg wash wastewater in an aerobic lagoon for 90 days significantly reduced the E_2Eqs contents. The treated wastewater could be safely used for crop irrigation (Shappell, 2013). Hauser and Hills (1981) studied the effects of aerobic treatments of egg processing wastewater in continuous flow stirred tank reactors operated at 20°C and different retention times (HRT) of 1, 2, 4 and 8 days. Aerobic treatment was effective in stabilizing the egg washing wastewater. The reduction in soluble COD was 78% and 96%, respectively, for HRT of 1 and 8 days. The volatile suspended solids increased from 580 mg/L in the influent, to approximately 6000 and 2000 mg/L at HRT of 1 and 8 days, respectively. This increase is due to bacterial growth.

5 Methods for treating waste and wastewater: thermochemical processes, burial, rendering and land application

5.1 Thermochemical processes

Thermochemical technologies have been applied for waste management. They use high temperatures at either ambient or high pressures to break the bonds of organic matter and produce heat, syngas and hydrocarbon fuels or ash depending on process operational conditions. Gaseous and liquid fuels produced from thermochemical processes can be further processed and upgraded (e.g. using catalytic thermochemical processes) to produce alcohols and light diesels (Cantrell et al., 2007). Thermochemical processes include direct combustion (i.e. incineration), pyrolysis, gasification and hydrothermal degradation. Compared with biochemical processes, thermochemical processes have the following advantages (Cantrell et al., 2007): (1) thermochemical reactors are more compact, (2) processing time is short (in the order of minutes), (3) the applied high temperatures destroy pathogenic microorganisms and most pharmaceutically active compounds, (4) these processes are flexible and can simultaneously process several materials such as manure and crop residues, and (5) possibilities of nutrient recovery. On the other hand, thermochemical processes, except for hydrothermal processes, require low moisture (<40%). Fresh manure and mortalities need to be dried before they can be used. More safety measures are required due to high temperatures and pressures used in these operations.

Incineration (combustion) is a complete oxidation of organic matter under high temperatures in the presence of air. It produces heat and flue gas that consists of carbon dioxide and water vapour. The heat can be used for the production of steam that subsequently can be used for electricity generation. Direct combustion is the simplest economical thermal technology for energy production from organic matter in large-scale installations (Flora and Riahi-Nezhad, 2006). Poultry manure has a heating value of 6–8 MJ/kg. Historically, it was believed that one of the limitations of direct combustion of manure is the high concentration of NO_x in the flue gas due to the low C/N ratio of manure. However, recent life cycle studies showed that electricity production from poultry manure via combustion reduces the emissions of GHG, SO_2, NO_x, nitrates and ammonia to the environment (Billen et al., 2015). The reduction of the emissions is partially due to the diversion of poultry manure from land application and partially due to reducing the consumption of electricity generated from fossil fuels. The efficiency of electricity production from poultry manure combustion is approximately 28%, which is lower than that of coal power plants due to the high ash content and composition of manure. Another limitation of the direct combustion of poultry manure is the high concentrations of potassium and phosphorus that could result in unwanted shutdown of power plants due to (1) agglomeration in the combustion bed, (2) deposits formation on freeboard and (3) fouling of boiler tubes (Billen et al., 2015). Incineration has been applied for the disposal of mortality to prevent environmental pollution and the spread of disease. It is considered the safest method for mortality disposal because it converts mortality carcass to ash. However, it has high cost and tends to cause environmental pollution (Damron et al., 2001). The main concern for the application of incineration is the emission of GHG and other gaseous compounds such as polycyclic aromatic hydrocarbons (Chen et al., 2003). To reduce the negative environmental impacts of incinerators, flue gases should be passed through after-burning devices to ensure complete combustion. With increasing concerns about the negative environmental impacts of the thermochemical processes, many countries have stringent regulations and permit requirements for installing and operating incinerators.

Pyrolysis is a thermochemical treatment of biomass materials under high temperature and ambient pressure in the absence of oxygen. The products of pyrolysis are gases (H_2, CO_2, CO), liquid (biooil) and solid (biochar). The biochar produced from pyrolysis and gasification can be used as a soil amendment and an energy source. There are several phases of pyrolysis, depending on the temperature, heating rate and residence time (Mohan et al., 2006). There is limited information in the literature on pyrolysis of laying hen manure and mortality. Using a fluidized bed reactor, the mass balance after fast pyrolysis of poultry litter at 450°C showed biochar, syngas and biooil yields of 41% (w.b.), 36% and 23%, respectively (Kim et al., 2009). Huang et al. (2015) found that it was technically and economically feasible to produce heat, power and biochar from poultry litter using pyrolysis/gasification processes.

Gasification is a partial oxidation, at the atmospheric pressure and a temperature ranging from 800 to 1100°C, of organic matter to produce syngas and biochar by-product. Oxygen, air or steam is usually used as an oxidative agent. Syngas is a mixture consisting primarily of H_2, CO and lesser amounts of CO_2 and CH_4. It could be used for electricity generation or upgraded to transportation fuels. Several endothermic and exothermic chemical reactions occur during gasification. The higher heating value of the gasification of poultry litter was 10.3 ± 0.70 MJ/m^3. The ash agglomeration occurred also during the gasification of poultry litter in a fixed bed gasifier due to higher levels (16.7%) of alkaline oxides ($Na_2O + K_2O$) in the ash (Priyadarsan et al., 2004). Blending poultry litter with low (6.1%) alkaline oxides content material such as feedlot biomass could prevent the occurrence of agglomeration.

Most of the thermochemical processes need low moisture content feedstocks. However, for wet materials such as laying hen manure and mortality, hydrothermal processing could be applied without the need for drying. Hydrothermal processing is a thermal treatment of wet biomass at elevated pressures to produce carbohydrate, liquid hydrocarbons or gaseous products. This process involves thermochemical conversion of biomass into liquid fuel, called biocrude, under a high temperature (250–374°C) pressurized (4–22 MPa) water environment (Elliott et al., 2015). At temperatures higher than 374°C, gasification reactions dominate producing syngas and the process is defined as a hydrothermal gasification. Vardon et al. (2011) applied hydrothermal liquefaction (300°C, 10–12 MPa and 30 min reaction time) to swine manure producing a biocrude oil yield of 30.2%. Biocrude oil can be directly combusted or upgraded to petroleum-grade oils.

5.2 Burial and rendering of laying hen carcasses

Burial is a common practice for disposal of dead laying hens and other animals. It has also been applied for disposal of spent layer hens, which retired from production and have low market value (Freeman et al., 2009). Usually, dead animals are put in trenches or large pits, after which the pits and trenches are covered with soil. If the burial sites are not well covered, predators and vectors (rats, snakes and flies) could increase around these sites. Burial of dead animals is a potential source of soil and groundwater contamination. To prevent the spread of disease, biosecurity measures should be taken carefully for the burial sites (NABCCCDWG, 2004).

Rendering is a process used to convert dead animals into biologically safe protein by-product meals. It has been applied to mortality and spent laying hens. Christmas et al. (1996) described rendering as a cooking process for dead carcasses under high pressure (9.81 Pa) for 40 min. The cooked material was then maintained at 120°C for at least two hours until the desired moisture level is attained. Cooked carcass is stabilized by adding ethoxyquin at 200 ppm of raw material. Rendered hen meals contain readily available nutrients for the growth of broiler chicks from hatch to six weeks of age. The hen meal is comparable to soya bean meal. The rendered whole-hen meal had 55% protein, 23% fat, 10% moisture content, and the remainder was amino acids, other nutrients and salts. Rendered hen meal is used as an additive for non-ruminant livestock feed. Compared with the control, supplementing a laying hen diet with 10% of rendered hen meal achieved similar or superior daily feed intake, egg production, feed conversion, egg specific gravity and body weight, but egg weight was reduced due to the low biodegradability of amino acids in the hen meal (Damron et al., 2001). Although rendering is a good option for disposal of dead animals, there is a risk of the spread of disease-causing microorganisms during handling and transportation to rendering facilities. Several methods and measures could be employed to reduce the spread of disease during handling of dead animals: central pick up of carcass, refrigeration, acid preservation, lactic acid fermentation and extrusion. Each of these methods has advantages and disadvantages from economical and energy consumption perspectives (Blake and Donald, 2001). Gwyther et al. (2011) pointed out that bioreduction and freezing could be promising on-farm storage for animal mortality.

5.3 Land application of laying hen manure

Land application of poultry manure is widely practised to improve soil fertility while recycling manure nutrients. Poultry manure contains macro and micro nutrients that are necessary for plant growth. On a cautionary note, poultry manure contains antibiotics, pesticides, larvicides, endocrine-disrupting compounds and coccidiostats (Sims and Wolf, 1994; Bolan et al., 2010). The effects of these compounds on crop growth and soil microorganisms and the impact of these compounds on the environment are not well understood (Bolan et al., 2010). In most cases, application rate of manure is determined based on the nitrogen needs of crop. After land application of poultry waste, organic nitrogen (uric acid and urea) is mineralized to ammonium (NH^{+}_{4}-N) that is quickly nitrified under proper conditions or converted to ammonia that is evaporated causing environmental pollution. The ammonia emissions depend on the production of ammonia from degradation of uric acid and urea (Lacey et al., 1981).The conversion rate of organic nitrogenous compounds depends on manure age, ambient temperature, precipitation and manure application method. Higher temperatures and moisture contents stimulate the enzymatic activity and hence the conversion rate (Groot Koerkamp, 1994). Incorporation of waste in soils results in more rapid conversion of organic nitrogen than surface application (Sims and Wolf, 1994). Ammonia volatilization from manure-applied fields depends on temperature, rainfall intensity, soil and crop types, and application rates and method. The application rates depend on the characteristics of soil and manure to be applied, and nutrient requirements of crops. The emission of ammonia could reduce the total nitrogen content of poultry manure by 50% (Reddy et al., 1980). Nitrogen is often a limiting nutrient for land application of manure. However, phosphorus can be a limiting factor for soils that have relatively high phosphorus content. Manure application rates should be determined based on the nutrient management plans of the fields.

6 Laboratory study of co-digestion of laying hen manure and mortality

A laboratory study was conducted to determine the biogas production potential and the stability of anaerobic digestion of laying hen manure and its mixtures with ground mortality, sewage sludge and dissolved air floatation (DAF) grease. Co-digestion of poultry manure with other organic wastes allows better use of nutrients in the manure and higher solids loading in anaerobic digesters.

The benefits of co-digestion of animal and other organic waste, such as food waste, were well documented by previous research publications (El-Mashad and Zhang, 2007, 2010; El-Mashad et al., 2008). This study was aimed at using the laboratory data to design a full-scale anaerobic digester for an egg production farm located in California. Manure was collected from belts in layer houses on two farms. Other organic wastes were provided by an industry collaborator. The composition of different test materials is shown in Table 4. Batch anaerobic co-digestion laying hen manure from the first farm (LHM 1) and its mixtures was carried out at 50 ± 2°C for 47 days. Biogas volume and composition were measured periodically. The pH and VS of the reactor contents were measured at the end of the digestion time.

Successful batch digestion tests were conducted. Biogas was produced quickly after starting the digestion tests. However, relatively low biogas production rates were noticed in the first five days of digestion. The low rates of biogas production might be due to the adaptation of the micro-flora present in digesters to the tested substrates, and the high contents of fats and proteins in ground mortality. Protein and fats had lower digestion rates than carbohydrates that were contained in manure. The biogas production rates increased after five days. After 47 days of digestion, the biogas yields of 753, 685, 708 and 823 mL/g VS were obtained for LHM1 + DAF, LHM1+ sewage sludge, LHM1 + sewage sludge + ground mortality and LHM1+ ground mortality + DAF, respectively. About 90% of the final biogas

Table 4 Composition of raw materials, biogas yield, VS reduction and final pH after 47 days of batch thermophilic digestion

Feedstock	Composition of materials				Biogas yield, L/gVS	Methane yield, L/gVS	VS reduction, %	Final pH
	TS, %	VS, %	MC, %	VS/TS, %				
Cooked mortality	35.0	31.9	65.0	91.0	0.883	0.699	68.7	7.08
Manure1	35.3	21.6	64.7	61.0	0.705	0.525	60.4	7.16
89% Manure1 + 11% DAF	24.8	15.9	75.2	64.0	0.753	0.567	62.0	7.09
LHM1 + sludge (85% manure + 15% sewage sludge)	22.4	13.9	77.6	62.0	0.685	0.503	62.4	7.04
LHM1 + sludge + mortality (82% manure + 15% sewage sludge + 3% mortality)	22.5	14.2	77.5	63.0	0.708	0.525	65.3	7.1
LHM1 + mortality + DAF (86% manure + 3% mortality + 11% DAF, wet weight base)	24.9	16.2	75.1	65.0	0.823	0.658	62.7	7.1
Manure2	25.5	16.8	65.7	75.0	Explained in the text			

yields were obtained during the first 14 days of digestion and little biogas was produced after 21 days.

Continuously fed anaerobic digester was used to treat laying hen manure from the second farm (LHM2). The digester was fed with a mixture of 95% LHM2 and 5% ground mortality. The digester was operated for 90 days at an organic loading rate (OLR) of 2.4 g VS/L/day, and a 20-day hydraulic retention time (HRT). The daily biogas production stabilized after 23 days. During the operational period of 23rd day to 90th day, the average biogas yield was 0.431 L/g VS, the pH was 7.88, the methane content of the biogas was 66.2% and the VS reduction was 62.9%. To increase the biogas yield, the OLR was reduced to 1.2 g VS/L/day and the HRT was increased to 40 days until 120th day. During this period, the average biogas yield was 0.503 L/g VS, the pH was 7.91, the methane content was 69.5% and the VS reduction was 68.6%. Thereafter, OLR increased to 1.8 g VS/L/day and the HRT was decreased to 30 days till day 129. The average biogas yield was 0.483 L/g VS and the pH was 7.85, the methane content was 69.3% and the VS reduction was 62.9%. The 1.8 gVS/L/day ORL and 30-day HRT are recommended from the standpoint of biogas yield and digester stability.

7 Summary and research needs

Laying hen manure and mortality, as well as wastewater from egg washing, are the main waste streams that should be properly managed and disposed of. Biochemical processes such as anaerobic digestion and aerobic degradation are the most common methods for treating laying hen manure and mortality, and egg washing wastewater because of high water content and biodegradability of the organic matter present in the waste and wastewater. Anaerobic digestion of laying hen manure and mortality produces bioenergy and nutrient-rich effluents. Ammonia concentration in the digester is a limiting factor for anaerobic digestion of hen manure. The mechanism of the adaptation of microorganisms to high ammonia concentration needs to be fully understood. The required time needed for microorganisms to adapt to different ammonia concentrations needs to be determined for different temperatures. There is also a need to develop high-ammonia-tolerant methanogens that could be used for the digestion of poultry manure and mortalities at thermophilic conditions. Co-digestion of manure and mortality with other organic materials that have high C/N ratio, such as food waste and crop residues, could be one of the best options. However, the logistics and economics of bringing egg farms wastes and other substrates together needs to be carefully analysed for a given project. There is a need to study the economics and effect of biochar on the stability of anaerobic digesters and biogas production. There is also a need to study the possibilities for concentration of the nutrients of the anaerobic digestion effluents to produce easily transferrable and storable organic fertilizers (Barzee et al., 2015).

Composting is a mature technology that has been applied for treating many waste materials such as food processing waste, organic fraction of municipal solid wastes, animal manure and mortality. However, composting system designs and management need to be designed and managed properly in order to obtain consistent characteristics of compost, reduce the emissions of GHG and ammonia, and the leachate of soil and water pollutants. Ammonia emission is a limiting factor for applying composting of poultry farm wastes due to its negative environmental impacts and reduction of the fertilizer value of the produced compost. Ammonia emissions could be reduced by treating the emitted ammonia by absorption, biofiltration, wet scrubbing; adsorbing the ammonia using materials such as clay, peat, clay and zeolite; and the formation of struvite. Biochar and acidic peat addition could also reduce the ammonia emissions from composting. More research is needed to determine the optimum amount and the economics of biochar and acidic peat as means of the reduction of ammonia emissions from different mixtures of mortality, laying hen manure and litter, and agricultural wastes. Moreover, the effect of biochar and acidic peat amended laying hen waste compost on crop yields need to be determined. There is a need to study the effect of stimulation of struvite formation on reducing ammonia emissions during composting of wastes from egg production farms.

Laying hen manure could be effectively processed using thermochemical processes such as combustion, gasification, pyrolysis or hydrothermal liquefaction for the production of energy and valuable products such as ash and biochar. Agglomeration is a limiting factor for applying thermochemical processes for poultry manure. Mixing manure with materials containing low alkaline oxides contents could alleviate the agglomeration problem. There is a need for research to optimize the pyrolysis and hydrothermal processes parameters for treating laying hen manure and mortality. The effects of different process parameters on the yields and characteristics of gaseous, liquid and solid products from these processes need to be studied. Novel and effective methods need to be developed to alleviate the sintering and agglomeration in thermochemical processing of poultry manure. The characteristics of biochar produced from the pyrolysis need to be determined and potential applications of that biochar need to be identified.

Land application is the most common disposal method for manure. However, there are several limitations that should be considered before land application of manure: (1) emissions of ammonia and GHG, (2) spread of diseases and (3) land overloading with nutrients that could negatively affect crop quality and cause soil and water pollution. There is a need to develop new technologies for the recovery of valuable products while reducing the negative environmental impact and the biosecurity impacts of laying hens mortality. Rendering is a common disposal method of mortality. It can produce protein-rich meals that can be used as animal feed. However, there are risks of environmental pollution and spread of diseases during farm storage of mortality and their transportation to rendering facilities. There are no publications of the applications of bioreduction and freezing methods for on-farm storage laying hen mortality.

Anaerobic and aerobic digestion have been proven to be effective technologies to reduce the organic contamination of egg washing wastewater. However, there are also other effective technologies such as electrocoagulation, precipitation/coagulation and ultrafiltration that could also be good alternatives to biochemical processes. There is limited information on the performance of these processes on the treatment of egg washing wastewater. The performance parameters of and the costs of these processes need to be optimized when they are applied to treat wastewater from egg washing. The effects of these technologies on the fate of hormones needs to be determined.

8 Where to look for further information

More information on biochemical and thermochemical processes for the treatment of waste streams can be obtained from the research groups of Professors Ruihong Zhang and Bryan Jenkins, respectively, in the Department of Biological and Agricultural Engineering at University of California, Davis. Mohan et al. (2006) presented a critical review of pyrolysis of woody biomass. This is a good article to understand the theory and applications of pyrolysis. Although it was written mainly for woody biomass, it provides the basic knowledge on the effect of biomass characteristics and process parameters on system performance and characteristics of process products. A report on different technical and economic aspects of several methods for the disposal of livestock mortality was presented by Sparks Companies, Inc. (2002). A comprehensive review of carcass disposal was published by NABCCCDWG (2004). The report summarizes the operational concepts, process applications, and social and economical considerations of selected carcass disposal technologies. The report also presents the knowledge gaps and the research needs in the areas of carcass disposal. There are several companies that are specialized in designing, installing and operating high rate anaerobic digesters for different bio-waste materials; among them are CleanWorld (http://www.cleanworld.com/) and Helee Bio-Energy (http://www.helee-bio-energy.com). *BioCycle* magazine, Biocycle.net and BioCycle Conferences are good sources of information on recovering and increasing the values of different wastes, including manure, by producing compost, renewable energy and engineered soils (http://www.biocycle.net/). The technical library of American Society of Biological and Agricultural Engineering (ASABE) has a large collection of articles related to poultry waste treatment and disposal technologies design and evaluation. The ASABE has an annual international meeting that usually has articles on new developments in the areas of different biological and agricultural engineering research areas such as manure treatment and disposal technologies. El-Mashad (2003) and Chen et al. (2008) reviewed the mechanisms, thresholds and the measures needed to alleviate the inhibition of anaerobic digestion by high ammonia concentrations. Collins et al. (1999) published a handbook on different technologies for the management of poultry waste. Bolan et al. (2010) published a review article on characteristics of poultry litter, its value as a fertilizer and alternative methods for producing valuable products from that material.

9 References

Ahring, B. K. (1995), 'Methanogenesis in thermophilic biogas reactors', *Antonie van Leeuwenhoek*, 67, 91–102.

Barzee, T., Zhang, R., Edalati, A., and El Mashad, H. M. (2015), 'Sustainable bio-fertilizer production from anaerobically digested organic wastes', ASABE Paper Number: 2190937.

Billen, P., Costa, J. B., der Aa, L. V., Van Caneghem, J. and Vandecasteele, C. (2015), 'Electricity from poultry manure: A clean alternative to direct land application', *Journal of Cleaner Production*, 95, 467–75.

Blake, J. P. and Donald, J. O. (1992), 'Alternatives for the disposal of poultry carcasses', *Poultry Science*, 71, 1130–5.

Bolan, N. S., Szogi, A. A., Chuasavathi, T., Seshadri, B., Rothrock Jr., M. J. and Panneerselvam, P. (2010), 'Uses and management of poultry litter', *World's Poultry Science Journal*, 66, 673–98.

Cantrell, K., Ro, K., Mahajan, D., Anjom, M. and Hunt, P. G. (2007), 'Role of thermochemical conversion in livestock waste-to-energy treatments: obstacles and opportunities', *Industrial & Engineering Chemistry Research*, 46, 8918–27.

Chen, S. J., Hsieh, L. T. and Chiu, S. C. (2003), 'Emission of polycyclic aromatic hydrocarbons from animal carcass incinerators', *Science of the Total Environment*, 313, 61–76.

Chen, T. H. and Huang, J. L. (2006), 'Anaerobic treatment of poultry mortality in a temperature-phased leach bed–UASB system', *Bioresource Technology*, 97, 1398–410.

Chen, Y., Cheng, J. J. and Creamer, K. S. (2008), 'Inhibition of anaerobic digestion process: A review', *Bioresource Technology*, 99, 4044–64.

Chowdhury, M. A., de Neergaard, A. and Jensen, L. S. (2014), 'Potential of aeration flow rate and bio-char addition to reduce greenhouse gas and ammonia emissions during manure composting', *Chemosphere*, 97, 16–25.

Christmas, R. B., Damron, B. L. and Ouart, M. D. (1996), 'The performance of commercial broilers when fed various levels of rendered whole-hen meal', *Poultry Science*, 75,536–9.

Collins, E. R., Barker, J. C., Carr, L. E., Brodie, H. L. and Martin, J. H. (1999), *Poultry Waste Management Handbook*, Natural Resource, Agriculture, and Engineering Service (NRAES), NRAES-132. ISBN 0-935817-42-5. Ithaca, New York, USA.

Cortus, E. L., Lin, X. J., Zhang, R. and Heber, A. J. (2010), *National Air Emissions Monitoring Study: Emissions Data From Two Layer Houses in California-Site CA2B*, Final Report. Purdue University, West Lafayette, IN.

Damron, B. L., Ouart, M. D. and Christmas, R. B. (2001), 'Rendered whole-bird layer mortality as an ingredient in layer diets', *Journal of Applied Poultry Research*, 10, 371–5.

Dolliver, H., Gupta, S. and Noll, S. (2008), 'Antibiotic degradation during manure composting', *Journal of Environmental Quality*, 37, 1245–53.

Elliott, D. C., Biller, P., Ross, A. B., Schmidt, A. J. and Jones, S. B. (2015), 'Hydrothermal liquefaction of biomass: Developments from batch to continuous process', *Bioresource Technology*, 178,147–56.

El-Mashad, H. M. (2003), *Solar Thermophilic Anaerobic Reactor (Star) For Renewable Energy Production*, PhD dissertation, Wageningen University, Netherlands.

El-Mashad, H. M., McGarvey, J. A. and Zhang, R. (2008), 'Performance and microbial analysis of anaerobic digesters treating food waste and dairy manure', *Biological Engineering*, 1(3), 235–44.

El-Mashad, H. M. and Zhang, R. (2007), 'Co-digestion of food waste and dairy manure for biogas production', *Transactions ASABE* 50(5), 1815–21.

El-Mashad, H. M. and Zhang, R. (2010), 'Biogas production from co-digestion of dairy manure and food waste', *Bioresource Technology*, 101, 4021–8.

Flora, J. R. V. and Riahi-Nezhad, C. (2006), 'Availability of poultry manure as a potential bio-fuel feedstock for energy production', http://dc.statelibrary.sc.gov/bitstream/handle/10827/15026/BCB_EO_Availability_of_PoultryManure_2006-8-31.pdf? sequence=1 (accessed 3 February 2016).

Freeman, S. R., Poore, M. H., Middleton, T. F. and Ferket, P. R. (2009), 'Alternative methods for disposal of spent laying hens: Evaluation of the efficacy of grinding, mechanical deboning, and of keratinase in the rendering processes', *Bioresource Technology*, 100, 4515–20.

Furuichia, T., Kannanb, K., Giesy, J. P. and Masunag, S. (2004), 'Contribution of known endocrine disrupting substances to the estrogenic activity in Tama River water samples from Japan using instrumental analysis and in vitro reporter gene assay', *Water Research*, 4491–501.

Glanville, T. and Harmon, J. (2007), 'Poultry & livestock disposal in Iowa', http://www.abe.iastate.edu/cattlecomposting/files/2013/05/Mortality-Disposal-FAQs11_2007. pdf (accessed 3 January 2016).

Groot Koerkamp, P. W. G. (1994), 'Review of emissions of ammonia from housing systems for layer hens in relation to sources, processes, building design and manure handling', *Journal of Agricultural Engineering Research*, 59, 73–87.

Gwyther, C. L., Williams, A. P, Golyshin, P. N., Edwards-Jones, G. and Jones, D. L. (2011), 'The environmental and biosecurity characteristics of livestock carcass disposal methods: A review', *Waste Management*, 31(4), 767–78.

Haug, R. T. (1993), *The Practical Handbook of Compost Engineering*, Lewis Publishers Press Inc., Boca Raton, Florida.

Hegelund, L., Sorensen, J. T. and Hermansen, J. E. (2006), 'Welfare and productivity of laying hens in commercial organic egg production systems in Denmark', *NJAS*, 54–2, 147–55.

Hills, D. J. and Hauser, J. R. (1979), *Egg Washing Wastewater Characteristics and Treatability*, Cooperative Extension, University of California.

Hauser, J. R. and Hills, D. J. (1981), 'Egg washing wastewater characteristics and treatability', *Poultry Science*, 60, 961–8.

Hobson, P. N., Bousfield, S. and Summers, R. (1981), *Methane Production from Agricultural and Domestic Wastes*, Applied Science Publishers Ltd., London, UK.

Huang, Y., Anderson, M., McIlveen-Wright, D., Lyons, G. A., McRoberts, W. C., Wang, Y. D., Roskilly, A. P. and Hewitt, N. J. (2015), 'Biochar and renewable energy generation from poultry litter waste: A technical and economic analysis based on computational simulations', *Applied Energy*, 160, 656–63.

Kawata, K., Nissato, K., Shiota, N., Hori, T., Asada, T. and Oikawa, K. (2006), 'Variation in pesticide concentrations during composting of food waste and fowl droppings', *Bulletin of Environmental Contamination and Toxicology*, 77, 391–8.

Kiepper, B. and Ritz, C. (2012), *Importance of Grit Removal From Commercial Shell Egg Processing Wastewater*, UGA Cooperative Extension Bulletin 1402.

Kim, S., Agblevor, F. A. and Lim, J. (2009), 'Fast pyrolysis of chicken litter and turkey litter in a fluidized bed reactor', *Journal of Industrial and Engineering Chemistry*, 15, 247–52.

Koster, I. W. (1986), 'Characteristics of the pH-influenced adaptation of methanogenic sludge to ammonia toxicity', *Journal of Chemical Technology and Biotechnology*, 36, 445–55.

Kroeker, E. J., Schulte, D. D., Sparling, A. B. and Lapp, H. M. (1979), 'Anaerobic treatment process stability', *Journal of the Water Pollution Control Federation*, 51(40), 718–27.

Lacey, R. E., Ross, I. J., Taraba, J. L. and Walton, L. R. (1981), 'Nutrient changes in poultry manure during batch liquid phase anaerobic fermentation', in *Livestock Waste*, Renewable Resource, *Proceedings of the 4th International Symposium*. Amarillo, TX, 1980, pp. 31–3.

Lay, Jr. D. C., Fulton, R. M., Hester, P. Y., Karcher, D. M., Kjaer, J. B., Mench, J. A., Mullens, B. A., Newberry, R. C., Nicol, C. J., O'Sullivan, N. P. and Porter, R. E. (2011), 'Emerging issues: social sustainability of egg production symposium. Hen welfare in different housing systems', *Poultry Science*, 90(1), 278–94.

Liang, Y., Xin, H., Li, H., Gates, R. S., Wheeler, E. F. and Casey, K. D. (2006), 'Effect of measurement interval on estimation of ammonia emission rates for layer houses', *Transactions of the ASAE* 49(1), 183–6.

Lin, X. J., Zhang, R., Jiang, S., El-Mashad, H. M. and Mitloehner, F. (2015), 'Nutrient flow and distribution in conventional cage, enriched colony, and aviary layer houses', *Poultry Science*, 95(1), 213–24.

Lin, X., Zhang, R., Jiang, S., El-Mashad, H. and Xin, H. (2017), 'Emissions of ammonia, carbon dioxide and particulate matter from cage-free layer houses in California', *Atmospheric Environment*, 152, 246–55.

Mahimairaja, S., Bolan, N. S., Hedley, M. J. and MacGregor, A. N. (1994), 'Losses and transformation of nitrogen during composting of poultry manure with different amendments: An incubation experiment', *Bioresource Technology*, 47, 265–73.

Martin, J. H. and Loehr, R. C. (1983), 'Enhancement of laying-hen manure as a feedstuff by aerobic stabilization', *Agricultural Wastes*, 8, 27–49.

Mohan, D., Pittman Jr., C. U. and Steele, P. H. (2006), 'Pyrolysis of wood/biomass for biooil: A critical review', *Energy & Fuels*, 20, 848–89.

NABCCCDWG, National Agricultural Biosecurity Center Consortium Carcass Disposal Working Group (2004), *Carcass Disposal: A Comprehensive Review*, http://amarillo.tamu.edu/files/2011/01/draftreport.pdf (accessed 5 January 2016).

Northcutt, J. K., Musgrove, M. T. and Jones, D. R. (2005), 'Chemical analyses of commercial shell egg wash water', *Journal of Applied Poultry Research*, 14(2), 289–95.

Ogejo, J. A. and Collins Jr., E. R. (2009), *Storing and Handling Poultry Litter*, Virginia Cooperative Extension Pub. No. 442-054. Blacksburg, VA: Virginia Tech.

Oliveira, P. A. V. de, Nicoloso, R. S., Angnes, G., Bellaver, C. and Higarashi, M. M. (2012), 'Greenhouse gas emissions in the treatment of laying hen farm residues by in-vessel composting with forced aeration', in Hassouna, M. and Guingand, N. (Eds), *Emissions of Gas and Dust from Livestock*, (EMILI 2012), 10–13 June 2012, in Saint-Malo, France.

Pan, J., Fadel, J. G., Zhang, R., El-Mashad, H. M., Ying, Y. and Rumsey, R. (2009), 'Evaluation of sample preservation methods for poultry manure', *Poultry Science*, 88, 1528–35.

Park, K. H., Jeon, J. H., Jeon, K. H., Kwag, J. H. and Choi, D. Y. (2011), 'Low greenhouse gas emissions during composting of solid swine manure', *Animal Feed Science and Technology*, 166–7, 550–6.

Patni, N. K., Fernandes, L., Zhan, W. and Jui, P. Y. (1992), 'Passively aerated composting of manure slurry', *ASABE Paper No. 924519.*

Priyadarsan, S., Annamalai, K., Sweeten, J. M., Mukhtar, S. and Holtzapple, M. T. (2004), 'Fixed – bed gasification of feedlot manure and poultry litter biomass', *Transactions of the ASAE*, 47(5), 1689–96.

Rahman, S., Xin, H., Roberts, S. A., Arthur, J. A., Burns, R. T., Li, H., Zhu, Z., Moody, L. B. and Bregendahl, K. (2012), 'Effects of laying-hen strain on manure properties and ammonia emissions', *Transactions of the ASABE*, 55(3), 1059–65.

Reddy, K. R., Khaleel, R. and Overcash, M. R. (1980), 'Nitrogen phosphorus, and carbon transformations in a coastal plain soil treated with animal manures', *Agriculture Wastes*, 2, 225–38.

Roberts, S. A., Xin, H., Kerr, B. J., Russell, J. R. and Bregendahl, K. (2007), 'Effects of dietary fiber and reduced crude protein on ammonia emission from laying-hen manure', *Poultry Science*, 86, 1625–32.

Rynk, R., van de Kamp, M., Wilson, G. B., Singley, M. E., Richard, T. L., Kolega, J. J., Gouin, F. R., Laliberty Jr., L., Kay, D., Murphy, D. W., Hoitink, H. A. J. and Brinton, W. F. (1992), *On Farm Composting*, Northeast Regional Agricultural Engineering Service, Ithaca, NY.

Safley Jr., L. M., Vetter, R. L. and Smith, L. D. (1987), 'Management and operation of a full-scale poultry waste digester,' *Poultry Science*, 66, 941–5.

Shappell, N. W. (2013), 'Egg wash wastewater: estrogenic risk or environmental asset?', *Integrated Environmental Assessment and Management*, 9(3), 517–23.

Shepherd, T. A., Zhao, Y., Li, H., Stinn, J. P. Hayes, M. D. and Xi, H. (2015), 'Environmental assessment of three egg production systems – Part II. Ammonia, greenhouse gas, and particulate matter emissions', *Poultry Science*, 94, 534–43.

Sims, J. T. and Wolf, D. C. (1994), 'Poultry manure management: agricultural and environmental issues', *Advances in Agronomy*, 52, 1–83.

Sivakumar, K., Kumar, V., Jagatheesan, P., Viswanathan, K. and Chandrasekaran, D. (2008), 'Seasonal variations in composting process of dead poultry birds', *Bioresource Technology*, 99(9), 3708–13.

Sparks Companies Inc. (2002), *Livestock Mortality: Methods of Disposal and their Potential Costs*, http://assets.nationalrenderers.org/mortality_final.pdf (accessed 5 January 2016).

Tiquia, S. M. and Tam, N. F. Y. (2002), 'Characterization and composting of poultry litter in forced-aeration piles', *Process Biochemistry*, 37, 869–80.

Van Horne, P. L. M. (1996), 'Production and economic results of commercial flocks with white layers in aviary systems and battery cages', *British Poultry Science*, 37, 255–61.

Vardon, D. R., Sharma, B. K., Scott, J., Yu, G., Wang, Z., Schideman, L., Zhang, Y., and Strathmann, T. J. (2011), 'Chemical properties of biocrude oil from the hydrothermal liquefaction of Spirulina algae, swine manure, and digested anaerobic sludge', *Bioresource Technology*, 102(17), 8295–303.

Zhang, R. and El-Mashad, H. M. (2007), 'Biodiesel and biogas production from seafood processing byproducts', in Shahidi, F. (Ed.), *Maximizing the Value of Marine By-Products*, Woodhead Publishing, Abington Hall, Abington, Cambridge, CB1 6AH, England.

Zhang, R., Mitloehner, F., Rumsey, T., El-Mashad, H. M., Li, C., Salas, W., Rotz, A., Hafner, S., Montes, F., Arogo, J., Xin, H. and Ni, J. (2009), 'Process based emission models for predicting gaseous emissions from animal feeding operations', White Paper Submitted to Agricultural Air Quality Task Force.

Zhang, W. and Lau, A. (2007), 'Reducing ammonia emission from poultry manure composting via struvite formation', *Journal of Chemical Technology and Biotechnology*, 82, 598–602. doi: 10.1002/jctb.1701.

Chapter 10

Assessing the sustainability of organic egg production

Jacqueline Jacob and Anthony Pescatore, University of Kentucky, USA

1 Introduction

Regardless of the management system chosen, the overall goal of all farmers is the production of a safe and nutritious food product at a reasonable price. Any system used must be sustainable. Sustainability is a multidimensional, complex concept. It has simply been defined as meeting the needs of the present generation without compromising the ability of the next generation to meet their needs. Any production system should be economically profitable, ecologically sound and socially acceptable. The evaluation should address human food needs, environmental preservation, economic feasibility and quality of life.

The basic premise of organic farming is that it is a systems approach to producing food in an environmentally and socially sustainable way. This led to the development of specific regulations dictating what can and cannot be used in organic farming. The European Union and the United States represent two different regulatory approaches. Within the EU, organic farming regulations are set out under Council Resolution 2029/91 and its amendments. Organic poultry production is dictated by EC 834/2007. For organic egg production in the EU, cages cannot be used. Poultry must be kept in free-range conditions. Each poultry house must have its own dedicated grazing space. At least one third of the poultry house floor must be solid, which does not include slatted or grid flooring. Perches must be provided as well as adequate exit/entry pop-holes for the chickens to access the grazing area. There are stipulations with regard to the floor space (6 hens/m^2) and pasture area (4 m^2/hen) required for each laying hen. The stocking density is based on 170 kg of nitrogen per year per hectare. The EU puts limits on flock size with no more than 3000 laying hens in a flock. Within the United States, organic farming regulations

http://dx.doi.org/10.19103/AS.2016.0012.31

are set out by the National Organic Program (NOP). As with the EU regulations, NOP regulations stipulate that cages cannot be used and that poultry must have access to the outdoors. Within the U.S. regulations there is no restriction on the size of flock. In December 2011, the National Organic Standards Board (NOSB), which advises the USDA on implementation of the NOP regulations, recommended stocking densities of 2 ft^2/hen (5.4 hens/m^2) indoors and 2–5 ft^2/hen (2.2–5.4 hens/m^2) outdoors (NOSB, 2011).

In Europe, pure amino acids are banned, but they have had a temporary variance to allow 5% of non-organic feedstuffs in a certified organic feed. This variance was to allow egg producers time to meet the dietary requirement for the amino acid methionine which is commonly added to conventional poultry diets. The removal of the variance has been delayed, while producers continue to research alternatives to pure methionine in poultry diets. In the United States, a temporary variance has been granted to allow egg producers to use a limited quantity of pure methionine in certified organic poultry diets. Again, this variance is temporary as research into alternatives continues. Both the EU and US ban the use of synthetic medications such as antibiotics, de-wormers, insecticides and coccidiostats in organic poultry production. Some areas allow vaccinations if they have not been produced by using biotechnology methods.

Regardless of the country, organic regulations require the feeding of diets made with organic feedstuffs. Therefore, an assessment of the environmental impact of organic egg production must take into account the effect of this part of the production system as well. All organic crop producers function under strict regulations which prohibit the use of agrichemicals such as synthetic pesticides, herbicides and fertilizers. In addition, crop varieties developed by using biotechnology (genetically modified organisms, or GMO) are not permitted. Watson et al. (2002) reported organic yields averaging 50–95% of conventionally produced crops, depending on the crop and the position in the organic rotation system commonly used by organic farms to control weeds and pests and to improve soil fertility. There are difficulties in accounting for differences in crop yields because of differences in types of farms compared. While there are specific regulations concerning organic crop production, there is no clear definition of what conventional agriculture is. What exactly is 'conventional agriculture', other than simply being non-organic?

2 Assessing environmental sustainability

There are several reports in the literature dealing with 'sustainability' of poultry production systems using different definitions as well as different methodologies for the evaluation. Schader et al. (2014) identified 35 different approaches to assessing sustainability developed in the last 15 years. A multi-criteria approach should be used to evaluate different food production systems. A one-size-fits-all evaluation system may not work. Any evaluation should provide a detailed definition of 'sustainability' as well as a description of the indicator tools used. Indicators should be relevant to something about the sustainability of different production systems, simple to understand, react to changes in the system and the data accessible (Mollenhorst et al., 2006). The other factor that affects the comparison between systems is the actual farms used in the comparison. The variety of conventional farming procedures and the experience of the farmers cause difficulty in attempting comparisons. Comparisons are often made with nearby farms to remove regional differences, but managerial quality is commonly ignored.

The Life Cycle Assessment (LCA) approach looks at all aspects of production to identify resource use and environmental burdens that result. This would include feed production, non-crop nutrient production, feed processing, maintaining breeder flocks, hatching and rearing of pullets for both breeders and egg layers, egg production, as well as egg processing and distribution. Aspects looked at include global warming potential (GWP), land occupation, acidification potential, energy use, fossil phosphorus (P) use and nitrogen (N) and P balances. The results are reported on the basis of a unit of production; for egg production, it is typically kg of eggs produced. The number of eggs represented by 1000 kg will depend on the production system. Leinon et al. (2012) reported that egg size varied between the production systems averaging 62 g for cage-raised hens and 63.5 g from hens raised in alternative production systems, which would include organics. Therefore, 1000 kg of eggs would be represented by 16 129 eggs for cage systems and 15 748 eggs for alternative production systems. Even with the reduced number of eggs represented, the lower performance for organic hens results in the number of hens required to produce 1000 kg of eggs to be highest in the organic system (56.3 hens) and lowest in the cage system (51.2 hens) (Leinon et al., 2012). This is based on 315 eggs per hen for cage systems and 280 eggs per hen in alternative production systems.

When looking at the LCA of egg production systems, feed production, processing and transportation represent 54–75% of the primary energy use and 64–72% of the GWP (Leinon et al., 2012). Two components of this assessment would include the amount of feed required by each production system and the yields for the crops used in poultry diets. Organic egg production systems are reported to have lower feed efficiency. Dekker et al. (2011) reported organic egg layers requiring 2.6 kg of feed for each kg of eggs produced. For hens kept in cage systems, the feed-conversion ratio was only 2.0. Leinon at al. (2012) reported higher overall feed-conversion ratios of 2.56 for caged layers and 3.10 for organic layers.

As previously indicated, organic crop yields can average 50–95% of conventionally produced crops (Watson et al., 2002). Organically produced crops also have more overheads for soil fertility building and cover crops. The differences in crop yield will have a major impact on the amount of land required to produce the feedstuffs necessary for poultry production (Williams et al., 2006). Dekker et al. (2011) estimated that organic wheat, a common grain used in poultry diets in Europe, requires twice the amount of land compared with conventionally grown wheat. Most organic cropping systems rely on a period in which a cover crop is grown and ploughed back in to provide the soil with nitrogen. No organic yield is obtained throughout this period, unless it is made into pastures and used for grazing animals. This will increase the total amount of land required for crop production. In a conventional cropping system, the same field can be used continuously (Trewavas, 2004). However, with conventional agriculture, there has been a major loss of the world's cropland because of erosion and degradation (Wood et al., 2006). Organic farmers, in general, pay particular attention to their soil since it is a fundamental factor that determines the success of their crops.

While organic crop production can require less energy compared to conventional agriculture, the larger reduction in energy use achieved by avoiding synthetic N sources is typically offset by lower yields and higher inputs into field work. Global warming potential was estimated to be only 2–7% less for organic compared to non-organic field crops (Williams et al., 2006). Fuel costs, however, go up when comparing egg production systems. In the winter, non-caged housing systems tend to be colder because of the lower stocking densities than in caged housing systems. This leads to greater feed and fuel energy use (Xin et al., 2011).

A major contributing factor in evaluating the ecological impact of transporting concentrates and manure will depend on the location. For example, when compared with Western and Northern Europe, the ecological impacts would be lower in Southern and Eastern Europe because of more favourable climate and soil conditions (Dekker et al., 2011).

Manure is another contributing factor in GWP. In addition to the different amounts of feed consumed, the composition of the feeds also varies between organic and non-organic egg production systems, resulting in different environmental burdens per equal amount of feed. Based on 1000 kg of feed, organic pullet production systems have a much higher GWP than conventionally reared pullets (Dekker et al., 2011; Leinon et al., 2012). The GWP was, however, lower for organic layers as compared to conventional. The eutrophication potential, expressed as PO_4 equivalent per 1000 kg of feed, was double for organic layers and more than double for organic pullets. This may reflect the phytate content of cereals. The use of commercial phytase enzymes could reduce this impact. In order to use phytase in organic poultry diets, it would be necessary to have an organically approved phytase enzyme that is not made by using GMO techniques. The acidification potential, expressed as SO_2 equivalents per 1000 kg of feed, was lower for organic layers compared with conventional systems, but higher for organic pullets (Dekker et al., 2011; Leinon et al., 2012). Manure has the greatest impact on the acidification and eutrophication potential because of ammonia emissions. With lower N excretion, there is a lower acidification and eutrophication potential. The better feed conversions seen with non-organic feeds reduce N excretion, thus reducing acidification and eutrophication potential. In addition, the manure in cage houses is frequently air-dried and removed on a weekly basis. The lower moisture content in air-dried manure makes it easier to transport and it emits less ammonia (Xin et al., 2011).

Organic crop production often uses poultry manure instead of synthetic fertilizer. However, the ammonia emissions from manure application are higher than from fertilizer application (Dekker et al., 2011). Loose-housed hens may need less amino acids since they produce fewer eggs. Reducing N excretion is hampered by the fact that pure methionine is not allowed in organic diets because it is produced by GMOs. The selection of breeds for high production potential has resulted in greater amino acid requirements than for the heritage breeds, especially with regard to the amino acid methionine (unpublished data). As a result, conventional egg producers include pure methionine in their feed allowing them to keep total nitrogen content low. Until alternatives to pure methionine can be found, the reduction in manure N from organic hens is not possible. There are reports of increased nitrogen leaching into the soil on organic poultry farms due to the ban on the use of pure animal acids (Tauson, 2005).

Aarnink et al. (2006) looked at the ammonia emission and nutrient load in outdoor runs of flocks of 3000 hens, the limit allowed by EU organic regulations. They reported that the emissions from the outdoor run were relatively small compared to the emissions from the house and that nutrient load in the outdoor area near the house far exceeded maximum acceptable levels by a factor of 15. As previously indicated, this may be due to the high protein diets used in organic feeds because of the need to meet the methionine requirement.

LCAs, however, largely ignore impacts on biodiversity and soil. Modern intensive crop production is believed to be a major threat to worldwide biodiversity. Over the last few decades, there has been a major decrease in many species associated with farmland. This would include birds, small mammals, invertebrates and arable flora (Hole et al., 2005).

Organic farming is seen as a potential solution to reduce, and perhaps reverse, this trend. It is unclear, however, whether organic farming has greater benefits for biodiversity than carefully maintained areas of natural habitat in specific areas of cropped and/or non-cropped habitats within conventional crop production.

Organic farming makes up only a small part of the total agricultural production in Europe and North America. Although the average price of organic products is twice that of conventionally grown food, the supply is not able to meet the demand. The shortfall of organically grown grains for poultry feeds has been met by importing feed from countries such as Romania and Turkey (Jaenicke and Demko, 2015).

3 Assessing economic sustainability

Yields for organic crop production were found to be significantly lower than the same crops grown under conventional management. However, these yield differences vary among crops as well as among countries and regions (Offermann and Nieberg, 2000). Organic cereals, a major constituent of poultry feed, can have yields 60–70% lower than those raised conventionally. Fortunately, farm gate prices for organically produced cereals can be 50–200% higher than for conventionally grown crops. As long as the opportunity for receiving higher farm gate prices for organically produced foods exists, organic farmers can remain economically viable. The economic stability depends on overcoming high input costs with high premium prices.

There are external factors that can impact production viability due to the risk of catastrophic loss. Concerns are raised regarding the risk of outdoor chickens becoming infected with avian influenza and having a major effect on the economic stability of the poultry industry in any region affected. While there are no reports comparing the viral poultry disease incidence between organic and conventionally reared chickens, contact with wild ducks may be acting as a silent reservoir for the avian influenza virus playing a role in the transmission of the virus to poultry flocks (Kijlstra and Eijck, 2006).

Eggs produced in alternative production systems receive a premium price. Research in the United States (Sumner et al., 2011) has shown that egg consumption tends to be relatively unresponsive to price change. It was estimated that a farm price increase of 40% for eggs would likely reduce egg consumption by less than 10%. Low-income consumers, however, would be hardest hit by such a price increase.

Interestingly, 75% of conventional crop producers in the United Kingdom believed that organic farming methods are better for the environment, but only 13% thought that such techniques could produce sufficient food and fibre for society. Conversely, only 73% of organic crop producers in the United Kingdom interviewed believed that organic production systems could produce the necessary food and fibre for the growing world population (Rigby and Cáceres, 2001).

4 Assessing welfare and health issues

While organic farming constitutes only a small part of the total agricultural food production, consumers have demonstrated an increasing demand for organic purchases. The reasons for purchasing organic products vary (Sylvander and Le Floch-Wadel, 2000). Some view organic products as an important symbol of their standard of living and as more socially acceptable. Many believe that organic production systems are better for the environment and animal welfare. While health and food safety concerns are the main motives for food purchases, ethical concerns, specifically animal welfare and environmental protection, play a significant role in the decision to purchase organic food (Harper and Makatoun, 2002).

Consumers assume that animals kept in organic systems are raised under higher animal welfare standards than those raised in conventional production systems (Sutherland et al., 2013). Evaluation of the animal welfare in any production system is difficult because of differences in defining what constitutes good animal welfare. The description most commonly used is that the animals should be able to lead natural lives, should be free from intense and prolonged fear and pain, and should have good health and physiology.

While organic poultry have freedom to express 'natural' behaviours, allowing them outdoor access makes maintaining a high standard of welfare difficult. A major challenge of organic animal production systems is the prevention and treatment of health-related issues. There is a lack of scientifically evaluated, organically acceptable, therapeutic systems. Organic poultry production systems can be plagued with ecto- and/or endo-parasites (Permin et al., 1999; Fossum et al., 2009; Sutherland et al., 2013; Hinrichsen, 2015). Outdoor access provides increased opportunities for chickens to become infected with parasites, especially intestinal parasites (Sherwin et al., 2013). Infections with internal parasites can impair the immune response elicited by vaccination and increase the severity of any bacterial infection. Hinrichsen (2015) reported that, on the farms studied, roundworm eggs appear in the faeces of laying hens 6–8 weeks after placement and maximum infection typically occurs 8–12 weeks later. A roundworm infection can cause weight loss, blockage of the intestinal lumen, reduced feed intake and feed efficiency, reduced movement and general weakness in the hens (Gauly et al., 2007). Roundworm infection can also affect plumage quality and decrease bone strength (Ramadan and Abou Znada, 1991). Some European regulations allow the use of Flubenol© to treat intestinal worms, but no approved commercial products are available in the United States.

A study of 18 organic egg producing flocks showed that mortality was mainly due to inappropriate behavioural patterns, disease and predatory attacks (Hegelund et al., 2006). Cannibalism is one of the main causes of death in laying hens kept in free-ranging systems in Sweden (Fossum et al., 2009). Variation in the prevalence of feather pecking among organically reared laying hens in Denmark, however, shows that there is room for improvement (Hegelund et al., 2006). While comparing 18 flocks, they found that only 6 showed little or no damage due to feather pecking, while 4 had indications of severe feather pecking.

Many consumers associate better food safety with organic food production. However, providing chickens with access to an outdoor area may increase the risk of poultry becoming infected with *Salmonella* and *Campylobacter*, two important food-borne pathogens. This is due to the increased contact with wild birds as well as other animals and their faeces. In a Swedish study, almost 100% of organically farmed chicken meat flocks were infected with *Campylobacter* compared with only 10% of the conventionally reared flocks (Engvall, 2001). There is very little data on *Campylobacter* infections in laying hens.

There is considerable variation in the incidence of disease between organic farms. There is also a higher incidence of disease on newly converted organic farms. This would suggest that management skills and experience make a significant contribution to animal health and welfare in organic production systems (Sutherland et al., 2013).

We need to solve apparent conflicts between organic farming objectives with regard to environmental protection and animal health and welfare. Nutritional imbalance can lead to increased disease susceptibility and behaviourally problems like cannibalism (Berg, 2001). Animal breeding programmes need to start selecting for reduced feather pecking in extensive production systems rather than just using the commercial strains currently available. There is only one study looking at the heritability estimates of cannibalism, but it suggests there is high heritability (Koene, 2001). The high feed-conversion ratios for organic egg production need to be decreased. Improved feed utilization can be achieved by changes in feed composition, farm management, genetic merit of hens, incidence of feather pecking or disease and the percentage of damaged eggs.

5 Conclusions

In order to produce poultry products organically, a specific set of rules must be followed. These rules vary among regions of the world. The aims of organic egg production are to ensure a high level of animal welfare, protection of the environment and the production of a healthy food item. While there is no universal standard for organic egg production, all standards require that organic feedstuffs be fed. Restrictions include no GMOs, no antibiotics or other drugs, no animal slaughter by-products, no chemically extracted feeds and no pure amino acids. European regulations go further requiring that feedstuffs be produced on the farm or at least in the same region.

Any poultry production system has scope for improvement and thus has the potential to reduce its environmental impact. The data available for comparison of the environmental impact of organic versus non-organic egg production systems are limited, especially with regard to North America. There are, however, some well-identified areas like parasite control and balanced feed ration formulation where efforts are needed to find solutions that meet organic standards while guaranteeing high levels of animal health and welfare.

If sustainable egg production means more small-scale, more labour-intensive operations, this would imply the need to return to the land and a stop to the ever-growing urbanization, since such large urban populations cannot be maintained in this framework (Rigby and Cáceres, 2001). We need to design productive systems that are sustainable on a farm level but have a high enough productivity to maintain a food supply for the growing human population.

6 Where to look for further information

Organic regulations that pertain to poultry production continue to evolve. The International Federation of Organic Agricultural Movements (IFOAM) is a 'worldwide umbrella organization for the organic agricultural movement that represents close to 800 affiliates in 117 countries'. They are a source of information on current regulations.

The National Sustainable Agriculture Coalition represents a group of 'grassroots organizations that advocates for federal policy reform to advance the sustainability of agriculture, food systems, natural resources and rural communities'.

The 31st symposium of the World Poultry Science Association, UK Branch (2014) was devoted to sustainability of egg production in general within Europe. A book based on the proceedings from this symposium can be downloaded from http://stikerdinding.net/download/file.php?asin=1780645309.

7 References

Aarnink, A. J. A., Hol, J. M. G and Beurskens, A. G. C. (2006), 'Ammonia emissions and nutrient load in outdoor runs of laying hens'. *NJAS – Wagen. J. Life Sci.* 54(2): 223–34.

Berg, C. (2001), 'Health and welfare in organic poultry production'. *Acta Vet. Scand. Suppl.* 95: 37–45.

Dekker, S. E. M., de Boer, I. J. M., Vermeij, I., Aarnink, A. J. A. and Groot Koerkamp, P. W. G. (2011), 'Ecological and economic evaluation of Dutch egg production systems'. *Livest. Sci.* 139: 109–21.

Engvall, A. (2001), 'May organically farmed animals pose a risk for Campylobacter infections in humans?' *Acta Vet. Scand. Suppl.* 95: 85–8.

Fossum, O., Jansson, D. S., Etterlin, P. E. and V gsholm, I. (2009), 'Causes of mortality in laying hens in different housing systems in 2001 to 2004'. *Acta Vet. Scand. Suppl.* 51: 3–11.

Gauly, M., Duss, C. and Erhard, G. (2007), 'Influence of *Ascaridia galli* infections and antihelminthic treatments on the behavior and social ranks of laying hens (*Gallus gallus domesticus*)'. *Vet. Parasitol.* 146: 271–80.

Harper, G. C. and Makatoun, A. (2002), 'Consumer perception of organic food production and farm animal welfare'. *Br. Food J.* 104: 287–99.

Hegelund, L., Sørensen, J. T. and Hermansen, J. E. (2006), 'Welfare and productivity of laying hens in commercial organic egg production systems in Denmark'. *NJAS – Wagen. J. Life Sci.* 54: 147–55.

Hinrichsen, L. K. (2015), 'Animal welfare in organic egg production with emphasis on mortality and helminth infections'. PhD Thesis. Aarhus University.

Hole, D. G., Perkins, A. J., Wilson, J. D., Alexander, I. H., Grice, P. V. and Evans, A. D. (2005), 'Does organic farming benefit biodiversity?' *Biol. Conserv.* 122: 113–30.

Jaenicke, E. C. and Demko, I. (2015), 'Report to the Organic Trade Association – Preliminary analysis of USDA's Organic Trade Data: 2011 to 2014'. Available online at http://ota.com/sites/default/files/indexed_files/OTAJaenickeMay2015_TradeDataReport.pdf. (accessed 29 February 2016).

Kijlstra, A. and Eijck, I. A. J. M. (2006), 'Animal health in organic livestock production systems: a review'. *NJAS – Wagen. J. Life Sci.* 54(1): 77–94.

Koene, P. (2001), 'Animal welfare and genetics in organic farming of layers: the example of cannibalism'. In the proceedings of the 4th NAHWOA workshop, Wageningen, The Netherlands. 24–27 March 2001, pp. 62–85.

Leinon, I., Williams, A. G., Wiseman, J., Guy, J. and Kyriazakis, I. (2012), 'Predicting the environmental impact of chicken systems in the United Kingdom through a life cycle assessment: egg production systems'. *Poult. Sci.* 91: 26–40.

Mollenhorst, H., Berentsen, P. B. M. and De Boer, I. J. M. (2006), 'On farm quantification of sustainability indicators: an application to egg production systems'. *Br. Poult. Sci.* 47: 405–17.

National Organic Standards Board (NOSB) (2011), 'Formal recommendation by the National Organic Standards Board (NOSB) to the National Organic Program (NOP): Animal welfare and stocking densities'. Available online at https://www.ams.usda.gov/sites/default/files/media/NOP%20Livestock%20Final%20Rec%20Animal%20Welfare%20and%20Stocking%20Rates.pdf (accessed 29 February 2016).

Offermann, F. and Nieberg, H. (2000), 'Economic performance of organic farms in Europe'. In *Organic Farming in Europe: Economics and Policy. Volume 5*. University of Hohenheim/ Dept of Farm Economics 410A. 220 pages.

Permin, A., Bisgaard, M., Frandsen, F., Preamna, M., Kold, J. and Nansen, P. (1999), 'Prevalence of gastrointestinal helminths in different poultry production systems'. *Br. Poult. Sci.* 40: 439–43.

Ramadan, H. H. and Abou Znada, N. Y. (1991), 'Some pathological and biochemical studies on experimental ascaridiasis in chickens'. *Die Nahrun* 35: 71–84.

Rigby, D. and Cáceres, D. (2001), 'Organic farming and the sustainability of agricultural systems'. *Agric. Syst.* 68: 21–40.

Schader, C., Grenz, J., Meier, M. S. and Stolze, M. (2014), 'Scope and precision of sustainability approaches to food systems'. *Ecol. Soc.* 19(3): 42–56.

Sherwin, C. M., Nasr, M. A. F., Gale, E., Petek, M., Tafford, K., Turp, M. and Coles, G. C. (2013), 'Prevalence of nematode infection and fecal counts in free-ranging laying hens: relations to housing and husbandry'. *Br. Poult. Sci.* 54: 12–23.

Sumner, D. A., Gow, H., Hayes, D., Matthews, W., Norwood, B., Rosen-Molina, J. T. and Thurman, W. (2011), 'Economic and market issues on the sustainability of egg production in the United States: analysis of alternative production systems'. *Poult. Sci.* 90: 241–50.

Sutherland, M. A., Webster, J. and Sutherland, I., (2013), 'Animal health and welfare issues facing organic production systems'. *Animals* 3: 1021–35.

Sylvander, B. and La Floch-Wadel, A. (2000), 'Consumer demand and production of organics in the E. U.' *AgBioForum* 3(2–3): 97–106.

Tauson, R. (2005), 'Management and housing systems for layers – effects on welfare and production'. *World's Poult. Sci. J.* 61: 477–90.

Trewavas, A. (2004), 'A critical assessment of organic farming and food assertions with particular respect to the UK and the potential environmental benefits of no-till agriculture'. *Crop Prot.* 23: 757–81.

Watson, C. A., Atkinson, D., Gosling, P., Jackson, L. R. and Rayns, F. W. (2002), 'Managing soil fertility in organic farming systems'. *Soil Use Manage.* 18: 239–47.

Williams, A. G., Audsley, E. and Sandars, D. L. (2006), *Determining the Environmental Burdens and Resource use in the Production of Agricultural and Horticultural Commodities. Main Report. Defra Research Project IS0205.* Bedford: Cranfield University and Defra. Available on www.silsoe.cranfield.ac.uk and www.defra.gov.uk.

Wood, R., Lenzen, Z., Dey, C. and Lundie, S. (2006), 'A comparative study of some environmental impacts of conventional and organic farming in Australia'. *Ag. Sys.* 89: 324–48.

Xin, H., Gates, R. S., Green, A. R., Mitloehner, F. M., Moore Jr., P. A. and Wathes, C. M. (2011), 'Environmental impacts and sustainability of egg production systems'. *Poult. Sci.* 90(1): 263–77.

Index